THE HISTORY
OF THE AMERICAN
SPACE
SHUTTLE

SCHIFFER
PUBLISHING

4880 Lower Valley Road · Atglen, PA 19310

DENNIS R. JENKINS

The original program logo (left) and the end-of-program commemorative logo (center) depicted a generic vehicle. The later program logo used beginning in the late 1990s (right) depicted, likely unintentionally, Columbia (note the black chines).

Designed by Dennis R. Jenkins
Cover design by Justin Watkinson
Type set in Neuropol/StempelGaramond

ISBN: 978-0-7643-5770-1
Printed in China

Published by Schiffer Publishing, Ltd.
4880 Lower Valley Road
Atglen, PA 19310
Phone: (610) 593-1777; Fax: (610) 593-2002
E-mail: Info@schifferbooks.com
Web: www.schifferbooks.com

For our complete selection of fine books on this and related subjects, please visit our website at www.schifferbooks.com. You may also write for a free catalog.

Schiffer Publishing's titles are available at special discounts for bulk purchases for sales promotions or premiums. Special editions, including personalized covers, corporate imprints, and excerpts, can be created in large quantities for special needs. For more information, contact the publisher.

We are always looking for people to write books on new and related subjects. If you have an idea for a book, please contact us at proposals@schifferbooks.com.

Contents

PREFACE

The space shuttle flight campaign began at 12:00:04 UTC on 12 April 1981 with the launch of STS-1 from the Kennedy Space Center (KSC). It ended at 09:57:51 UTC on 21 July 2011 with wheels stop of STS-135, also at KSC. During the thirty years in between, the program experienced triumphs and tragedies, amazed the world with its orbital exploits, and was frequently the subject of adMiration, condemnation, pride, and despair. The men who created the American space shuttle did so with grand ambitions, but the exercise must be looked at in the context of time. Although the concept of space travel was ancient, its practice was little more than a decade old. In fact, when the initial studies that led to space shuttle began, the United States had completed fewer than two dozen manned space flights; indeed, the maiden voyage of *Columbia* was only the thirtieth American manned orbital flight. It was a small experience base upon which to begin an ambitious program.

Space shuttle was born of the age-old dream to fly into space, an expectation put on hold during the 1960s due to the limited throw-capability of the early ballistic missiles that were hastily converted into launch vehicles at the beginning of the space race. Mostly because of time constraints, the capsules continued when John Kennedy committed the United States to landing on the moon by the end of the 1960s. Only after that goal was satisfied did engineers return to the dream of flying into space. The United States was riding high on its successes, and visions of space stations and the space shuttles that serviced them were firmly planted in the minds of engineers, science fiction writers, and the public. Almost religiously, Arthur C. Clarke and Stanley Kubrick furthered these dreams to the music of the *An der schönen blauen Donau* in the movie *2001: A Space Odyssey*.

Unfortunately, some within NASA seriously oversold space shuttle, promising unrealistic economies and capabilities. Various mission manifests were predicting as many as one flight per week.

In retrospect, it was obvious the vehicle could never live up to the hype, but somehow it seemed appropriate at the time. Regardless, space shuttle was an engineering triumph; a reusable spacecraft that could carry heavy payloads up and bring back satellites or other heavy downmass. The vehicle was an incredible achievement.

The most important, and oft-overlooked, downside was that NASA had not evolved, organizationally or culturally, to manage an operational program. The agency, as the National Advisory Committee for Aeronautics (NACA), had been established in 1917 to conduct fundamental research and had morphed into the organization that put men on the moon. But that did not mean it was capable of managing a complex operational vehicle on a sustainable basis. The development culture led to a standing army that would never be economically efficient, while inexperience and lapses in judgment contributed to two fatal accidents. Much of the space shuttle legacy, unfortunately, will be in how NASA operated it rather than the capabilities of the vehicle itself.

Regardless of the expectations and circumstance, space shuttle accomplished remarkable things. These included launching significant payloads such as the Hubble Space Telescope and assembling the International Space Station. During its thirty-year flight campaign, space shuttle carried more crew members to orbit than all other launch systems, worldwide, combined. It carried more than 3.5 million pounds of cargo up and essentially everything that has ever been brought back down. Perhaps more importantly, space shuttle taught us invaluable lessons about how to operate spacecraft on the ground and in space. We learned about inspection, maintenance, and refurbishment; about extravehicular activities; and about troubleshooting and repair. Unfortunately, we also learned about tragedy. Many of these were hard lessons, and most will likely be forgotten before whatever follows becomes operational. But we must continue to keep the dream alive.

NOTES ABOUT DATA

This book is not intended as the definitive work and, therefore, does not use source citations or endnotes. In general, the vast majority of the operational data used here came from two program documents: Shuttle Crew Operations Manual, USA007587, OI-33 release, 15 December 2008, and Shuttle Flight Data and In-Flight Anomaly List, JSC-19413, 30 October 2011. These documents are derived from engineering sources, and the data may disagree with information released by public affairs during or immediately following a mission.

In addition, dates and times for each mission are given in coordinated universal time (UTC), mostly to avoid dealing with daylight savings time and because engineering data tend to be in UTC. As a result, dates and times may appear to disagree with other published sources. Florida is +4 or +5 hours from UTC (depending on daylight savings time), so launches or landings near midnight may have a different date than expected.

ACKNOWLEDGMENTS

Over the past thirty years, many people helped with the collection of data and played sounding board to some of my crackpot ideas of what should, or should not, be in my various space shuttle books. The list is far too extensive for the space available here and any mistakes that remain are entirely mine.

The stack that was used for STS-1 rolled out to Launch Complex 39A on 31 December 1980 and would spend 102 days at the pad, including the first flight readiness firing (FRF). Note the unique black lightning ring around the top of the white external tank. (NASA)

The last stack, STS-135, rolled out to LC-39A on 31 May 2011 and spent 38 days at the pad. By this time, Discovery was already being prepared for transfer to the Smithsonian's National Air and Space Museum, and the Space Shuttle Program was well into efforts to close down. (NASA)

EVOLUTION

Since the beginning, most visions of manned space flight involved the concept of a "spaceplane," a seemingly logical extension of the airplane that had become commonplace during the twentieth century. Scientists, engineers, science fiction writers, and artists all predicted a future featuring spaceplanes. Therefore, it came as a major disappointment to many when the first men ventured into space aboard ballistic capsules. Test pilots were known to call astronauts "spam in a can," and many engineers were not much kinder in their critiques of the simple, blunt-body capsules used by Mercury, Gemini, and Apollo.

However, the fact was there was no other way to get men into space at the time. The early Redstone and Atlas boosters, repurposed intermediate and intercontinental ballistic missiles, respectively, were not powerful enough to launch a winged vehicle, and political realities dictated the United States could not wait. Spam in a can or not, the early astronauts became instant celebrities and proved it was possible to survive and, eventually, work in the space environment. Nevertheless, the Soviets maintained a lead in the space race for several years as the Americans tried to catch up with the impressive missiles developed by Sergei Pavlovich Korolev.

The second-generation American intercontinental ballistic missile, Titan, could probably have launched a small winged vehicle, perhaps like the one proposed by John Becker from the National Advisory Committee for Aeronautics (NACA, a predecessor of NASA) Langley Aeronautical Laboratory at the Conference on High-Speed Aerodynamics in March 1958. But by the time Titan flew, John Kennedy had committed the United States to a moon landing before the end of the decade, so the momentum of the capsules continued. They served their purpose, with Neil Armstrong landing on the moon on 20 July 1969, well ahead of the Soviets, who had largely given up after a series of failures and the realization that the United States had an unbeatable lead.

Still, visions of winged vehicles were abundant. The Air Force and NACA/NASA had begun studying vehicles that resembled space shuttles during the late 1950s and continued throughout the 1960s. Engineers intended most of these efforts to determine if the technologies were available to develop a large spaceplane, not to actually build one. Along the way, researchers and engineers developed improved rocket engines, investigated new thermal protection systems, and conceived advanced vehicle automation systems and ground processing techniques.

The Air Force committed to building the first reusable, lifting-reentry spaceplane when they began developing Dyna-Soar. Men had not yet ventured beyond the atmosphere, so the investigations started from scratch. It was a long, winding process that went through several iterations of names and missions. Initially led by the Bell Aircraft Company, the requirements eventually converged on a potentially useful concept and Boeing won the development contract that included the production of ten X-20 spacecraft.

The resulting sleek, one-man glider benefited little from the NASA capsules, since its development was concurrent and the entire concept was far more advanced. Instead of throwing away the vehicle after every flight, like the capsules, engineers intended Dyna-Soar to fly ten times with only minor refurbishment of its superalloy metallic heatshield. It would be maneuverable during entry and able to land on conventional runways. It was also incredibly controversial, and secretary of defense Robert McNamara ultimately canceled the program before Boeing completed the first vehicle. Regardless of whether there was a valid military mission for the diminutive spacecraft, its development and flight test would have provided a much-needed body of knowledge when it came time to build space shuttle. But it was not to be, although Dyna-Soar became part of aerospace folklore amid much wild speculation about what the Air Force had really intended to use the vehicle for.

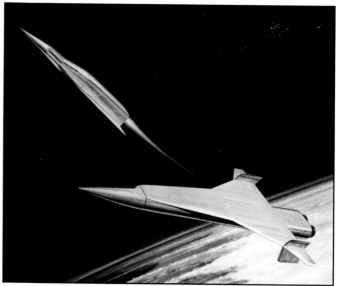

The Boeing X-20 Dyna-Soar glider represents the largest missed opportunity on the road to space shuttle. Canceling the X-20 deprived future programs of the data that would have come from testing and operating the small reusable glider. (US Air Force)

This 1962 Lockheed Reusable Aerospace Passenger Transport (RAPT) used a large reusable flyback first stage, an expendable second stage, and a reusable winged orbital stage. The vehicle, strictly a design exercise, carried ten passengers and a crew of two. (National Archives)

Almost all of these early studies concentrated on fully reusable designs. A few were single-stage-to-orbit concepts, but engineers soon found that technology did not yet (or even now) exist for a vehicle to carry sufficient propellants to deliver a meaningful payload to orbit. Therefore, most designs were two-stage vehicles that used large, winged boosters to carry smaller, winged orbiters to about 6,800 mph (Mach 10) and 200,000 feet before releasing them. Conceptually it was not much different from how the X-1 or X-15 research airplanes operated, just on a vastly larger, and faster, scale. It was a grand plan.

Even early on, however, there were some researchers and engineers who believed this was a step too far, and recommended a partially expendable configuration, usually throwing away either the booster or the propellant tanks. Nobody was listening. Yet.

PHASE A

Many of these early studies concluded it would be possible to develop a large, winged space shuttle, and several senior NASA officials, particularly George Mueller, began to make incredible predictions about its potential capabilities and economics. Perhaps it was part of a grander scheme to secure funding, or maybe it was just irrational exuberance. In retrospect, it was pure science fiction.

The development of the actual vehicle we call space shuttle began officially on 30 October 1968 when the Manned Spacecraft Center (MSC, now the Johnson Space Center, JSC) and Marshall Space Flight Center (MSFC) released a joint request for proposals (RFP) for an eight-month study of an Integral Launch and Reentry Vehicle (ILRV). Five companies, Convair, Lockheed, McDonnell Douglas, Martin Marietta, and North American Rockwell, participated in the study, which became Phase A of the space shuttle design and development cycle. In the then-new four-phase process, Phase A was called advanced studies, Phase B was project definition, Phase C was vehicle design, and Phase D was production and operations. Given how long it seems to take to develop complex systems in the twenty-first century, space shuttle moved remarkably quickly.

Combined with independent, concurrent Air Force studies, Phase A confirmed that cross-range, the ability to maneuver off the orbital track during entry, was the major sticking point when trying to reconcile Air Force and NASA requirements. The Air Force wanted to be able to land at the launch site after only a single orbit during a mission that deployed or retrieved an American reconnaissance satellite (contrary to many reports, this was not to "steal" a Soviet satellite). During this time, the earth had rotated

Of all the art depicting rockets drawn before the Space Age, perhaps those by artist Chesley Bonestell (1888–1986) are the best known. Bonestell influenced at least two generations of scientists, engineers, and science fiction aficionados. (NASA)

approximately 1,100 nm, meaning the returning spacecraft had to fly at least that distance from its nominal flight path during entry. On the other hand, NASA simply wanted an opportunity to land back at the launch site once every twenty-four hours, requiring a relatively

The Convair T-18 triamese concept from 1965 was a major player in the early Air Force space shuttle studies. The three vehicles were aerodynamically largely identical, with the outer pair being boosters that flew back to the launch site after the center orbiter separated. (Convair)

During the NASA Phase A study, Lockheed conceived this two-stage concept. The booster was 237 feet long, about the same as a 747, and had a top speed of nearly 9,000 mph. The orbiter could carry a 15x60-foot payload and had a cross-range of 1,500 nm. (Lockheed)

Max Faget was a true believer in the blunt-body concept and described his orbiter as "falling" from orbit until it was well within the atmosphere and transitioning to horizontal flight. This concept did not thrill the Air Force and many others. (University of Houston Clear Lake Archives)

modest 265 nm cross-range, although various abort scenarios raised this to 450 nm in most studies. Still later, the need for more abort options increased the NASA requirement to 1,100 nm, conveniently coinciding with the Air Force desire. Given the need for continuing support from the national security establishment for funding, the convergence of requirements presented a more unified appearance.

Throughout the Phase A studies, there was considerable technical controversy within NASA and its contractors about the size and configuration of the orbiter. These studies, supported by more than 200 man-years of engineering effort backup up by thousands of hours of wind tunnel testing, materials evaluation, and structural design, resulted in four basic configurations. These included lifting-bodies, stowed-wing, straight-wing, and delta-wing concepts.

Despite a certain romance and a good public-relations campaign, engineers found the lifting body would make a poor space shuttle, mostly because the shape did not lend itself to efficient packaging and installation of a large payload bay, propellant tanks, and major subsystems. The complex double curvature of the body resulted in a vehicle that would be difficult to fabricate, and the airframe could

not easily be divided into subassemblies to simplify manufacture. In addition, its large base area yielded a relatively poor subsonic lift-to-drag (L/D) ratio, resulting in a limited cruise capability. Although lifting bodies continued to be studied for another year, the concept was a dark horse, at best. The stowed-wing designs, where a set of wings resided in the fuselage for most of the flight and deployed just before landing, had many attractive features, including a low burnout weight and the high hypersonic L/D needed to meet the Air Force cross-range requirements. In addition, the stowed-wing approach allowed the wing to be optimized for low-speed flight, providing better landing characteristics. Drawbacks included a high vehicle-weight-to-planform-area ratio that resulted in higher average base temperatures relative to either straight- or delta-wing designs. In addition, the structure and mechanisms needed to operate the wing resulted in significantly increased design and manufacturing complexity. The maintenance required between flights was expected to be high and insufficient data existed to reliably determine potential failure modes and effects.

Maxime Faget at MSC, arguably the father of the Mercury capsule, was not a believer in lifting reentry. Instead, Faget held to the idea of a high-drag blunt body that had shaped all of the capsules. In 1952, Julian Allen and Alfred Eggers, two researchers at the NACA Ames Aeronautical Laboratory, developed the blunt-body theory to support the early ballistic missile programs, notably Atlas.

As the range of the missiles increased from the several hundred miles of the German A4 (V-2) to the intercontinental distances of Atlas, engineers discovered the sleek, pointy warheads were burning up as they reentered the atmosphere. Allen and Eggers developed a theoretical solution after they deduced about half the heat generated by aerodynamic friction was being transferred to the warhead, quickly exceeding its structural limits. The obvious solution was to deflect the heat away from the warhead. In place of the traditional sleek missile with a sharply pointed nose, the researchers proposed a blunt shape with a rounded bottom. The blunt body, when reentering the atmosphere, created a powerful detached shock wave that deflected the airflow, and its associated heat, outward and away from the vehicle. As Allen and Eggers observed, "not only should pointed bodies be avoided, but the rounded nose should have as large a radius as possible." The blunt-body theory allowed the development of the first successful ICBM warheads, and Faget soon applied it to the design of the Mercury capsule.

Adapting the blunt-body theory to a winged vehicle, engineers at MSC penned a straight-wing spaceplane usually called the DC-3

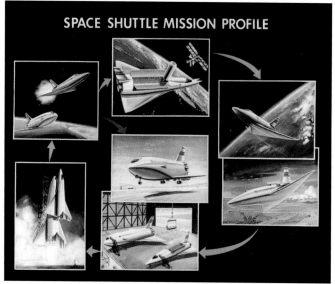

A typical mission profile for the fully reusable two-stage vehicles. The mated pair launched vertically and the orbiter separated from the booster at approximately Mach 10. The booster landed as the orbiter continued into space and ultimately back to the launch site. (NASA)

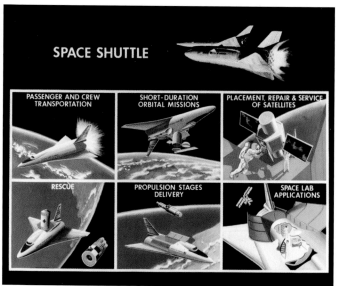

NASA envisioned a wide variety of missions for its new space shuttle, but the primary rationale for its existence was to build a space station. Nevertheless, as early as 1971 NASA foresaw the Spacelab concept, which allowed science experiments aboard the orbiter. (NASA)

(more officially, the MSC-001, the first in a series of about fifty designs carrying the MSC prefix). To operate the DC-3 as a blunt body, Max Faget proposed to enter the atmosphere at an extremely high angle of attack with the broad lower surface of the vehicle facing the direction of flight. This would create a shock wave that would carry most of the heat around the vehicle instead of into it. The vehicle would maintain this attitude until it got below 40,000 feet and about 200 mph, when the nose would come down and it began diving to pick up sufficient speed for level flight. The DC-3 would then head toward the landing site, touching down at a modest 140 knots. Since the only "flying" was at low speeds during the landing phase, the wing design could be optimized for subsonic cruise and landing; hence the simple straight wing proposed by Faget. The design did have one major failing, at least in the eyes of the Air Force: since it did not "fly" during entry, it had almost no cross-range. Max Faget convinced many within NASA that his simple straight-wing concept would be more than adequate. But others disagreed. In particular, Charles Cosenza and Alfred Draper at the Air Force Flight Dynamics Laboratory (AFFDL) did not accept the idea of building a space shuttle that would come in nose high, then dive to pick up flying speed. With its nose so high, the vehicle would be in a classic stall, and the Air Force, as well as most pilots, disliked both stalls and dives, regarding them as preludes to crashes. Draper preferred to have the vehicle enter a glide at hypersonic velocities (Mach 5 or above), thus maintaining much-better control while still avoiding much of the severe aerodynamic heating.

However, if the vehicle were going to glide across a broad Mach range, from hypersonic to subsonic, it would encounter another aerodynamic phenomenon; a shift in the center-of-lift. At supersonic speeds, the center of lift is located about midway along the wing chord (the distance from the leading to the trailing edge); at subsonic speeds, it moves much closer to the leading edge. Keeping an aircraft in balance requires aerodynamic forces that can compensate for this shift. Another MSC design, the Blue Goose, accomplished this in an extreme manner by translating the entire wing fore and aft as the center of lift changed. Nobody believed the idea was worth the mechanical complexity, not to mention the control and stability issues before the advent of workable fly-by-wire flight-control systems.

The development of supersonic combat aircraft had provided the Air Force with extensive experience regarding this phenomenon, and engineers had already determined a delta planform readily mitigated most of the problem. Al Draper proposed that any space

This McDonnell Douglas Phase B concept shows the immense size of the fully reusable two-stage concepts then under consideration. The booster weighed slightly less than 3,000,000 pounds, and staging took place at nearly 6,000 miles per hour. (National Archives)

shuttle should use delta-wings instead of straight ones. Max Faget disagreed, pointing out that his design did not "fly" at any speed other than low subsonic—at other speeds it "fell" and was not subject to center-of-lift changes since it was not using lift at all. This, of course, brought the discussion full-circle to stalls and crashes.

The Air Force researchers argued that delta-wings had other advantages. Since it was relatively thick where it joined the fuselage, a delta wing offered more room for landing-gear and other systems that could be moved out of the fuselage, allowing a larger payload bay. Its sharply swept leading edge produced less drag at supersonic speeds and its center-of-lift changed slowly compared to a straight wing. But the delta offered one other advantage, one that became increasingly important as the military became more interested in using space shuttle. Compared to a straight wing, a delta produces considerably more lift at hypersonic velocities, allowing a returning space shuttle to achieve substantially greater cross-range.

Payload size and weight were major considerations when sizing the vehicle, but the dimensions of the payload bay was uncertain

Initially, the Phase B study required each of the contractors to examine low-cross-range straight-wing concepts (left) and high-cross-range delta-wing vehicles (right). In each case, the vehicles were fully reusable with large flyback boosters that usually (but not always) shared the same wing planform as the orbiter. Various numbers of space shuttle main engines powered both stages, which were equipped with air-breathing engines for landings and ferry flights. These North American concepts had gross liftoff weights of more than 4,000,000 pounds. (NASA)

During the initial studies, engineers in Houston were not convinced the vehicles could successfully glide to landings on a concrete runway, so almost all of the designs used air-breathing engines of some sort. These were deployed from the payload bay just before landing. (NASA)

given conflicting requirements. The Air Force wanted to carry 40,000-pound payloads up to 60 feet long since that was the projected size of the next generation of reconnaissance satellites. NASA wanted to carry 15-foot-diameter payloads since that was the expected diameter of modules for some future space station. Usually not mentioned, NASA also had a few payloads, particularly the planned planetary probes, that could benefit from the 60-foot payload bay.

PHASE B

To help resolve these controversies, NASA baselined two concepts for the follow-on Phase B study, including a Faget-style straight-wing low-cross-range orbiter and a Draper-supported delta-wing high-cross-range orbiter. The straight-wing orbiter would be configured to provide design simplicity, minimal weight, decent handling, low cross-range, and good landing characteristics. The vehicle would enter at a high angle of attack to minimize heating and use of a heat shield fabricated from materials (ablators) available in the early 1970s. The delta-wing orbiter would provide

the capability to trim over a wide angle of attack range, allowing initial entry at a high angle of attack to minimize the severity of the heating environment, and then transition to a lower alpha during a hypersonic glide to achieve a high cross-range. Its more challenging heat shield became the subject of several intense research projects in academia and industry. The study contractors would investigate payload bays ranging from 25 to 65 feet long and 10 to 25 feet in diameter, carrying payloads between 15,000 and 65,000 pounds.

NASA issued two Phase B contracts on 6 July 1970, one to a team of McDonnell Douglas and Martin Marietta, and the other to North American Rockwell, who was joined by Convair as a risk-sharing subcontractor. This phase included a more detailed analysis of mission profiles and the vehicles needed to complete them. NASA entered Phase B with the goal of developing a vehicle to fulfill an unrealistically high flight rate that everybody had repeated so many times they were beginning to believe its validity.

Phase B resulted in some ambitious two-stage vehicles designed to meet the stated preference for a fully reusable space shuttle. All of the concepts would have been expensive and contained large development risks, even if the contractors were unwilling to admit it fully. In retrospect, it is questionable if any could have been built given the technology of the era. To put it in perspective, in 1968, Pete Knight managed to take the X-15A-2 research airplane to 4,520 mph—the fastest manned flight to-date (a feat not surpassed until *Columbia* flew back from STS-1). The X-15A-2 weighed about 50,000 pounds fully loaded. The Phase B boosters, which were supposed to fly twice as fast, would weigh more than 3,000,000 pounds. It did not bode well for the ambitious two-stage concepts.

Because of the high speeds during ascent and entry, both the booster and orbiter would need robust thermal-protection systems. But the entire issue of thermal protection was unsettled since it was not apparent that an adequate method of protecting either vehicle was at hand. The X-15 flights, and indeed the returning Apollo capsules, spent considerably less time in high thermal environments than anticipated for the orbiter—and neither the X-15 nor Apollo had a truly reusable thermal protection system. Charles Donlan, the NASA space shuttle program director at NASA Headquarters, later commented that all of the proposed reusable thermal protection systems had significant problems. For one, imagine covering an aircraft larger than a 747 with metallic shingles made of exotic superalloys that had previously only been used in small turbine blades or with ablators that had to be applied by hand after every flight.

Then there was the issue of separating two winged vehicles at 6,000 mph. The experience base consisted entirely of four launches

Lockheed had been suggesting using external propellant tanks since the original Starclipper study for the Air Force in 1965. These tanks were decidedly different than the ones ultimately used, wrapping around the sleek delta-wing orbiter to form a large vee. (NASA)

This influential concept from Grumman was one of the first in the later NASA studies to move the liquid hydrogen out of the orbiter and into external tanks, but it still used a large, fully reusable booster. At this point, the orbiter carried its liquid oxygen internally. (NASA)

during 1966 of a D-21 reconnaissance drone from a Lockheed M-21 Blackbird at 2,000 mph. Three of the launches were successful; the fourth had resulted in the loss of both vehicles and the death of one crew member. This brought up how to provide meaningful escape for the pilots of the booster during ascent, a problem Charlie Donlan did not believe was ever satisfactorily resolved. The development of the main engines was also problematic since they were meant to power both the booster and orbiter and, therefore, could not be optimized for either. They also needed to be, by weight, the most powerful engines ever developed.

Hindsight indicates the technical risks of developing a fully reusable two-stage space shuttle were tremendous, even if the money could be found—and that was in serious doubt. Funding issues were beginning to force NASA to pay greater attention to criticism of the fully reusable concept, but at the beginning of 1971, the ultimate configuration of the space shuttle and any possible expendable components remained an open question.

ALTERNATE SPACE SHUTTLE CONCEPTS

One of the drawbacks of fully reusable two-stage configurations studied during Phases A and B was their extremely high development cost. In a 1968 AIAA research paper, Charles Cosenza and Al Draper had argued that concepts using expendable external propellant tanks would allow more internal volume for payload and not require the development of vehicles large enough to carry their engines, propellant, and payload within a fully reusable airframe. This would significantly shrink the development costs in exchange for some incremental increase in operational costs. It was not a new idea, but perhaps it was time to pay more attention to it.

At the same time, the White House and Congress were telling NASA it would not continue to receive the high level of funding it had become accustomed to during the race to the moon. In response, concurrent with the issuance of the Phase B contracts, the agency initiated the Alternate Space Shuttle Concepts (ASSC) study to investigate various alternatives to the fully reusable Phase B concepts that potentially cost much less to develop, but somewhat more to operate. It also allowed NASA a fallback position if space shuttle funding was reduced, which was looking more and more likely.

MSFC awarded ASSC study contracts to the Chrysler Space Division and the Lockheed Missiles & Space Company, and, subsequently, MSC issued an ASSC contract to a team of Grumman and Boeing. The initial matrix of alternate concepts included "fractional" stages (stage and a half), partially reusable vehicles (expendable booster / reusable orbiter), and fully reusable two-stage alternatives similar to the Phase B vehicles. Chrysler, however, took "alternate" literally and did not feel bound by these choices. The company pursued an unconventional single-stage-to-orbit vehicle with an aerospike propulsion system that was different from any of the other competitors. NASA quickly dismissed the idea as too radical, although the effort continued for some time.

Ultimately, the ASSC study concluded that the fully reusable vehicles being investigated in Phase B were the "best" since they provided the lowest cost per flight and would have the lowest total program costs (nonrecurring development costs plus recurring operational expenses), at least if one believed the high flight rates depicted by the mission models. However, if NASA wanted to minimize peak yearly funding or development risk or fly fewer missions per year, the best option was a phased development approach using a reusable orbiter with an expendable booster, with the option of developing a recoverable booster later.

REALITY INTERVENES

On 17 May 1971, the Office of Management and Budget (OMB) told NASA that its budget would remain essentially constant for the next five years. This was the third major budget blow for the agency within eighteen months. In late 1969, the Bureau of the Budget (BoB, the predecessor to the OMB) had cut the NASA budget request by more than $500 million, forcing then administrator Thomas Paine to abandon all hopes of a manned mission to Mars and to concentrate on a space station and space shuttle. During the summer of 1970, the agency received more cuts, effectively canceling the space station. Now the space shuttle was in jeopardy.

The new budget guidance was a drastic blow because it meant the agency could not develop the fully reusable space shuttle it had been investigating for the previous two years. If limited to the $3,200 million budget approved for FY72, the most NASA could hope to put into space shuttle development and still maintain a balanced science and application program was roughly $1,000 million annually for five years. At the same time, an in-house analysis had led the OMB to conclude a fully reusable space shuttle was simply not cost competitive with the existing Titan III launch vehicle. All of this brought a new urgency to some of the partially expendable concepts examined during the ASSC studies.

In essence, the OMB funding left NASA with enough to develop an orbiter, but not the booster to go with it, so engineers began investigating replacing the reusable flyback booster with some sort of

A view of the North American orbiter on the ground. Note the four retractable air-breathing engines under the fuselage that were used during landing. The extremely simple maintenance facility and minimal ground crew were recurring features of the early concepts. (NASA)

Without a doubt, the most unusual design was this Chrysler concept submitted as part of the ASSC study. The Single-stage Earth-orbital Reusable Vehicle (SERV) used an innovative aerospike propulsion system arranged around the bottom outer edge. (Scott Lowther Collection)

One of the first missions envisioned for space shuttle was to boost *Skylab* into a higher orbit during early 1979. Unfortunately, development delays postponed the first flight of *Columbia* until well after 11 July 1979, when *Skylab* entered the atmosphere. (NASA)

expendable stage. This did not appear technically feasible, however, given the large size of the Phase B orbiters. NASA and its contractors began investigating orbiters with small payload bays, but this seemed to defeat the main rationale to build the vehicle in the first place.

The problem was resolved in June 1971, when NASA finally decided to endorse some variation the "external tank" concept advocated by Charles Cosenza and Al Draper at the AFFDL during 1968, and initially investigated in detail by both Lockheed and McDonnell Douglas for the Air Force as early as 1965. This concept moved the large liquid hydrogen tanks outside the orbiter airframe and made them expendable, allowing a much-smaller orbiter. This resulted in a significant reduction in development costs, but with a corollary increase in costs every time the orbiter was launched.

Engineers at Grumman had revisited this concept during the ASSC study, marking the first time that NASA had seriously sanctioned a partially reusable concept. Larry Mead, a Grumman vice president, vigorously pursued the external-tank concept by arguing it was the only way to meet the budget limitations. The economics, however, depended on what mission model one believed. Some mission models, essentially guesses at how many times a space shuttle would fly, showed one flight per week, while others showed only a couple dozen per year and a couple showed even fewer. Mead thought the low end of the estimates was more plausible, perhaps 15–25 missions annually. Obviously if part of the vehicle was not reusable, costs would increase as more missions were flown. But if only a relatively modest number of missions were flown, the development savings would make up for the increased per-mission cost. Ultimately, it would be economics, not technology, which convinced NASA to listen more closely to what the AFFDL and Grumman were telling it.

Initially, engineers only moved the liquid hydrogen into external tanks. Since the hydrogen tanks were large, moving them outside the airframe resulted in a much-smaller orbiter, significantly reducing its weight and the amount of thermal protection needed to protect it. Eventually, several studies also moved the liquid oxygen into external tanks, further reducing the size of the orbiter. Some designs used separate LO_2 and LH_2 tank; others carried all the propellants in a single large tank. Perhaps as significant as the changes to the orbiter itself, its smaller size allowed a much lower staging velocity, meaning a smaller booster that did not need to fly as fast, eliminating much, if not all, of its thermal protection. This allowed a smaller orbiter to still carry the 15x60-foot payloads that were so important to NASA and the Air Force, a seemingly adequate, if not ideal, solution.

However, the configuration of the booster was more in doubt. A few studies attempted to marry an external-tank orbiter with a smaller version of the flyback boosters studied during Phase B, but this still proved expensive. Other studies looked at using modified Saturn V stages to lift the orbiter, but this too was an expensive solution and was fraught with technical risk. This left a variety of liquid and solid "strap-on" boosters as possibilities. Economics were rapidly pushing the choice toward solid-propellant stages since they were much less expensive to develop, but the final decision would not come until after Richard Nixon approved the program.

Some engineers within NASA had been listening to the debates, and one of the in-house designs, the MSC-040, was a delta-wing orbiter with a 15x60-foot payload bay that relied on a single expendable external propellant tank and recoverable boosters. Most of the studies investigated liquid- and solid-propellant boosters in both throwaway and recoverable configurations. On 12 September 1971, both Phase B contractors, along with Grumman/Boeing and Lockheed, were told to reevaluate their studies using the MSC-040 orbiter and an external tank. The Chrysler aerospike concept was simply too far out of the mainstream and the company elected not to continue participating in the airframe competition.

An early Phase C concept from North American. Note the reaction control system pods on the top of the vertical stabilizer and wingtips. The air-breathing engines were in the back of the payload bay, and there was a docking airlock in the nose. (NASA)

Easy maintenance was a selling point of all the space shuttle proposals. This is a 1975 illustration showing what became the Orbiter Processing Facility, albeit missing the large maintenance stands and equipment actually used to service the orbiter between missions. (NASA)

Officially termed the Phase B Extension (even for the ASSC contractors), this was usually called Phase B Prime and evolved into Phase B Double Prime as the studies attempted to find an economical compromise. This represented a major change from the original approach of reducing the number of contractors as the phases progressed; four contractors would now compete on a theoretically equal basis for the Phase C development contract.

PHASE C/D

It appears to be the fate of winged spacecraft that their development is shrouded in controversy and political maneuvering. So it had been for Dyna-Soar and so it was with space shuttle. There were significant battles between the various NASA field centers, between the Air Force and NASA, and between everybody and the OMB. Nevertheless, when the dust finally settled, on 5 January 1972, President Richard Nixon approved the development of space shuttle, albeit with some significant funding limitations.

To save time and treasure, NASA opted to combine Phases C and D. On 15 March, the agency announced the decision to use recoverable solids instead of liquid boosters. NASA administrator James Fletcher explained the solids could be developed quicker and for $700 million less than equivalent liquid boosters, bringing the total development costs within the $5,150 million (FY71 dollars) ceiling imposed by the OMB and Richard Nixon. The change increased the cost per mission, but "not enough to adversely affect the economical use of the shuttle." The decision had apparently been made well in advance of this announcement, since the Phase C/D request for proposals, which would be released two days later, already contained requirements to use solid rocket boosters. The agency released the request for proposals on 17 March 1972 to Grumman, Lockheed, McDonnell Douglas, and North American Rockwell for the production and initial operations of the space shuttle system. The companies had sixty days to respond.

The statement of work required each orbiter have a useful life of ten years and be capable of 500 missions, but asked each contractor to provide information on lowering this to only 100 missions, a figure that was subsequently adopted. There were three reference missions: (1) 65,000 pounds into a 100 nm due-east orbit from the Kennedy Space Center (KSC) in Florida, (2) 25,000 pounds into a 270 nm 55-degree orbit from KSC while carrying a set of air-breathing engines, and (3) 40,000 pounds into a 100 nm polar orbit from Vandenberg AFB in California. The first and last missions excluded the use of air-breathing engines that were intended to provide more landing options at the end of the mission.

Numerous space shuttle models were tested in a variety of wind and arc-jet tunnels around the country. This shows how the shockwaves form around the orbiter at a 40-degree angle of attack, carrying much of the heat of entry away from the vehicle. (NASA)

The payload bay was to have a clear volume 15 feet in diameter and 60 feet long. This allowed the orbiter to carry all of the foreseeable national security payloads, the approved planetary probes and large space telescopes, and the expected modules for some future space station. The vehicle also needed to be able to carry 45,000 pounds back to Earth, allowing satellites and other high-value objects to be returned for servicing and refurbishment.

The crew cabin needed to accommodate four astronauts and support them for up to a week on-orbit. An additional six astronauts were to be accommodated for shorter periods as needed. Maximum acceleration during ascent or entry was to be limited to 3-g, a much-gentler ride than any of the earlier capsules. The vehicle was to be capable of being held in a standby status for up to 24 hours and launched within two hours from that condition.

Two solid rocket boosters (SRB) were ignited on the ground in parallel with three space shuttle main engines (SSME). The SRBs were to include a thrust termination system, and the vehicle was to be capable of intact aborts (safely landing the orbiter with the crew) even while the SRBs were thrusting. The SRBs were to be designed

Desktop computer terminals, and computer-aided design, were just becoming a reality as space shuttle was being designed. Small parts of the orbiter were done in CAD software, but the majority of the vehicle was designed with slide rules and drawn on vellum. (NASA)

Thousands of hours of wind tunnel time was used to understand the aerodynamics of the space shuttle stack. This highly detailed model is being prepared for engine-out testing in the 10x10-foot supersonic wind tunnel at the NASA Langley Research Center. (NASA)

for water recovery, refurbishment, and subsequent reuse. The only expendable element was the external tank (ET) that held the liquid oxygen and liquid hydrogen propellants for the SSMEs.

Unsurprisingly, the four contractors proposed vehicles that looked remarkably similar, since all were based on the MSC-040C concept. Grumman and North American designed their orbiters for the baseline 500-mission service life; Lockheed and McDonnell Douglas opted for the 100-mission alternate. Given this was likely the last major manned space effort for the remainder of the twentieth century, it was an intense competition.

CONTRACT AWARD

NASA administrator James Fletcher, deputy administrator George Low, and associate administrator for organization and management Richard McCurdy met the morning of 26 July 1972 for the final review of the space shuttle proposals. The source evaluation board had ranked Lockheed and McDonnell Douglas significantly below the other two, so they concentrated on reviewing the Grumman and North American results. After reviewing the mission suitability scores, the three men determined the advantage went, slightly, to North American. Since North American also presented the lowest probable cost, they deemed the company the winner.

After the stock market closed that afternoon, NASA announced it had awarded a $2,600 million ($18,500 million in 2016) contract to the Space Transportation Systems Division of North American Rockwell to design the orbiter and operate the space shuttle system. The contract included a main propulsion test article (MPTA-098), a full-scale structural test article (STA-099), and two operational orbiters (OV-101 and OV-102).

Previously, on 13 July 1971 NASA had awarded a contract to the Rocketdyne Division of North American Rockwell to develop and manufacture the space shuttle main engine, although a protest by Pratt & Whitney delayed the signing until 14 August 1972. The external-tank competitors included Chrysler, McDonnell Douglas, and Martin Marietta, with NASA selecting the latter on 16 August 1973. Interestingly, at the direction of NASA Headquarters, MSFC split the solid rocket booster procurement into two parts. The solid rocket motor contract went to Thiokol on 20 November 1973 over competitors Aerojet General, Lockheed Propulsion, and United Technology, but again a protest delayed the signing until 24 June 1974. A new company, United Space Boosters, Inc. (USBI), would provide the nonmotor parts of the solid rocket booster as well as operational support. USBI subcontracted much of the actual hardware manufacturing to McDonnell Douglas.

North American constructed a full-scale mockup of the orbiter at their Downey facility. Normally, the mockup was housed inside, but it apparently was moved outside at least once, as shown here. (NASA)

For the first two years of the flight campaign, Rockwell handled most operational aspects as part of the original space shuttle systems contract while the other three major contractors (Martin Marietta, Rocketdyne, and USBI) supported their specific flight elements. In January 1984, NASA and the Air Force consolidated all ground operations at KSC and Vandenberg into the shuttle-processing contract (SPC) awarded to the Lockheed Space Operations Company (LSOC). Similarly, NASA consolidated all flight planning and mission operations at JSC into the space transportation system operations contract (STSOC) awarded to the Rockwell Shuttle Operations Company (RSOC).

In August 1995, NASA decided to further consolidate the space shuttle contracts under a single prime contractor. Although originally to be procured as a competitive bid, in the end NASA awarded the space flight operations contract (SFOC) as a sole source to a joint venture of Lockheed and Rockwell called United Space Alliance (USA). In April 1996, USA assumed management responsibility for both the SPC and STSOC efforts, establishing the company as the prime contractor for the Space Shuttle Program. Other contracts, such as the USBI effort at MSFC and the former-IBM flight software effort at JSC, were subsequently brought into the SFOC contract. Although the initial intent was to consolidate all space shuttle contracts into the SFOC, this never happened and the Lockheed Martin (ET) and ATK (SRM) contracts (and several others) remained intact through the end of the flight campaign. When the SFOC contract expired in 2006, NASA awarded the non-competitive space processing operations contract (SPOC) to USA, and this effort ran through the end of the program in 2013.

LAUNCH AND LANDING SITES

Initially, NASA and the Air Force intended to launch equatorial-orbit missions from two launch pads (LC-39A and LC-39B) at the Kennedy Space Center in Florida and polar-orbit missions from a space launch complex (SLC-6) at Vandenberg AFB, California. Technical concerns delayed the Vandenberg complex, and waning Air Force interest after the *Challenger* accident caused the complex to be first mothballed and then canceled.

For most of the flight campaign, the preferred end-of-mission landing site was the Shuttle Landing Facility (SLF) at the Kennedy Space Center, although early concerns about the orbiter brakes and tires caused many pre-*Challenger* missions to land on the dry lakebeds at Edwards AFB in California. Bad weather in Florida often caused the program to land at Edwards and bad weather at both sites once caused a single diversion to a gypsum runway at Northrup Strip in New Mexico (later called White Sands Space Harbor).

There were eighty-two launches from LC-39A, fifty-three from LC-39B, and none from SLC-6. As defined by NASA, 101 of these were day launches and thirty-four were night launches (within fifteen minutes of nautical sunrise or sunset). Of the 133 landings, seventy-eight were at the Shuttle Landing Facility, fifty-four were at Edwards AFB, and one was at White Sands. Again, as defined by NASA, twenty of the KSC landings were at night as were six at Edwards. Twenty-six of the landings were at a location other than originally intended. Two flights never landed.

NASA reports the orbiters flew 542,398,878 miles during 21,068 orbits, although this is an imprecise measurement based on a standard orbit without factoring in maneuvering, altitude changes, etc. The shortest mission was STS-2 at two days, six hours, thirteen minutes, two seconds while the longest was STS-80 at seventeen days, fifteen hours, fifty-three minutes, seven seconds; interestingly, both were on *Columbia*.

At approximately 3,500,000 pounds, the amount of payload carried by space shuttle accounts for more than half the cargo launched since the dawn of the space age. The orbiters also returned roughly 250,000 pounds of material from orbit, particularly from the International Space Station, marking essentially all of the downmass returned. It was an enviable record.

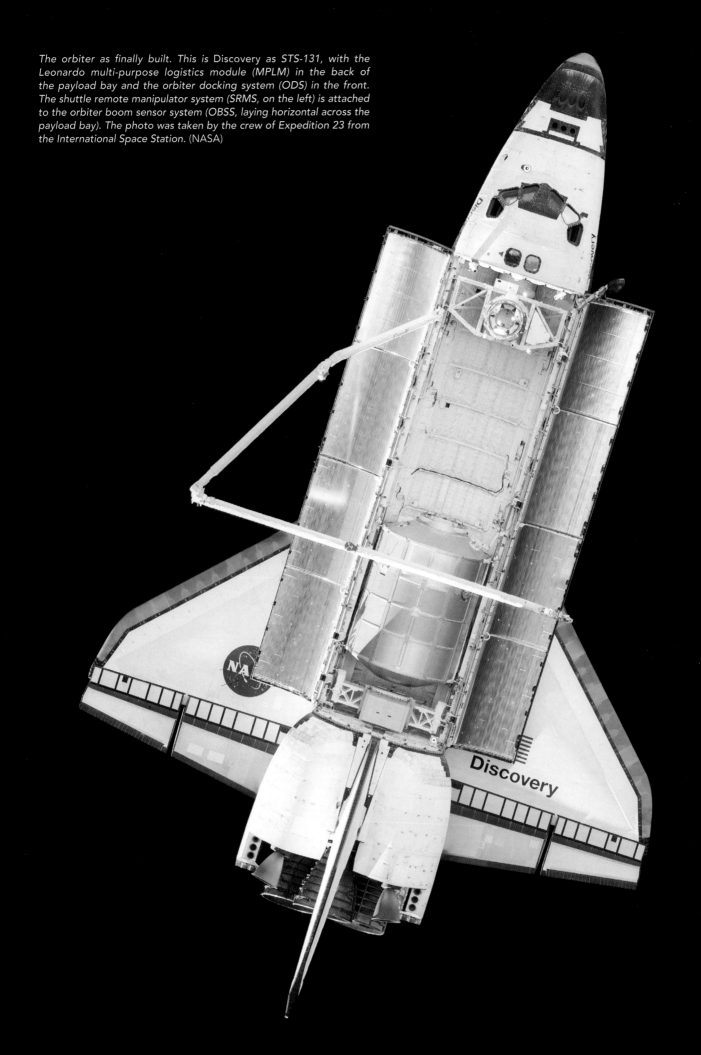

The orbiter as finally built. This is Discovery as STS-131, with the Leonardo multi-purpose logistics module (MPLM) in the back of the payload bay and the orbiter docking system (ODS) in the front. The shuttle remote manipulator system (SRMS, on the left) is attached to the orbiter boom sensor system (OBSS, laying horizontal across the payload bay). The photo was taken by the crew of Expedition 23 from the International Space Station. (NASA)

VEHICLE DESCRIPTION

The space shuttle system consisted of four primary flight elements: the orbiter, an external tank (ET), two solid rocket boosters (SRB), and three space shuttle main engines (SSME). Of these, only the external tank was not recovered and reused.

Rockwell manufactured six orbiters along with several partial airframes for various test regimens and a seventh partial set of spares. The orbiter was 122.17 feet long, had a 78.06-foot wingspan, and was 56.58 feet high. The double-delta wing provided 2,690 square feet of area. The orbiter was constructed primarily of conventional aluminum alloys, with limited use of advanced composites. The orbiter had an empty weight of approximately 150,000 pounds and each SSME weighed about 7,480 pounds, for an empty mission weight of approximately 175,000 pounds. The payload bay could accommodate payloads 15 feet in diameter and 60 feet long, weighing up to 65,000 pounds depending on the altitude and inclination of the desired orbit. Rockwell International Corporation assembled the orbiters at Air Force Plant 42 in Palmdale, California.

Each orbiter was required to be capable of 100 flights, but this was a theoretical limit and no orbiter approached that number. For most of their careers, each orbiter underwent a major overhaul (called an orbiter maintenance and down period, OMDP, or an orbiter major modification, OMM) every eight missions or three years. No major structural issues were identified during any of these overhauls, and it was possible the orbiters had a nearly indefinite structural life given the extensive maintenance philosophy used.

The forward fuselage was of conventional aircraft manufacture and contained the crew module. It was manufactured in upper and lower sections to allow the crew module to be inserted during final assembly. Rockwell manufactured the forward fuselage and forward reaction control system (FRCS) module in Downey, California.

The pressurized crew module contained three levels with a volume of 2,325 cubic feet if the internal airlock was installed and 2,625 otherwise. The upper level was the flight deck and provided normal seating for up to four crew members. The forward flight deck, which included the commander and pilot seats, occupied approximately 24 square feet, while the aft flight deck had approximately 40 square feet of usable area. The middeck was located directly beneath the flight deck and was accessible through two 26x28-inch openings. The middeck contained passenger seating, living accommodations (potty, galley, sleeping), avionics equipment, and storage. Up to three passengers could be accommodated on the middeck except on *Columbia*, which accommodated four. The completely stripped middeck was approximately 160 square feet while the gross mobility area was nominally 100 square feet. The lower deck contained environmental control equipment and was accessible through removable floor panels on the middeck. The crew module was pressurized to 14.7±0.2 psia to provide a "shirtsleeve" environment for the crew.

A hatch on the port side of the middeck was used for normal crew ingress/egress and could be operated from within the crew module or externally. Modifications made after the *Challenger* accident allowed the hatch to be explosively jettisoned in an emergency. The side hatch was 40 inches in diameter, had a small window in its center, and weighed 294 pounds.

Two different airlocks were used during the flight campaign. Originally, each orbiter except *Enterprise* was equipped with an internal airlock at the rear of the middeck. This airlock had a hatch into the crew module and another into the payload bay and supported EVAs while preserving the maximum payload bay volume. The internal airlock had an inside diameter of 63 inches, was 83 inches long, and had an interior volume of approximately 150 cubic feet. The airlock was equipped with two 40x36-inch D-shaped hatches.

To facilitate docking with Mir and the ISS, NASA replaced the internal airlock with a 185-cubic-foot external airlock in the forward payload bay as part of the orbiter docking system (ODS). This airlock had three hatches—one into the crew module, one aft

Discovery under construction at the North American Rockwell facility in Palmdale. The lower forward fuselage has been mated to the mid-fuselage and wings. Note the conventional aluminum aircraft-type construction and the green Koropon® primer. (NASA)

This is Discovery after the crew module and upper forward fuselage had been installed. The pressurized crew module was a separate structure that floated inside the forward fuselage to isolate it from vibration and thermal effects. Note the sign at right. (NASA)

into the payload bay, and one upward for docking. The ODS was 15 feet wide, 6.5 feet long, 13.5 feet high and weighed 4,016 pounds. *Columbia* retained her internal airlock until she was lost, but the other three flight orbiters used external airlocks to support ISS operations.

Normally, two extravehicular mobility unit (EMU) spacesuits were stowed in the airlock, although the internal airlock could accommodate up to four. The EMU was an integrated spacesuit and life support system that allowed crew members to leave the pressurized crew compartment and work near the orbiter. The airlock was sized to accommodate two suited crew members at the same time, although the first flight of *Endeavour* (STS-49) proved that three suited astronauts could fit in the airlock during a final attempt to rescue the Intelsat VI communications satellite.

During the Approach and Landing Tests (ALT), *Enterprise* was equipped with modified Lockheed SR-71 zero-zero ejection seats for the two pilot positions, and no other seating was provided. During the orbital flight tests (OFT, STS-1 through STS-4), *Columbia* was also equipped with the same ejection seats for use on the launch pad and during landing. The ejection seats were disabled after STS-4 and removed at Palmdale following STS-9. No crew escape system was provided during STS-5 through STS-33/51L.

After the loss of *Challenger*, NASA reexamined the issue of crew escape systems and, after again rejecting various proposals for ejection seats, escape pods, etc., selected a system that was useful only below 200 knots and 30,000 feet during controlled gliding flight. A 9.8-foot-long telescopic pole extended from the orbiter side hatch, and crew members would slide down the pole by using special attachments on the parachute harness. The escape pole curved sharply downward and slightly aft, directing the crew members under the left wing. Tests showed a crew of eight could evacuate the orbiter in approximately 90 seconds. It was more about public perception than crew safety.

The crews of the first four missions used David Clark Company S1030A ejection escape suits (EES) to satisfy the OFT emergency-egress requirements up to Mach 2.7 and 80,000 feet. The S1030A was a derivative of the Air Force S1030 pilot's protective assembly (PPA) worn by SR-71 crew members, and included an integrated g-suit. When the ejection seats were disarmed after STS-4, ascent was considered a "shirtsleeve" environment, and no protective suits were worn, although all crew members wore a clamshell Gentex launch-entry helmet (LEH) to provide supplemental oxygen in the event of a cabin depressurization during ascent or entry below 50,000 feet.

NASA abandoned the shirtsleeve concept after the *Challenger* accident and initiated an effort to develop a pressure suit to provide hypobaric protection during ascent and entry, as well as cold-water immersion protection in the event of a bailout over water. The David Clark Company developed a partial-pressure suit, using as many already proven concepts as possible. The resulting S1032 launch-entry suit (LES) was ready in time for return to flight of STS-26R.

During 1990, the David Clark Company developed the S1035 advanced crew escape suit (ACES), which drew heavily from the Air Force S1034 pilot's protective assembly used on the SR-71 and U-2R. The S1035 was a full-pressure suit that first flew on STS-64. Because it took a while for David Clark to manufacture the ACES suits in a complete range of sizes, LES suits continued to be used as needed through STS-98. Beginning with STS-102, all crew members wore ACES suits through the end of the flight campaign.

More than 2,100 displays and controls were located on the flight deck instrument panels, and many additional functions were controlled through keypads and CRTs that interface with the five onboard general-purpose computers (GPC). Dual heads-up displays (HUD) were included on *Challenger* and subsequent orbiters, and were retrofitted to *Columbia* after STS-9. Rotational hand controllers (RHC) controlled vehicle rotation about all three axes. During ascent, the controllers could be, but never were, used to gimbal the SSMEs and SRBs. The RHCs were used to gimbal the OMS engines and to command thrusting of the RCS during orbital insertion and deorbit. On-orbit, the controllers were used to command the RCS jets. During entry, the hand controllers provided normal flight control inputs, commanding either the RCS jets or aerodynamic flight controls as appropriate. Normal rudder pedals controlled the rudder during atmospheric flight and the nose wheel steering system and main wheel brakes during ground operations.

As originally delivered, and flown for almost twenty years, the orbiters were equipped with the multifunction CRT display system (MCDS). This used three 5x7-inch green monochrome CRTs and a large selection of conventional round and tape-style flight instrumentation. In 1992, Rockwell developed a "glass cockpit" called the multifunction electronic display system (MEDS) that used color flat-panel displays measuring 6.71 inches square. Although seemingly obsolete by personal computer standards, MEDS was powered by 16 MHz 80386DX processors.

The mid-fuselage contained the payload bay and connected with the forward fuselage, aft-fuselage, wings, and payload bay doors. Convair manufactured the mid-fuselage in San Diego, California. The assembly was 60 feet long, 17 feet wide, 13 feet high and weighed 13,502 pounds. It contained payload-mounting provisions along each longeron sill and the centerline keel. Each longeron sill could also support a shuttle remote manipulator system (SRMS) arm.

The internal airlock being removed from Discovery during 1995. Note the large opening in the forward payload bay bulkhead, where the airlock was originally mounted. Removing the internal airlock freed up a considerable amount of volume in the middeck. (NASA)

The internal airlock was replaced by an external airlock and orbiter docking system (ODS) that sat in the front of the payload bay and enabled docking with the Russian Mir space station and the International Space Station. This unit is being installed in Discovery. (NASA)

Each payload bay door was made up of five segments that were interconnected by circumferential expansion joints and supported by thirteen external hinges attached to the midfuselage longeron sill. Thermal seals on the doors provided a relatively airtight payload compartment, although the payload bay was not pressurized and leaked water when the orbiter was exposed to rain. The right door had to be opened first and closed last because of the arrangement of the centerline latch mechanisms and seal overlaps. Rockwell manufactured the lightweight composite doors in Tulsa, Oklahoma.

The shuttle remote manipulator system used an arm that was 50.25 feet long 15 inches in diameter and weighed 905 pounds. The "Canadarm" had six joints that corresponded roughly to those in the human arm, with shoulder yaw and pitch joints, an elbow pitch joint, and wrist pitch, yaw and roll joints. The two-piece boom was made of reinforced graphite-epoxy and weighed 93 pounds. The SRMS was carried on the left payload bay longeron sill on many (but not all) missions. Structural provisions and controls existed to carry a second SRMS on the right longeron sill, but this capability was never used.

After the *Columbia* accident, NASA developed the orbiter boom sensor system (OBSS) to allow detailed inspection of the entire orbiter thermal-protection system. This 50-foot boom was carried on the right payload bay longeron sill beginning on STS-114, using the mounting locations that had been provided for the second SRMS arm. The OBSS doubled the SRMS length to 100 feet and used an instrumentation package of cameras and lasers to scan the leading edges of the wings, nose cap, and crew compartment for damage. The boom was essentially identical to the SRMS arm, except the articulating joints were fixed. A trio of booms were assembled relatively quickly, largely using spare SRMS parts.

The aft fuselage provided the primary support for the main propulsion system, orbital maneuvering system (OMS) pods, and vertical stabilizer. Rockwell fabricated the 18-foot-long, 22-foot-wide, and 20-foot-high structure in Downey. The internal thrust structure that supported the three SSMEs was composed of twenty-eight machined and diffusion-bonded titanium truss members in *Challenger* through *Atlantis*, and of built-up titanium forgings in *Endeavour*. In diffusion bonding, titanium strips are bonded together under heat and pressure that fuses the strips into a single hollow, homogeneous mass that is lighter and stronger than a forged part.

After the *Challenger* accident, a drag chute compartment was added to the upper thrust structure at the base of the vertical stabilizer. The 40-foot-diameter drag chute was a production feature in *Endeavour* and was subsequently incorporated in *Columbia*, *Discovery*, and *Atlantis*. It was one of the few outer moldline changes made to the orbiters during the flight campaign.

The original multifunction CRT display system (MCDS) instrumentation on the flight deck of Columbia while under construction in Palmdale. Most of the flight instruments were conventional round dials or tape-style indicators, with three green-phosphorous stroke-written CRTs connected to the general-purpose computers. Note the ejection seats and lack of heads-up displays, which had not yet been added to the design. (NASA)

The blended double-delta wings used a modified NACA-0010 section with an 81-degree sweep on the inner leading edge and 45 degrees on the outer. The wing trailing edge had a 3° 31' dihedral. Each of the main wing assemblies were 60 feet long and had a maximum thickness of approximately 5 feet at the fuselage. Grumman manufactured the main wing assemblies in Bethpage, Long Island, New York and the completed wings were shipped using a commercial freighter via the Panama Canal to the Port of Long Beach, California and then transported overland to Palmdale.

The vertical stabilizer had a leading-edge sweep of 45 degrees with a trailing-edge rudder that split into two to serve as a speedbrake. This also provided the classic hypersonic wedge shape pioneered on the X-15 research airplane. Fairchild Republic manufactured the vertical stabilizers for the original orbiters in Farmingdale, New York. However, after the demise of that company, Grumman manufactured the vertical stabilizer for *Endeavour* in nearby Bethpage.

The orbital maneuvering system (OMS) provided thrust for orbital insertion, orbit circularization, orbit transfer, rendezvous, and deorbit. The system consisted of two pods, one on each upper aft-fuselage shoulder, flanking the vertical stabilizer. Each OMS pod was 21.8 feet long, 11.37 feet wide at its aft end, and 8.41 feet wide at the forward end, with a surface area of approximately 435 square feet. Each pod could carry 7,773 pounds of nitrogen tetroxide (N_2O_4)

oxidizer and 4,718 pounds of monomethyl hydrazine (MMH) fuel, the same propellants used by the reaction control system. Each OMS engine produced 6,000 lbf in a vacuum. The OMS pods also contained the aft reaction control system. McDonnell Douglas Astronautics fabricated the OMS pods and propellant tanks while Aerojet-General manufactured the OMS engines.

The reaction control system (RCS) had thirty-eight primary jets and six vernier jets to provide attitude control and three-axis translation during the orbit insertion, on-orbit, and entry phases of flight. The RCS consisted of three propulsion modules, one in the forward fuselage, and one in each OMS pod. The forward RCS module contained fourteen primary and two vernier jets while each of the aft RCS modules had twelve primary and two vernier jets.

Electrical power was generated by the three fuel cells using cryogenic oxygen and hydrogen reactants. The combined fuel cell system provided 14 kW continuous and 24 kW peak at 28 Vdc. The fuel cells were located under the forward portion of the payload bay. The only byproduct of operating the fuel cells was water. During missions to the ISS, this water was transferred to the station and provided its primary water source. Since each gallon of water weighs 8.35 pounds, it is expensive to transport from Earth, and using the fuel cell byproduct saved considerable transportation costs. On non-ISS missions, the crew dumped the water overboard at regular intervals.

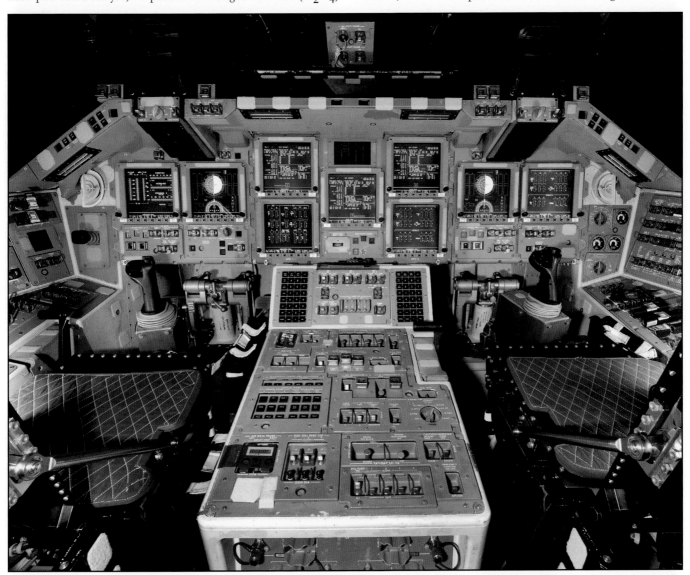

The multifunction electronic display system (MEDS) added a "glass cockpit" to the orbiters, using LCD flat-panels based on the units used in the Boeing 777. Note the HUDs on the top of the instrument panel cowl and the conventional rudder pedals. Most of the switch panels were unchanged from the MCDS configuration. This photo was taken in one of the mission simulators in Houston but is identical to how the orbiters looked. (NASA)

To better support Spacelab science missions, Rockwell developed the extended-duration orbiter (EDO) cryo kit. A 3,500-pound, 15-foot-diameter pallet provided a support structure for tanks that stored 3,124 pounds of liquid oxygen and 368 pounds of liquid hydrogen. When filled with cryogens, the pallet weighed about 7,000 pounds. The EDO pallet attached vertically at the back of the payload bay on *Columbia* and *Endeavour*. The additional reactants allowed the orbiter to stay on-orbit up to sixteen days.

Three Sundstrand auxiliary power units (APU) and three 3,000 psi hydraulic pumps in the aft fuselage provided power for the triple-redundant hydraulic systems. Fuel tanks had a maximum capacity of 350 gallons of hydrazine (N_2H_4). The three APU exhaust ports were located on the top of the aft-fuselage, beside the leading edge of the vertical stabilizer.

The thermal-protection system (TPS) protected the aluminum orbiter, maintaining skin temperatures less than 350°F. The TPS also established the outer moldline of the orbiter and was capable of withstanding all aero-loads.

Reinforced carbon-carbon (RCC) was used on the wing leading edges, nose cap, immediately aft of the nose cap on the lower surface (chin panel), and the immediate area around the forward orbiter-ET structural attachment. RCC protected areas where temperatures exceeded 2,300°F during entry. By itself, RCC was not truly an insulative material, but rather a high-temperature material used to preserve the outer moldline and cover other insulation materials.

Black high-temperature reusable surface insulation (HRSI) tiles were used where temperatures were between 1,200 and 2,300°F. This included the entire underside of the vehicle, base heat shield, around the forward fuselage windows, portions of the OMS/RCS pods, leading and trailing edges of the vertical stabilizer, wing glove areas, elevon trailing edges, adjacent to the RCC on the upper wing surface, behind the RCC on the nose and wing leading edge, the upper body flap surface, and around the SILTS pod on *Columbia*. These tiles had a black reaction-cured glass (RCG) for thermal emittance.

Black fibrous refractory composite insulation (FRCI) tiles were used instead of HRSI in selected areas of the orbiter. Toughened unipiece fibrous insulation (TUFI) began replacing HRSI tiles on the base heat shield around the SSMEs during 1993. The black-colored FRCI and TUFI tiles were more resistant to damage than the HRSI tiles but never gained widespread use.

White low-temperature reusable surface insulation (LRSI) tiles were initially used on the forward fuselage, midfuselage, aft-fuselage, vertical stabilizer, upper wing, and OMS/RCS pods. These tiles protected areas where temperatures were below 1,200°F and had a white surface coating to provide better thermal characteristics while on-orbit. As the flight campaign progressed, most LRSI tiles were replaced by AFRSI blankets.

White advanced, flexible, reusable surface insulation (AFRSI) blankets offered improved producibility and durability, reduced fabrication and installation costs, and a weight reduction over LRSI tiles. The AFRSI blankets protected areas where temperatures were below 1,200°F and aerodynamic flight loads were minimal. Late in the flight campaign, NASA started calling these flexible insulation blankets (FIB) instead of AFRSI.

White-coated Nomex felt reusable surface insulation (FRSI) was used on the upper payload bay doors, portions of the mid-fuselage and aft-fuselage sides, portions of the upper wing surface, and parts of the OMS/RCS pods. The FRSI protected areas where temperatures were below 700°F.

Three Rocketdyne RS-25 space shuttle main engines (SSME) were located in the extreme rear of the orbiter. Each main engine consumed 889 pounds of LO_2 and 146 pounds of LH_2 per second at 100 percent power while developing 375,000 lbf at sea level and 470,000 lbf in a vacuum. Each engine had a rated specific impulse of 452 seconds. Almost all missions used a 104 percent power level that corresponded to 393,800 lbf at sea level and 488,800 lbf in a vacuum, consuming 933 pounds of liquid oxygen and 155 pounds of liquid hydrogen per second. Each engine was throttleable in one percent increments from 67 (originally 65) to 104 percent. The main engines were 13.9 feet long 7.8 feet in diameter at the nozzle exit, and weighed approximately 7,480 pounds.

By the end of 1982, the amount of maintenance required between each flight and a series of ground failures during testing intended to certify the engines at 109 percent power led NASA to initiate a program to improve the SSME. There were three goals: Phase I would decrease the maintenance required between flights, Phase II would provide the 109 percent power level required for Vandenberg launches, and Phase III would provide long-term margin for flight at 109 percent.

After the *Challenger* accident, NASA canceled the 109 percent certification effort, but the other improvements developed during the effort were incorporated into so-called Phase II flight engines. Nevertheless, continuing issues with the SSME turbopumps led NASA, in August 1986, to contract with Pratt & Whitney to develop and produce forty-four alternate turbopumps that would be supplied to Rocketdyne as government-furnished equipment. The alternate

Space shuttle technician Tim Keyser prepares to re-install one of the three fuel cells in Discovery after the end of the flight campaign. The fuel cells were removed and drained of all fluids in preparation for the orbiter's transfer to the National Air and Space Museum. (NASA)

Technicians install a new auxiliary power unit in the aft compartment of Discovery at LC-39B during STS-31. The first launch attempt on 10 April 1990 was scrubbed because APU-1 did not start correctly; the second attempt on 24 April was successful. (NASA)

high-pressure oxidizer turbopumps (HPOT), a two-duct powerhead, and a single-coil heat exchanger would be incorporated into a new configuration known as the Block I engine. A follow-on Block II configuration would add the alternate high-pressure fuel turbopump (HPFT) and a large-throat main combustion chamber (LTMCC).

The Block I configuration first flew in the center position on STS-70, with Phase II engines occupying the other two positions. STS-77 was the first flight to use three Block I engines. The Block I engines provided essentially the same power as the Phase II engines and were slightly heavier but offered a significant decrease in the risk of losing a vehicle during ascent. With three Phase II engines installed, the computed loss-of-vehicle probability was 1 in 262 flights; three Block I engines reduced this to 1 in 335 flights.

Continued difficulties in certifying the Pratt & Whitney high-pressure fuel turbopump delayed the introduction of the Block II engine. But the tests of the LTMCC proved so successful at lowering risk during ascent that NASA decided not to wait until the alternate high-pressure fuel turbopump was ready. An interim configuration, dubbed Block IIA, included all the improvements originally intended for Block II except the alternate high-pressure fuel turbopump. Less than a year passed from the decision to create the interim Block IIA configuration to its first flight.

Rocketdyne assembled fourteen Block IIA engines, using components from earlier engines. The Block IIA first flew on STS-89, with all three engines being the new design. The Block IIA engine further reduced the ascent risk of the vehicle, to 1 in 438. The full-up Block II engine incorporated all of the Block IIA improvements plus the alternate HPFT and reduced the ascent risk to 1 in 483. Single Block II engines flew on STS-104 and STS-108, while STS-110 was the first to use three Block II engines.

The external tank (ET) contained the liquid oxygen and liquid hydrogen propellants for the space shuttle main engines. Martin Marietta (later, Lockheed Martin Space Systems Company) manufactured the tanks at the Michoud Assembly Facility (MAF) near New Orleans, Louisiana. In all, they manufactured 139 tanks, including three test articles (ET-MPTA, ET-GVTA, and ET-STA) that were used during various test efforts. At the conclusion of the flight campaign, a single unflown external tank remained at Michoud (ET-94 had been used during the *Columbia* accident investigation), as well as parts of the uncompleted ET-139 and ET-140.

Each ET was 153.8 feet long 27.58 feet in diameter and consisted of a forward LO$_2$ tank, an unpressurized intertank, and an aft LH$_2$ tank. Three different types of external tanks were manufactured. The first six flight tanks were called standard-weight tanks (SWT) and varied in weight from 77,100 pounds to 74,800 pounds. However, the Space Shuttle Program needed to increase the performance of the vehicle, particularly for polar-orbit launches from Vandenberg. Since the ET was carried almost all of the way to orbit, every pound that could be removed from its empty weight translated into almost a pound of payload. The next production batch of ETs, called lightweight tanks (LWT), weighed some 6,000 pounds less than the standard-weight tanks.

However, the decision to build the International Space Station in a 51.6-degree orbital inclination cost space shuttle 13,500 pounds of payload capability. Since NASA had already canceled the advanced solid rocket motor (ASRM), this left it to the external tank to make up a significant portion of the payload loss. In 1991, NASA asked Martin Marietta to reduce the weight of the external tank by an additional 7,500 pounds. Improvements in the manufacturing of the last production batch of lightweight tanks had already decreased their weight to approximately 66,000 pounds. The first superlightweight tank (SLWT) weighed 58,039 pounds, a rather substantial 7,961-pound reduction from the last lightweight tank. Approximately 4,500 pounds of the weight reduction was the direct result of using a new aluminum-lithium (Al-Li) alloy, and the remainder came from various design changes.

The ET was covered with a thermal-protection system that maintained propellant quality and minimized ice formation on its outer surface. Internal insulation was not possible because of the monocoque construction chosen to provide a lightweight structure. Instead, Lockheed applied several varieties of spray-on foam insulation (SOFI) and lightweight ablators to the outside of the ET. During ascent, the material had to endure temperatures as high as 1,200°F yet allow the tank to break-up during reentry.

Unfortunately, neither the ascent environment nor the materials used in the thermal protection system were as well understood as initially thought. During the launch of STS-107, a piece of foam separated from the ET and struck the wing leading edge of *Columbia*, causing a breach in a piece of reinforced carbon-carbon. During entry, superheated air flowed through this breach, destroying the internal wing structure and causing the orbiter to breakup over Texas, killing her crew of seven. After the accident, many detailed design changes were made to the ET thermal protection system to minimize the chance of debris during ascent.

Approximately eight minutes thirty seconds after launch, just short of orbital velocity, the general-purpose computers shut down

Workers in the Space Shuttle Main Engine Processing Facility move a Block II main engine to the Orbiter Processing Facility for installation on Discovery as STS-105. Each SSME was 14 feet long 7.5 feet in diameter, and weighed about 7,500 pounds. (NASA)

ET-119, used by STS-121, shows the protuberance air load (PAL) ramps (the flat area at the lower center of the photo and a similar area farther aft) that shed foam during the first mission after the Columbia accident. NASA removed the PAL ramps for STS-121 and later flights. (NASA)

A crawler-transporter delivers Discovery atop a mobile launch platform to LC-39A in preparation for STS-128. The crawler is moving over the flame trench, which diverted the exhaust of the main engines and solid rocket boosters away from the vehicle during liftoff. The 3.5-mile trip was made at 0.8 mph. (NASA)

the main engines (main engine cutoff, MECO) and jettisoned the external tank. This time varied slightly mission to mission based on the ascent profile and the actual performance of the SRBs and SSMEs. Nominal velocity at MECO was 17,489 mph for a standard-insertion trajectory or 17,625 mph for a direct-insertion mission. The ET continued on a ballistic trajectory and broke up during reentry. On the original standard-insertion 28.5-degree missions, the few remnants of the tank impacted in the Indian Ocean. Direct-insertion flights targeted a desolate area of the Pacific Ocean near Hawaii. On 57-degree standard-insertion flights, the impact was south of Australia, with direct-insertion tanks impacting in the south or mid-Pacific.

The two solid rocket boosters (SRB) provided the primary thrust to lift the space shuttle off the pad and to about 150,000 feet. The SRBs provided 71 percent of the total vehicle thrust during first-stage ascent and were the largest solid-propellant rocket motors yet flown. In addition, the two SRBs carried the entire weight of the ET and orbiter transmitted the load to the mobile launch platform. Alliant Techsystems (ATK) used cases provided by Ladish Forging to manufacture the solid rocket motors. The nonmotor parts (forward and aft skirts, separation motors, frustum, parachutes, and the nose cap) were originally manufactured by McDonnell Douglas for United Space Boosters, Inc. (USBI), with other portions being made in-house by the MSFC Science and Engineering Directorate. On 1 October 1999, the functions performed by USBI were rolled into the space flight operations contract (SFOC) under United Space Alliance (USA).

Each solid rocket motor case consisted of eleven individual cylindrical D6AC steel sections that were assembled into four segments using tang-and-clevis joints held together by 177 1-inch-diameter steel pins. The joints mated in Utah were called factory joints, while the joints mated at the launch site were called field joints. The empty booster weighed approximately 150,000 pounds. The segments were loaded with 1,106,640 pounds of TP-H1148 propellant in matched pairs, using the same batches of ingredients to minimize thrust imbalance. Each segment was shipped to the launch site on a heavy-duty railcar with a special cover. Each assembled booster was 149.16 feet long and 12.17 feet (146 inches) in diameter. The original boosters used for STS-1 through STS-7 produced 2,800,000 lbf in a vacuum, and later boosters generated just over 3,000,000 lbf.

TP-H1148 was a solid propellant formulated of aluminum powder (16 percent) as a fuel; ammonium perchlorate (69.6 percent) as an oxidizer; iron oxidizer powder (0.4 percent) as a burning rate catalyst; polybutadiene acrylic acid acrylonitrile terpolymer (PBAN, 12.04 percent) as a rubber-based binder; and an epoxy curing-agent (1.96 percent). The binder and epoxy were also burned as fuel, adding a small amount of thrust, and, unfortunately, a significant amount of pollutants. The propellant was cast as an eleven-point star-shaped perforation in the forward motor segment and a double-truncated-cone perforation in each of the aft segments. This provided high thrust at ignition and reduced the thrust by approximately thirty percent fifty-five seconds after liftoff to prevent overstressing the vehicle during the period of maximum dynamic pressure.

Due to its part in the *Challenger* accident, engineers redesigned the field joint during 1986. In the STS-33/51L design, the application of actuating pressure to the upstream face of the O-ring was essential for proper joint performance because large sealing gaps were created by pressure-induced case deflections ("joint rotation"), compounded by significantly reduced O-ring sealing performance at low temperature. The major change in the motor case was a tang capture feature that provided a positive metal-to-metal interference fit around the circumference of the tang and clevis ends of the mating segments. The interference fit limited the deflection between the tang and clevis O-ring sealing surfaces caused by motor pressure and structural loads. The post-*Challenger* solid rocket motors, first flown on STS-26R, were originally known as redesigned solid rocket motors (RSRM) but by 1995 had been renamed reusable solid rocket motors (still RSRM).

Each SRB was attached to the ET intertank at the forward skirt by a single thrust attachment. The recovery parachute system was attached to the same structure. Three non-thrust-bearing sway braces connected the aft portion of the SRB with the bottom of the ET. Four "hold-down" studs, 28 inches long and 3.5 inches in diameter, secured each SRB to the MLP with a frangible top nut that was explosively initiated at SRB ignition.

The SRBs provided the primary thrust for the first two minutes of ascent before being jettisoned at approximately 4,472 fps and 150,000 feet. SRB apogee occurred at approximately 220,000 feet some seventy-five seconds after separation, and the boosters impacted the Atlantic Ocean approximately 141 nm downrange from the launch site. The SRBs were equipped with three 136-foot-diameter parachutes and were retrieved by a pair of purpose-built ships based at the Cape Canaveral Air Force Station. After being disassembled in Hangar AF, the SRM segments were shipped by rail back to Utah to be refurbished and reused.

One of the solid rocket boosters from STS-119 being disassembled in Hangar AF at the Cape Canaveral Air Force Station. Each SRB was retrieved from the Atlantic Ocean, disassembled, and the cases shipped to Utah to be refilled with propellant. (NASA)

Each SRB was secured to the mobile launch platform by four hold-down studs like this one on STS-131. The upper nut was explosively fractured at T-0, using two NASA standard initiators (NSI) at the end of the wires. The silver device captured the fragments as the nut fractured. (NASA)

ENTERPRISE

Enterprise, OV-101, named after the starship in the science-fiction television series *Star Trek*. This airframe was originally to be named *Constitution* in honor of the American Bicentennial, and several famous naval vessels have also carried the name. However, a write-in campaign during 1976 produced over 100,000 requests for the White House to override NASA, which it did. Nevertheless, *Enterprise* was more than fitting, having been used by eight US Navy ships, including the first nuclear-powered aircraft carrier (CVAN/CVN-65).

Structural assembly of the airframe began on 4 June 1974, only thirty months after Richard Nixon formally approved the space shuttle and less than six years since the Phase A studies had begun. The vehicle was initially used in a variety of tests.

During the summer of 1976, before the vehicle was formally rolled out, *Enterprise* was used for the horizontal ground vibration tests (HGVT) in Palmdale. This was the first opportunity for engineers to obtain structural dynamic data from an actual flight vehicle for math model verification. After rollout, *Enterprise* was used for the Approach and Landing Tests (ALT) at nearby Edwards AFB.

Following the ALT flights, on 13 March 1978, *Enterprise* was ferried to Redstone Arsenal, next to the Marshall Space Flight Center, for the Mated Vertical Ground Vibration Tests (MVGVT). These used a set of exciters and sensors placed on the skin of the orbiter, ET, and SRBs to create vibrations and resonances similar to those that would be encountered during powered ascent.

After these tests, *Enterprise* was supposed to have gone back to Palmdale to be refitted for space flight. But several things convinced NASA to change its plan. As *Enterprise* was being built, numerous lessons were learned regarding the design and the materials used in its construction. All subsequent airframes used wings and a mid-fuselage significantly stronger than those installed on *Enterprise*, and some aluminum castings were changed to titanium to save weight. The need to retrofit these changes on *Enterprise* led to a decision to modify the structural test article instead. Since STA-099 was never completely assembled, there would not be as much rework required, and most of the structural and materials changes had been incorporated during production.

Instead, *Enterprise* was taken to the Kennedy Space Center on 10 April 1979 to check out facilities and procedures that would support STS-1. On 23 July 1979, the vehicle was rolled back to the Vehicle Assembly Building and demated from the SRBs and ET. After a one-day stop at Vandenberg AFB, *Enterprise* was flown to Edwards and moved overland to Palmdale on 30 October, where selected parts were removed and refurbished for use on later orbiters.

During May and June 1983, NASA ferried *Enterprise* to Europe for an appearance at the Paris Air Show and also visited Germany, Italy, England, and Canada. It marks the only time an orbiter was outside the United States. On 5 April 1984, *Enterprise* was floated down the Mississippi River on a barge from Mobile, Alabama, to the World's Fair in New Orleans, Louisiana.

Enterprise was used as a pathfinder at the Vandenberg Launch Site during late 1984 and early 1985 and then was put on display at KSC for several months while a place at Dulles Airport was prepared. *Enterprise* was officially transferred to the National Air and Space Museum (NASM) on 18 November 1985. For eighteen years, the orbiter would be stored in a small hangar at Dulles.

In 2003, the Smithsonian opened the new $153 million Steven F. Udvar-Hazy Center with a refurbished *Enterprise* as one of the centerpiece exhibits. Also that year, various parts of *Enterprise* were borrowed by the Columbia Accident Investigation Board (CAIB) to aid in their efforts to determine the cause of the STS-107 accident. *Enterprise* was moved out of the Udvar-Hazy Center on 19 April 2012 to make room for *Discovery* and was then ferried to the Intrepid Sea, Air and Space Museum for permanent display.

Perhaps not surprisingly, the cast of the television show Star Trek *were guests of honor at the rollout of* Enterprise *at Air Force Plant 42 in Palmdale, California. Although impressive looking, OV-101 was not much more than an extremely large glider.* (National Archives)

OV-101 SIGNIFICANT DATES

Contract Award	26 Jul 1972
Start Assembly of Crew Module	04 Jun 1974
Start Assembly of Aft Fuselage	26 Aug 1974
Wings Arrive in Palmdale	23 May 1975
Start Final Assembly	24 Aug 1975
Complete Final Acceptance Test	12 Mar 1975
Rollout at Palmdale	17 Sep 1976
Overland Transport to Edwards	31 Jan 1977
First ALT Captive Flight	18 Feb 1977
First ALT Free Flight	12 Aug 1977
Delivered to MSFC for MVGVT	13 Mar 1978
Delivered to KSC for Pad Validation	10 Apr 1979
Paris Air Show and European Tour	May/Jun 1983
Delivered to New Orleans World's Fair	05 Apr 1984
Delivered to Vandenberg Launch Site	11 Nov 1984
Delivered to National Air and Space Museum	18 Nov 1985
Moved to the Udvar-Hazy Center	20 Nov 2003
Delivered to Intrepid Sea, Air, and Space Museum	06 Jun 2012

COLUMBIA

*C*olumbia, OV-102, named after a Boston-based sailing frigate launched in 1836 that was one of the first ships to circumnavigate the globe, under the command of Robert Gray. *Columbia* was also the name of the Apollo 11 command module that carried Neil Armstrong, Mike Collins, and Buzz Aldrin to the moon for the first lunar-landing mission on 20 July 1969.

Ordered at the same time as *Enterprise*, *Columbia* shared many of the same structural limitations as its slightly older sister. Because it was built to an early specification, Columbia weighed more than later orbiters, making it unsuitable for ISS assembly missions. Instead, the vehicle was used primarily to support science missions, where its ability to carry the extended-duration orbiter (EDO) pallet provided particularly valuable.

Also like *Enterprise*, *Columbia* was initially equipped with ejection seats for its two-man crew during the orbital flight test series. A modification period after the fourth flight disabled the ejection seats, installed a seat behind the center console on the flight deck and another on the middeck, and removed portions of the developmental flight instrumentation pallet in preparation for STS-5.

Originally, *Columbia* had been scheduled for an extensive overhaul following STS-9. However, further analysis and events during STS-5 convinced NASA to split the modifications into two phases. The "Spacelab only" modifications took place after STS-5, and contrary to their name, involved more items than necessary to support the STS-9 Spacelab mission. The changes included adding one more seat on the flight deck and three on the middeck, improving the communications systems, strengthening the mid-fuselage, adding crew sleepstations, strengthening the landing-gear and brakes, adding provisions for Spacelab, and incorporating various enhancements to the thermal protection system.

The second phase of the modifications took place after STS-9. Primarily these involved removing the ejection seats (they had been disabled after STS-5) and incorporating the orbiter experiments (OEX) package that included the shuttle infrared leeside temperature sensing (SILTS), shuttle entry air data system (SEADS), and shuttle upper atmospheric mass spectrometer (SUMS) packages. The top section of the vertical stabilizer was replaced with a 20-inch-diameter SILTS pod that housed infrared sensors that monitored the temperature of the left wing. The SEADS experiment included a new nosecap with fourteen holes that fed pressure transducers, while SUMS used a small hole just aft of the nose cap to feed a mass spectrometer. The OEX packages were removed from *Columbia* during a "J1 mod" modification period after STS-40, although the SILTS pod remained (without the sensors) on the vertical stabilizer until the vehicle was lost on STS-107. Other modifications included installing new auxiliary power units, carbon brakes, a drag chute, and new AP-101S general-purpose computers. In addition, the thermal-protection system was brought up to mostly the same standard used by the later orbiters.

After STS-65, *Columbia* entered her first full-blown OMDP, a midlife refurbishment that also incorporated more enhancements to the thermal-protection system. Measures were taken to control corrosion on the wing leading-edge spar that occurred because the orbiter spent a lot of time on LC-39A before the weather protection system was installed.

Four years later, following STS-93, the orbiter entered its second OMDP, which uncovered more than 4,500 discrepancies with the Kapton wiring used throughout the vehicle. During the extended down period, *Columbia* also received the MEDS glass cockpit, several structural enhancements, and lightweight crew seats.

Columbia was lost during STS-107 on 1 February 2003 when a hole in the reinforced carbon-carbon on the leading edge allowed superheated air to destroy the left wing. The remains of the orbiter are currently stored in one of the towers in the VAB at KSC.

OV-102 Significant Dates

Contract Award	26 Jul 1972
Start Assembly of Crew Module	28 Jun 1976
Start Assembly of Aft Fuselage	13 Sep 1976
Wings Arrive in Palmdale	26 Aug 1977
Start Final Assembly	07 Nov 1977
Complete Final Acceptance Test	03 Feb 1979
Rollout at Palmdale	08 Mar 1979
Overland Transport to Edwards	12 Mar 1979
Delivered to Kennedy Space Center	24 Mar 1979
Flight Readiness Firing	20 Feb 1981
First Launch (STS-1)	12 Apr 1981
Start J1 Mod (Palmdale)	10 Aug 1991
Complete J1 Mod	09 Feb 1992
Start OMDP-J2 (Palmdale)	08 Oct 1994
Complete OMDP-J2	14 Apr 1995
Start OMDP-J3 (Palmdale)	24 Sep 1999
Complete OMDP-J3	06 Mar 2001
Destroyed during Entry (STS-107)	01 Feb 2003

Columbia completing her overland move from Palmdale to the Dryden Flight Research Center at nearby Edwards AFB, California. Note the orbiter does not yet have her distinctive black chines, something that would be added at KSC before her first flight. (NASA)

CHALLENGER

Challenger, OV-099, was named after HMS *Challenger*, a British corvette that, from 1872 to 1876 led the *Challenger* expedition, conducting pioneering global marine research in the Atlantic and the Pacific oceans. It was also the name of the Apollo 17 lunar module that carried Eugene Cernan and Harrison Schmitt to the surface of the moon. The name was doubly appropriate since the basic airframe was rebuilt from the structural test article.

The third airframe in the original order, STA-099, was the structural test article. This airframe was manufactured in parallel with *Columbia*, and the two vehicles were generally similar structurally and somewhat more advanced than *Enterprise* since they benefited from lessons learned during the construction of *Enterprise*.

On 14 February 1978, Rockwell delivered the nearly complete airframe named *Challenger* to the Lockheed-California Company, located across the Plant 42 runway. The vehicle was installed in a 430-ton steel test rig that contained 256 hydraulic jacks that distributed loads across 836 application points to simulated various stress levels under control of a computer. These stress levels duplicated the launch, ascent, on-orbit, entry, and landing phases of flight. Three 1,000,000 pounds-force hydraulic cylinders were used to simulate the thrust from the SSMEs. Heating and cooling simulations were conducted along with the stress tests up to 120 percent of the design loads. The data from these tests closely matched data from a computer model that showed the orbiter could withstand the required 140 percent of the design loads. Testing was stopped at 120 percent to ensure the airframe was not damaged and could be converted into an operational orbiter.

In the meantime, on 5 January 1979, NASA ordered Rockwell to modify STA-099 into a space-rated orbiter (OV-099) and followed this on 29 January 1979 with an order to manufacture two additional orbiters (OV-103 and OV-104). After testing was completed, STA-099 was returned to Rockwell on 7 November 1979 for conversion into OV-099. This conversion, while easier than it would have been to modify *Enterprise*, still involved a major disassembly of the vehicle. Within a month of arriving back from Lockheed, the payload bay doors, elevons, and body flap had been removed so they could be returned to the original vendors for modifications. By 18 January 1980 the vertical stabilizer had been removed and shipped back to Fairchild-Republic in New York for rework. *Challenger* had been built with a simulated crew module, so technicians separated the forward fuselage halve to remove it. The upper forward fuselage was sent to Downey for rework, and the lower fuselage was modified in Palmdale. Additionally, the wings and aft fuselage were removed and modified. *Challenger* would end up some 2,889 pounds lighter than *Columbia*, in spite of having additional equipment installed and a more robust structure.

The orbiter arrived at KSC in July 1982 and began her maiden voyage, STS-6, on 4 April 1983. That mission saw the first spacewalk of the flight campaign, as well as the deployment of the first satellite in the tracking and data relay satellite (TDRS) system constellation. The orbiter carried the first American woman, Sally Ride, into space on STS-7 and was the first to carry two American female astronauts on STS-17/41G. *Challenger* was the first orbiter to launch and land at night on STS-8 and also made the first landing at the KSC Shuttle Landing Facility at the end of STS-11/41B.

Because *Challenger* was lost on only her tenth flight, the orbiter had not been through any significant modifications or maintenance periods. However, minor changes to her thermal-protection system and systems had been made during normal postflight maintenance.

Challenger was destroyed about seventy-three seconds after launch as STS-33/51L on 28 January 1986 when a solid rocket booster field joint seal failed. The remains of the orbiter are currently stored in an abandoned Minuteman ICBM silo (Complex 31) on the Cape Canaveral Air Force Station, adjacent to KSC.

Challenger *was the first orbiter to carry her name on the forward fuselage instead of the payload bay doors. This made her much easier to identify when the doors were open on-orbit. Note the extensive use of white LRSI tiles instead of the later blankets. (NASA)*

OV-099 SIGNIFICANT DATES

Contract Award (STA-099)	26 Jul	1972
Start Assembly of Aft Fuselage	14 Jun	1976
Wings Arrive in Palmdale	16 Mar	1977
Start Final Assembly	30 Sep	1977
Complete Final Assembly	10 Feb	1978
Rollout at Palmdale	14 Feb	1978
STA-099 Delivered to Lockheed for Testing	14 Feb	1978
Contract Award (OV-099)	05 Jan	1979
Start Assembly of Crew Module	28 Jan	1979
Start Final Assembly	03 Nov	1980
Complete Final Acceptance Test	23 Oct	1981
Rollout at Palmdale	30 Jun	1982
Overland Transport to Edwards	01 Jul	1982
Delivered to Kennedy Space Center	05 Jul	1982
Flight Readiness Firing	18 Dec	1982
First Launch (STS-6)	04 Apr	1983
First Space Shuttle EVA (STS-6)	07 Apr	1983
Destroyed during Ascent (STS-33/51L)	28 Jan	1986

DISCOVERY

Discovery, OV-103, named after two ships: Henry Hudson's, which, in 1610–11, attempted to search for a northwest passage between the Atlantic and Pacific oceans and instead discovered Hudson Bay, and James Cook's, which discovered the Hawaiian Islands and explored southern Alaska and western Canada. *Discovery* was also the name of two ships used by the Royal British Geographical Society to explore the North Pole and Antarctica.

On 29 January 1979, NASA issued a change order for North American Rockwell to fabricate two additional orbiters (OV-103 and OV-104) at a cost of approximately $1,500 million. *Discovery* benefited from lessons learned and weighed approximately 6,870 pounds less than *Columbia* when she was rolled-out in Palmdale on 16 October 1983. Initially, NASA planned on basing *Discovery* at the Air Force launch site at Vandenberg AFB, but this was at first delayed because the launch site was not ready, and finally canceled when the DoD withdrew from the program after the *Challenger* accident. *Discovery* was flown from Edwards AFB to KSC on 9 November 1983 and spent seven months in the OPF for final checkout before her flight readiness firing on 2 June 1984.

The first *Discovery* OMDP followed STS-42. Unlike previous efforts, this OMDP was performed at KSC instead of Palmdale. Seventy-eight modifications were incorporated into the orbiter, with the most noticeable being the installation of the drag chute at the base of the vertical stabilizer. Other work included a complete structural inspection and the refurbishment of the thermal protection system.

Following STS-70, *Discovery* was ferried to Palmdale for her second OMDP. This time, 96 modifications were incorporated, including removing her internal airlock and added provisions for the orbiter docking system to support ISS missions. This was the first installation of the definitive ODS since *Atlantis* was still using the interim ODS-Mir hardware. Other modifications included the installation of improved payload bay lighting and replacing the

shutters over the star trackers. In addition, the airframe, wiring, and thermal protection system were inspected and repaired as needed.

Discovery entered her third and final OMDP following STS-105. This effort was again performed at KSC instead of Palmdale. During the modifications, *Discovery* received the MEDS glass cockpit and various other changes to support the ISS assembly flights. In addition, the now-normal Kapton wiring inspection and repair activity was performed. All told, *Discovery* received 368 modifications and underwent 88 special tests. This OMDP was taking significantly longer than expected, but the impact was largely mitigated by the stand-down following the *Columbia* accident. Since the vehicle was already undergoing major modifications, NASA extended the OMDP to incorporate the changes identified during the accident investigation.

Although each orbiter contributed equally to the success of the Space Shuttle Program, there is a perception that *Discovery* was the most significant vehicle because the luck of the draw allowed her to fly some of the more famous missions. For instance, *Discovery* carried the Hubble Space Telescope into space during STS-31R in April 1990, and provided both the second and third Hubble servicing missions (STS-82 and STS-103). In addition, *Discovery* hosted the first Russian cosmonaut (Sergei Krikalev) to fly aboard a space shuttle, made the first fly-around of Mir, the last docking with Mir, the first docking with the International Space Station, and flew the most missions (39) of the flight campaign. *Discovery* also has the distinction of twice being chosen as the return-to-flight orbiter. The first was STS-26R in 1988 after the *Challenger* accident and the second was STS-114 in 2005 following the *Colombia* accident.

Based on its perceived notoriety, the National Air & Space Museum selected *Discovery* to replace *Enterprise* as the orbiter on display at the Steven F. Udvar-Hazy Center near Dulles International Airport, Virginia.

OV-103 SIGNIFICANT DATES

Contract Award	29 Jan 1979
Start Assembly of Crew Module	25 Jun 1980
Start Assembly of Aft Fuselage	03 Feb 1981
Wings Arrive in Palmdale	30 Apr 1982
Start Final Assembly	06 Jul 1982
Complete Final Acceptance Test	11 Aug 1983
Rollout at Palmdale	16 Oct 1983
Overland Transport to Edwards	05 Nov 1983
Delivered to Kennedy Space Center	09 Nov 1983
Flight Readiness Firing	02 Jun 1984
First Launch (STS-14/41D)	30 Aug 1984
Start OMDP-J1 (KSC OPF-2)	17 Feb 1992
Complete OMDP-J1	17 Aug 1992
Start OMDP-J2 (Palmdale)	27 Sep 1995
Complete OMDP-J2	29 Jun 1996
Start OMDP-J3 (KSC OPF-3)	01 Sep 2002
Complete OMDP-J3	01 Apr 2004
Last Landing (STS-133)	09 Mar 2011

Like all aircraft, each orbiter had a data plate that described it. For the orbiters this data plate (along with one describing the wings) was on the forward bulkhead of the right main landing-gear well. Difficult to see is the "VO" prefix ahead of the part number. (Dennis R. Jenkins)

ATLANTIS

tlantis, OV-104, was named after a two-mast ketch designed and built by Wermeister & Wein shipyard in Copenhagen, Denmark. The Woods Hole Oceanographic Institute operated the vessel from 1930 to 1966, traveling more than half a million miles. The 460-ton ketch was the first American vessel specifically designed for oceanographic research. *Atlantis* had a seventeen-member crew and accommodated up to five scientists who worked in two onboard laboratories, examining water samples and marine life. The crew also used the first electronic sounding devices to map the ocean floor. In 1967, the ship was transferred to the Argentine Naval Prefecture under the name *Dr. Bernardo Houssay*. The experience gained during the construction of the first four airframes (STA-099, OV-101, OV-102, and OV-103) allowed Rockwell to complete *Atlantis* with 49.5 fewer man-hours compared to *Columbia*. A rather significant part of the decrease can probably be traced to the greater use of thermal-protection blankets on *Atlantis*, which required less manpower to install than LRSI tiles. Much like her near-sister *Discovery*, *Atlantis* was nearly 7,000 pounds lighter than *Columbia*.

The first flight of *Atlantis*, STS-28/51J, carried a classified DoD payload (a pair of Defense Satellite Communications System satellites), and the orbiter subsequently carried out four more DoD missions. The orbiter also carried the planetary probes Magellan (to Venus, STS-30R) and Galileo (to Jupiter, STS-34), as well as the Compton Gamma Ray Observatory (STS-37). The first OMDP for *Atlantis* was in Palmdale after STS-46. During the twenty month maintenance period, 165 modifications were made, including adding improved nosewheel steering, configuring the orbiter to carry the extended-duration orbiter (EDO) pallet, structural modifications to the airframe, changes to the landing-gear, and more than 800 new TPS tiles and blankets. In addition, provisions were incorporated to allow adding the ODS-Mir external airlock.

Beginning with STS-71 in 1995, *Atlantis* made seven flights to the Russian space station Mir as part of the Shuttle-Mir Program. When linked, *Atlantis* and Mir together formed the largest spacecraft on-orbit at the time. The missions to Mir included the first on-orbit U.S. crew exchanges, and on STS-79, the fourth docking mission, *Atlantis* ferried astronaut Shannon Lucid back to Earth after her record-setting 188 days on-orbit aboard Mir.

After STS-86, *Atlantis* again headed to Palmdale for her second OMDP, which consisted of 130 modifications. These included the installation of the MEDS glass cockpit, adding a single-string GPS navigation capability, and installing the definitive ODS to replace the ODS-Mir originally installed. Several weight reduction modifications were also performed on the orbiter, including replacing most of the AFRSI on the upper surfaces with FRSI blankets. Lightweight crew seats were installed and the EDO package installed during OMDP-J1 was removed to lighten the orbiter to allow better payload performance for the ISS assembly flights.

Following the 2004 decision was made to retire the space shuttles in 2010, NASA originally planned to withdraw *Atlantis* from service in 2008. This was mostly because it would be time for her third OMDP (one was required every eight flights or three years) and the program did not want to expend the funds. The intent was to use *Atlantis* as a "hangar queen" that could be scavenged for parts to keep the other orbiters flying. In the end, NASA waived the OMDP requirement so that *Atlantis* could continue flying until the end of the program. Instead, during the stand-down that followed the *Columbia* accident, *Atlantis* underwent a third maintenance period called an orbiter major modification (rather than an OMDP). Modifications included the installation of the station-shuttle power transfer system (SSPTS), provisions to carry the OBSS, installation of thicker side window outer panes, and the new wing leading-edge sensors mandated after the *Columbia* accident.

Workers put the finishing touches on the NASA "meatball" insignia on the upper left-wing surface. The orbiter markings were applied using a special high-temperature paint that could withstand the thermal environment during ascent and entry. (NASA)

OV-104 SIGNIFICANT DATES

Contract Award	29 Jan 1979
Start Assembly of Crew Module	05 Feb 1982
Start Assembly of Aft Fuselage	20 Nov 1981
Wings Arrive in Palmdale	17 Jun 1983
Start Final Assembly	31 Oct 1983
Complete Final Acceptance Test	14 Feb 1984
Rollout at Palmdale	06 Mar 1985
Overland Transport to Edwards	09 Apr 1985
Delivered to Kennedy Space Center	13 Apr 1985
Flight Readiness Firing	12 Sep 1985
First Launch (STS-28/51J)	03 Oct 1985
Start OMDP-J1 (Palmdale)	18 Oct 1992
Complete OMDP-J1	29 May 1994
Start OMDP-J2 (Palmdale)	11 Nov 1997
Complete OMDP-J2	27 Sep 1998
Start OMM-3 (KSC OPF-1)	01 Jun 2003
Complete OMM-3	30 Sep 2006
Last Landing (STS-135)	21 Jul 2011

*E*ndeavour, OV-105, named after the British HMS *Endeavour*, the ship that took James Cook on his first voyage of discovery. This is why the name is spelled in the British English manner, rather than American English. In August 1768, on *Endeavour*'s maiden voyage, Cook sailed to the South Pacific on a mission to observe a seldom-occurring event when the planet Venus passes between the Earth and sun. *Endeavour* was also the name of the Apollo 15 command module, which carried David Scott, Al Worden, and James Irwin to the Moon in July 1971.

The orbiter was named through a national competition involving students in elementary and secondary schools. They were asked to select a name based upon an exploratory or research sea vessel. *Endeavour* was a popular entry, and in May 1989, George H. W. Bush announced the winners as Senatobia Middle School in Senatobia, Mississippi, (elementary division), and Tallulah Falls School in Tallulah Falls, Georgia (secondary division).

During the construction of *Discovery* and *Atlantis*, NASA had opted to have the various contractors manufacture a set of "structural spares" to repair an orbiter if one was damaged during an accident. The $389 million spares consisted of an aft fuselage, midfuselage, forward fuselage halves, vertical stabilizer and rudder, wings, elevons, and a body flap. Following the destruction of *Challenger*, on 31 July 1987 NASA awarded Rockwell a $1,372 million contract to assemble these spares into a new orbiter. Interestingly, the total number of man-hours required to assemble *Endeavour* was considerably greater than those required for the final assembly of *Atlantis*. This was attributed to the fact that the orbiter assembly line had been shut down for almost two years when *Endeavour* was ordered, and many of the skilled and experienced workers had found other jobs, mostly on the Rockwell B-1B program (which was also built in Palmdale).

Endeavour was built with all the improvements incorporated on the other orbiters since the beginning of the program, including

the drag chute. *Endeavour* was also equipped as the first extended-duration orbiter (EDO) and updated avionics, including AP-101S general-purpose computers. An improved version of the auxiliary power unit was also installed. These modifications would be incorporated into the rest of the orbiter fleet as funds permitted.

Endeavour was the only orbiter that did not start her career with an overland move from Palmdale to Edwards. Instead, the SCA ferried her straight from Palmdale to KSC on 7 May 1991, and she made her maiden flight, STS-49, a year later on 7 May 1992.

The first OMDP for *Endeavour* took place in Palmdale following STS-77. The primary modifications were the removal of the internal airlock, the installation of the external airlock and orbiter docking system, and a general weight reduction effort to maximize payload to the International Space Station. In all, sixty-three major modifications were incorporated into the orbiter in Palmdale, and a further thirty-three at KSC either before or after the Palmdale visit. As part of the weight reductions, the EDO capability was removed, leaving *Columbia* as the only extended-duration orbiter. As with *Discovery* and *Atlantis*, most of the AFRSI blankets were replaced with lighter FRSI. The weight reductions went so far as to replace the aluminum foil tape that covered the inside of the landing-gear doors with lighter Kapton tape.

Between December 2003 and September 2005 *Endeavour* underwent her second OMDP. The primary changes were the installation of the MEDS glass cockpit and a three-string GPS navigation capability, in addition to various changes recommended in response to the *Columbia* accident. Workers spent 900,000 hours installing 491 modifications, replacing more than 1,000 TPS tiles and inspecting more than 150 miles of Kapton wiring. *Endeavour* also received the station-shuttle power transfer system (SSPTS), which allowed the orbiter to be powered from the ISS solar arrays while she was docked. *Discovery* received a similar modification.

OV-105 SIGNIFICANT DATES

Contract Award	31 Jul 1987
Start Assembly of Crew Module	30 Jul 1984
Start Assembly of Aft Fuselage	17 Aug 1984
Wings Arrive in Palmdale	22 Dec 1987
Start Final Assembly	01 Aug 1987
Complete Final Acceptance Test	06 Jul 1990
Rollout at Palmdale	25 Apr 1991
Overland Transport to Edwards	n/a
Delivered to Kennedy Space Center	07 May 1991
Flight Readiness Firing	06 Apr 1992
First Launch (STS-49)	07 May 1992
Start OMDP-J1 (Palmdale)	30 Jul 1996
Complete OMDP-J1	27 Mar 1997
Start OMDP-J2 (KSC OPF-2)	01 Dec 2003
Complete OMDP-J2	30 Sep 2005
Last Landing (STS-134)	01 Jun 2011

Workers paint the delivery markings on Endeavour. Note that the NASA "worm" logo was gold, not red as often shown on models. These markings were later replaced on all the orbiters with revised livery that used the traditional NASA "meatball." (NASA)

ATMOSPHERIC TESTS

On 4 June 1974, workers at Air Force Plant 42 in Palmdale started structural assembly of the first orbiter, designated OV-101 (orbiter vehicle 101). The vehicle was to carry the name Constitution in honor of the American bicentennial, but some people had other ideas about the name for the first reusable manned spacecraft. Reportedly, nearly 100,000 fans of the TV science fiction show *Star Trek* staged a write-in campaign urging the White House to rename OV-101. The "Trekkers" had their way, and when the doors opened on 17 September 1976, the name on the side of the orbiter was *Enterprise*.

For the atmospheric flight tests, *Enterprise* would not carry all of the systems required for space travel. The orbiter did not include any main propulsion system plumbing, and the SSMEs and OMS pods were mockups. Lead ballast was used to maintain the vehicle weight and balance within the limits anticipated for space-rated orbiters. The fuel cells were fed hydrogen and oxygen from high-pressure tanks instead of cryogenic dewars. The payload bay doors did not have radiators, and the crew module was bare inside except for the flight deck. The fragile thermal-protection-system tiles were mostly simulated by white and black polyurethane foam (there were a few real tiles used for aero evaluations), while the majority of the exotic carbon-carbon nose cap and wing leading edges were made of fiberglass.

On the flight deck, most of the navigation, guidance, and propulsion controls were missing, and the heads-up displays were not installed. A small amount of instrumentation was added just below the left display electronics unit for the air data system on the long flight test boom on the nose. Three cameras recorded crew actions, and Lockheed SR-71 zero-zero ejection seats were provided in the event escape was necessary. Two blowout panels were installed above the pilots to facilitate ejection, or rapid egress on the ground. The two windows looking from the aft flight deck into the payload bay were absent, covered by aluminum panels, as were the overhead rendezvous windows.

Engineers wanted to see of the orbiter could fly and land, despite thousands of hours of wind tunnel testing that said it could. Since the orbiter did not have any engines, the only way to accomplish this was to drop the orbiter from a much-larger airplane. NASA evaluated the Lockheed C-5A Galaxy and the Boeing 747 as possible Shuttle Carrier Aircraft (SCA). Various wind-tunnel tests showed that either airplane would work, but the Air Force was reluctant to give up one of its few C-5s, and there were plenty of used 747s available. Once the 747 was selected as the SCA, a great deal of wind tunnel data needed to be gathered to prove the vehicles could be separated in flight—it was these tests that led to the addition of the endplates to the horizontal stabilizers. As initially envisioned, the Approach and Landing Tests (ALT) would consist of four separate evaluations.

Taxi tests would verify the very low-speed dynamics of the mated vehicles using the concrete runway (04/22) at Edwards AFB. The first two runs would use normal braking from fairly low speed (25 knots), while the last run would use full braking and thrust reversers to simulate an aborted takeoff from about 50 knots.

Initially, NASA planned to conduct six captive-inert flights with an unmanned, unpowered orbiter. These would verify the performance, stability and control, flutter margin, and buffet characteristics of the mated configuration in flight patterns similar to the manned free flights. All of these would use the orbiter tailcone.

A crew would be aboard *Enterprise* during six captive-active flights to determine the optimum separation profile based on the results from the captive-inert flights, refine orbiter and SCA crew procedures, and evaluate orbiter integrated systems operations. Five of the flights were scheduled with the orbiter tailcone attached, and one flight without.

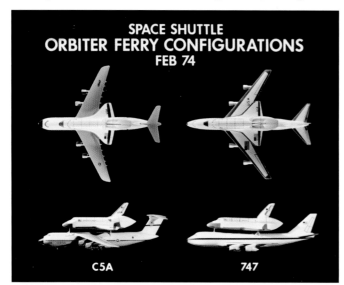

NASA evaluated the Lockheed C-5A Galaxy and Boeing 747 as possible Shuttle Carrier Aircraft to ferry the orbiter from landing sites to the launch sites at Kennedy Space Center and Vandenberg AFB. Ultimately, the agency chose the 747, based mostly on availability. (NASA)

Enterprise, during its overland move from the assembly facility in Palmdale to nearby Edwards AFB. Each of the orbiters except Endeavour was moved in this manner since Palmdale did not have a Mate/Demate Device. (Dennis R. Jenkins)

Enterprise *and the original Shuttle Carrier Aircraft (N905NA) over the High Desert around Edwards AFB during the first captive-inactive flight on 18 February 1977. The five captive-inactive flights demonstrated that the 747 could safely carry an orbiter aloft. (NASA)*

Enterprise *lifts away from the SCA, starting her second free flight on 13 September 1977. The large tailcone attached to the rear of the orbiter significantly reduced the drag from the orbiter base area and lessened aerodynamic stress on the 747 vertical stabilizer. (NASA)*

APPROACH AND LANDING TESTS (ALT)

Taxi Test #	Date	OV Crew	Maximum Speed	Braking Speed
1	15 Feb 77	--	78	23
2	15 Feb 77	--	122	20
3	15 Feb 77	--	137	50

Captive-Inert Fight #	Date	OV Crew	Duration	Tailcone	Speed	Altitude
1	18 Feb 77	--	2 hr 05 min	On	249	16,000
2	22 Feb 77	--	3 hr 13 min	On	285	22,600
3	25 Feb 77	--	2 hr 28 min	On	369	26,600
4	28 Feb 77	--	2 hr 11 min	On	369	28,565
5	02 Mar 77	--	1 hr 39 min	On	412	30,130

Captive-Active Flight #	Date	OV Crew	Duration	Tailcone	Speed	Altitude	Combined Weight
1	18 Jun 77	H/F	55 min 46 sec	On	180	14,970	470,123
2	28 Jun 77	E/T	62 min 00 sec	On	270	23,020	538,593
3	26 Jul 77	H/F	59 min 53 sec	On	270	30,292	537,423

Free flight Flight #	Date	OV Crew	Duration	Tailcone	Launch Speed	Launch Altitude	Landing Speed	OV Inert Weight	OV Landing Weight
1	12 Aug 77	H/F	5 min 21 sec	On	269	24,100	185	127,144	149,574
2	13 Sep 77	E/T	5 min 28 sec	On	269	26,000	196	127,144	149,574
3	23 Sep 77	H/F	5 min 34 sec	On	252	24,700	183	127,144	149,971
4	12 Oct 77	E/T	2 min 34 sec	Off	248	22,400	200	127,459	150,876
5	26 Oct 77	H/F	2 min 01 sec	Off	246	17,600	190	127,459	150,846

NOTES:

OV crew H/F was Fred W. Haise Jr. and C. Gordon "Gordo" Fullerton
OV crew E/T was Joe H. Engle and Richard H. "Dick" Truly

SCA crew included Fitzhugh L. Fulton Jr. (P), Thomas C. McMurtry (P), Louis E. Guidry Jr. (FE), and Victor W. Horton (FE) for all flights except Captive-Inert 5 where A. J. Roy replaced McMurtry and Captive-Active 3 where Vincent A. Alvarez replaced Guidry

Speeds are in knots, altitudes in feet agl, weights in pounds

Joe Engle and Dick Truly pilot Enterprise *during its second free flight on 13 September 1977. The orbiter was aerodynamically optimized for high-speed flight and essentially fell like a brick at low speeds, although the pilots found it easy enough to fly. (NASA)*

Finally, a series of up to eight free flights would verify orbiter subsonic airworthiness, integrated system operations, and both piloted and automated landing capabilities. The first five free flights would be conducted with the tailcone on. Following the successful completion of these, the last tailcone-off captive-active flight would be conducted. If successful, NASA would fly the last three free flights with the tailcone off.

During the tailcone-on free flights, *Enterprise* would glide for about 25 miles on its way to a landing, which would be either on the lakebed or the concrete runway at Edwards. Because of the increased drag from the exposed orbiter base area, the tailcone-off flights would be conducted at 18,000 feet, and *Enterprise* would only glide about 12 miles. Landings would all be made on the lakebed since it had more overrun available.

On 31 January 1977, workers towed *Enterprise* 36 miles through the California desert from Palmdale to Edwards. The trip took the better part of the day, using public roads that had been cleared of traffic and a specially constructed gravel road as a shortcut around one of the dry lakebeds. *Enterprise* was mated to the SCA on 7–8 February 1977 and the mated vehicles underwent weight, balance, and vibration checks during the morning of 15 February 1977. During the taxi tests, the 747 weighed approximately 400,000 pounds, including the 144,000-pound orbiter. These test validated the 747 steering and braking systems, and no unexpected problems were uncovered.

NASA successfully flew five captive-active flights between 18 February and 2 March 1977, and the results were sufficiently encouraging that the agency canceled the sixth flight. The pilots of the SCA were pleasantly surprised to find the orbiter had little adverse effect on the handling of the 747. None of the five captive-inert flights revealed any stability or flight control problems.

The first captive-active flight was scheduled for 17 June 1977, but one of the general-purpose computers was "voted out" by the other three during a preflight check. The flight was postponed twenty-four hours while the computer was replaced. On 18 June veteran astronaut Fred Haise (Apollo 13) and test pilot Gordon Fullerton were aboard *Enterprise* when the SCA took off just after 08:06 hours. The flight was once around an oval that measured 78 miles on the straight sections and 10 miles through the curves. This test was the first in which the orbiter was powered up, and Haise and Fullerton practiced moving the aerosurfaces and rudder/speedbrake for the first time. No problems were uncovered and Fullerton later said that the positioning of the orbiter on the SCA resulted in the orbiter crew not being able to see any part of the 747, which made it feel as if they were flying alone.

The next captive-active flight, with Joe Engle and Dick Truly at the controls of the orbiter, was designed to simulate the separation maneuver that would be used for the first free flight. Take-off was at 08:52 hours, and the vehicles climbed to 22,030 feet for the separation maneuver test. The SCA crew performed a pushover and descended at approximately 3,000 feet per minute. Following the separation test, the vehicles climbed to 19,300 feet and established a 6-degree glide slope to simulate an orbiter automated landing (autoland). The orbiter crew monitored the MSBLS (microwave scanning beam landing system) to ensure it was performing as expected (it was). After waving off the autoland approach, the vehicles performed a normal landing.

The third captive-active flight was flown on 26 July 1977 with Fred Haise and Gordo Fullerton in the orbiter for a 07:47 takeoff. This was a dress rehearsal for the first free flight and consisted of once around an oval measuring 84 miles on the straights and 24 miles through the curves. During the climb-out, Fullerton shut down APU-1 after it malfunctioned. The trouble was later traced to a faulty sensor, and the shutdown had no impact on the remainder of the flight. The 747 pitched down from 30,292 feet, performing

Enterprise *rolls-out on the Edwards lakebed after the second free flight. The speedbrake is slightly opened and the elevons are trailing edge up. Note the T-38 chase plane on the far side of the orbiter. The photo was taken from the back seat of another T-38. (NASA)*

The crew of the SCA during the ALT flights included Fitz Fulton, Thomas McMurtry, Louis Guidry, and Victor Horton. This was after the third captive-active flight. Note the orbiter landing-gear is deployed, a test to see if it worked after being cold-soaked during the flight. (NASA)

the maneuver that would be flown during an actual separation. The mated vehicle then continued to descend on essentially the same profile that would be flown by the orbiter in free flight. The pair landed on Runway 22 with no problems. After landing, while still mated to the SCA, Haise deployed the orbiter landing-gear as a final check prior to the first separation. These flights revealed no reason not to proceed with the free flights, and NASA canceled the last two tailcone-on captive-active flights.

Shortly after 08:00, 12 August, Fitzhugh Fulton and Thomas McMurtry guided the SCA down Runway 22, with Fred Haise and Gordo Fullerton back at the controls of *Enterprise*. At 08:48, while flying at 269 knots, Fulton pushed the nose of the 747 down 7 degrees, and Haise detonated the seven explosive bolts holding the pair together. *Enterprise* separated cleanly and Haise put the orbiter into a right-hand turn and pitched up. Haise held 2 degrees of pitch for three seconds, and then banked 20 degrees to the right and toward lakebed Runway 17. Descending at 1 foot per second, the main gear tires hit the lakebed at 185 knots, and the orbiter rolled 11,000 feet with minimal usage of the wheel brakes. The first flight of *Enterprise* had lasted just over five minutes, with no serious problems.

Joe Engle and Dick Truly were aboard when *Enterprise* flew again on 13 September 1977 for a series of more extensive maneuvers. *Enterprise* was stable as it began a wide 55-degree turn with its nose angled up 3 degrees. Truly took control and flew the vehicle down to 2,000 feet, handing off to Engle for a final series of maneuvers with the speedbrake open 40 percent. Five minutes and 28 seconds after leaving Fitz Fulton and Tom McMurtry in the 747, the orbiter settled onto the lakebed, rolling out 10,037 feet in just over a minute.

During the third free flight, on 23 September 1977, Fred Haise and Gordo Fullerton tested the autoland system, letting the computers fly the orbiter down to the 900-foot preflare. The crew hard-braked the vehicle on lakebed Runway 17 without any problems. The first three free flights were deemed so successful that NASA canceled the final tailcone-on flight. The time had come to see how *Enterprise* would fly in a return-from-space configuration.

The fourth free flight took place on 12 October 1977, and this time *Enterprise* flew like the brick it was, with a flight time less than half of what it was with the tailcone on. The buffeting in the SCA as the orbiter pulled away was moderate but acceptable, and the handling qualities of the orbiter were basically the same, except for the more rapid descent. Engle and Truly experienced some problems with the TACAN system, and some minor error messages from the GPCs, but the flight cleared the way for the final test—landing *Enterprise* on the concrete runway.

Without its tailcone, Enterprise *falls toward the Edwards runway during the fourth free flight. Note the Northrop T-38 Talon, with its landing-gear extended to allow a slow and steep dive, flying formation with the orbiter.* Enterprise *touched-down at 200 knots. (NASA)*

The fifth, and last, free flight was on 26 October 1977 with Haise and Fullerton at the controls of *Enterprise*. The orbiter separated 51 minutes after takeoff at an altitude of 17,600 feet. As before, the crew put the orbiter through a series of maneuvers, finding again that the gliding performance of the orbiter was better than predicted. This time, however, as the crew set up their final approach, trouble began. Coming out of the preflare, *Enterprise* was dropping at 290 knots, considerably quicker than planned. In an attempt to slow the orbiter, Haise opened the speedbrake early, but instead of slowing down, speed increased, so Haise deployed the landing-gear and pitched the nose down to make the desired touchdown point. Since *Enterprise* was unpowered, there was no option of making a second pass. The wings dipped and Haise struggled to correct as the rear wheels hit the tarmac hard. Instead of dropping the nose, *Enterprise* bounced 20 feet back into the air. After a few anxious seconds, the orbiter smoothed out for the remainder of the rollout, with the nose finally dropping onto the runway—a classic case of pilot-induced oscillation (PIO).

Regardless, NASA announced that the ALT program had met its objectives, and no further flights would be necessary.

Fred Haise and Gordon Fullerton bring Enterprise *onto the concrete runway for the fifth and final free flight. Haise deployed the speedbrake a bit too early and caused a pilot-induced oscillation that raised everybody's heart rate, but landed successfully. (NASA)*

Enterprise *certainly looked the part, but this was as close as the orbiter would come to being operational. In the days before overly paranoid security, several dozen employees and photographers stood alongside the concrete runway during the ALT free flights. (Dennis R. Jenkins)*

MISSION DESCRIPTION

A space shuttle mission began in the Vehicle Assembly Building (VAB) with a pair of solid rocket boosters being stacked on one of the three mobile launch platforms (MLP) while the orbiter was serviced in one of three Orbiter Processing Facilities (OPF). The external tank was then mated to the stacked SRBs. The orbiter was moved from the OPF to the VAB (called "roll-over") and mated to the ET. Electrical and fluid connections were made, the integrated vehicle checked out, and the range safety system ordnance installed while still in the VAB.

One of the two crawler-transporters moved the MLP (called "rollout") with the entire space shuttle stack to one of the two launch pads approximately 3.5 miles away, where servicing and checkout activities began. The move typically took place at night due to weather concerns and happened at a stately 1 mph (early in the flight campaign) or 0.8 mph (later). Nondeployable payloads, such as Spacelab and some of the early satellites, were installed horizontally while the orbiter was in the OPF. Deployable payloads, such as most components of the International Space Station (ISS), were installed vertically at the launch pad, using the payload changeout room (PCR) in the rotating service structure (RSS).

On launch day, the external tank was filled with cryogenic propellants. The crew boarded the orbiter approximately three hours prior to launch. At T-20 minutes (twenty minutes prior to launch), there was a built-in ten-minute hold that allowed the launch team and flight crew to catch up on any outstanding preparations and to make final consultations before proceeding. There was another built-in hold at T-9 minutes.

At T-6.6 seconds, the orbiter general-purpose computers (GPC) began sending ignition commands to the three main engines, one at a time. Because of their asymmetric location, the SSME thrust build-up resulted in a "twang," where the top of the stack moved several feet laterally. Assuming all three engines were generating at least 90 percent thrust, the GPCs commanded SRB ignition at T-0. The signal to ignite the SRBs was timed so the stack was vertical when the eight hold-down bolts were explosively released. Compared to the lumbering liftoff of the Saturn V, the space shuttle jumped off the pad due to the rapid thrust buildup of the SRBs.

The SSMEs throttled up, usually to 104 percent (104.5 percent for Block II engines), as the velocity reached 40 mph at 73 feet altitude (T+4 seconds). As soon as the velocity exceeded 87 mph (about T+7 seconds), the vehicle rolled to a heads-down attitude to align its velocity vector with the desired orbital plane. During first-stage (SRB) ascent, heads-down made it easier for the stack to maintain a negative angle of attack that reduced wing loading. As the vehicle approached the period of maximum dynamic pressure (max-q), the main engines throttled-down to between 65 and 72 percent, depending on the mission and atmospheric conditions, and the shape of the SRB propellant minimized acceleration for a few seconds to avoid overstressing the orbiter. Max-q was reached early in the ascent, normally twenty to thirty seconds after liftoff, while the stack was traveling about 475 mph at 9,000 feet.

Approximately 123 seconds into ascent, the two SRBs separated from the ET. This was also accomplished explosively, and eight small rocket motors fired to ensure the boosters separated cleanly from the vehicle. This took place at a nominal 3,050 mph and 150,000 feet. At a predetermined altitude, the SRBs deployed parachutes that lowered the boosters into the Atlantic Ocean approximately 141 nm downrange from the launch site. The boosters were recovered by a pair of retrieval ships, *Freedom Star* and *Liberty Star*, operating from the Cape Canaveral Air Force Station (CCAFS), adjacent to KSC.

Meanwhile, the orbiter and ET continued using the thrust provided by the SSMEs. At this point, the stack was flying above most of the atmosphere, so wing loading was less critical. Originally, the

Technicians install the last main engine in Discovery in preparation for STS-131. The base heat shields had not yet been installed around the engines, allowing a limited view inside the aft fuselage. The two red covers protected the orbital-maneuvering-system nozzles. (NASA)

A view looking down on the left solid rocket booster of STS-135 as it sat waiting for launch. Each booster was secured to the mobile launch platform by four large hold-down studs. Note the gray plumbing for the sound suppression water system. (NASA)

vehicle stayed heads-down to allow S-band communications through the Bermuda tracking station. After that station was closed, most flights beginning with STS-87 executed a roll-to-heads-up (RTHU) maneuver that allowed communications through the Tracking and Data Relay Satellite (TDRS) system. This occurred at approximately T+6 minutes when the vehicle was traveling at 8,310 mph. The roll maneuver was performed at about 5 degrees per second.

Approximately 8.5 minutes after launch, and just short of orbital velocity, the SSMEs were shut down (main engine cutoff, MECO) and the ET was jettisoned. The forward and aft reaction control system (RCS) thrusters provided the translation away from the ET at separation and returned the orbiter to the proper attitude prior to the orbital maneuvering system (OMS) burn. Nominal velocity at MECO was approximately 17,489 mph for a standard-insertion trajectory, or 17,625 mph for a direct-insertion mission.

The ET continued on a mostly ballistic trajectory and entered the atmosphere, where it broke-up during reentry. On the original standard-insertion 28.5-degree missions, the ET impacted in the Indian Ocean; for direct-insertion flights this moved to a desolate area of the Pacific Ocean near Hawaii. On 57-degree standard-insertion flights, the impact area was south of Australia, with direct-insertion tanks impacting in the south or mid-Pacific.

For the early missions, two thrusting maneuvers were made using the orbital maneuvering-system engines. The first OMS burn (OMS-1) raised the maximum altitude and occurred a few minutes after MECO. The second burn (OMS-2) raised the lowest point of the orbit (perigee) to circularize the orbit. This became the "standard-insertion" mission profile since it was standard practice early in the flight campaign, when there was less information on the exact performance of the main engines. The last mission to use a standard-insertion profile was STS-30R, and STS-38 was the last mission to use an OMS-1 burn. Later missions flew "direct-insertion" trajectories that used a single OMS burn to circularize the orbit (the OMS-1 burn was omitted); the main engines placed the orbiter directly at the correct apogee. The OMS-2 burn typically lasted between one and three minutes and added between 50 and 340 mph. Most of these missions included an OMS-assist burn, but these were different than the earlier OMS-1 burns. An OMS-assist burn using 4,000 pounds of propellant boosted payload capacity by about 250 pounds.

Orbital altitude was highly dependent on mission objectives. In theory, the vehicle could fly anywhere between 100 and 400 nm. However, the lowest orbit flown during the flight campaign was 120 nm during STS-68, while the highest was 335 nm during STS-82.

At the completion of orbital operations the orbiter was oriented into a tail-first attitude by the RCS and a deorbit burn by the OMS engines reduced the velocity 135 to 340 mph, depending on orbital altitude. This was sufficient to deorbit the vehicle. The RCS turned the nose forward for entry interface (EI), arbitrarily defined as descending through 400,000 feet, slightly less than 5,200 miles from the landing site. The velocity at entry interface was approximately 16,773 mph, with the orbiter descending wings level at a 40-degree angle of attack.

The forward reaction control system jets were inhibited immediately prior to entry. Aerosurface control began when dynamic pressure reached 2 psf, although in reality the surfaces were ineffective in the rarefied atmosphere, and the aft reaction control system continued to provide most control authority. The aft RCS roll jets were deactivated at 10 psf, which was when the ailerons actually became effective. At 40 psf, the elevators became effective and the aft RCS pitch jets were deactivated. The speedbrake was used below Mach 10 to induce positive downward elevator trim. At approximately Mach 5, the rudder was activated, and the aft RCS yaw jets were deactivated at Mach 1 and 54,000 feet, leaving only the aerosurfaces to maneuver the vehicle.

The descent rate and range were controlled by roll angle; the steeper the angle, the greater the descent rate and the greater the drag. Conversely, the minimum drag attitude was wings level. Cross-range was controlled by roll reversals. Once the velocity was below 13,000 mph, the orbiter entered equilibrium glide flight where the flight path angle—the angle between the local horizontal and the local velocity vector—remained constant. Equilibrium glide flight provided the maximum downrange capability.

The orbiter had a theoretical design cross-range capability of 1,100 nm. In practice, the actual cross-range varied from 2 nm on STS-61 and STS-113 to 813 nm on STS-116.

The constant drag phase began at this point, with an initial 40-degree angle of attack, ramping down to 36 degrees. In the transition phase, the angle of attack continued to ramp down to about 14 degrees as the orbiter reached the terminal area energy management (TAEM) interface some 6 minutes, 34 seconds prior to touchdown. The orbiter was flying Mach 2.5 (1,700 mph) at about 81,000 feet, some 69 miles from the runway. The vent doors opened as the orbiter slowed below Mach 2.4 at 80,000 feet.

Control then transferred to TAEM guidance that steered the orbiter to the nearest of two 18,000-foot-diameter heading-alignment cones (HAC) located tangent to, and on either side of, the approach

Discovery, as STS-131, shows the external airlock and orbiter docking system (ODS) in the payload bay and the Ku-band antenna on the right payload bay sill. The Ku-band antenna was used for communications and as radar during rendezvous and docking. (NASA)

A drag chute was added to the orbiter fleet after the Challenger accident to lessen wear and tear on the brakes and provide an extra margin of safety during landing. Like the landing-gear and air data probes, the chute could be deployed only by the crew. (NASA)

end of the runway centerline. Excess energy was dissipated in a final S-turn, increasing the ground track range as the orbiter turned away from the nearest HAC until sufficient energy was dissipated to allow a normal approach and landing. The vehicle slowed to subsonic velocity at approximately 49,000 feet, about 30 miles from the runway.

The approach and landing phase began at 10,000 feet while the orbiter was traveling about 345 mph, either 6.7 or 7.3 miles from touchdown for –20- or –18-degree glideslopes, respectively. These are more than seven times those normally used by commercial airliners. Approximately 10 seconds before touchdown, the crew deployed the landing-gear at 300 feet altitude, and a flare maneuver lowered the sink rate to 3 fps. Touchdown occurred approximately 2,500 feet past the runway threshold at roughly 225 mph for lightweight orbiters or 235 mph for heavy vehicles.

Ground crew wearing self-contained atmospheric protective ensembles (SCAPE) approached the orbiter as soon as it stopped rolling, and took air samples to ensure the atmosphere in the vicinity of the orbiter was not hazardous. In the event of a hypergolic or ammonia leak, a wind machine truck carrying a large fan was moved into the area to create a turbulent airflow that reduced the potential for an explosion. Technicians connected an air-conditioning purge unit to the right T-0 umbilical so cool air could be directed through the aft fuselage, payload bay, forward fuselage, wings, vertical stabilizer, and OMS pods to dissipate the remaining heat of entry. A second ground unit connected to the left T-0 umbilical Freon-21 loops provided cooling for the flight crew and avionics during the postlanding shutdown activities. The orbiter fuel cells remained powered-up until the flight crew exited the vehicle, and then the ground crew powered-down the orbiter.

If the landing was at KSC, the orbiter and its support convoy were moved to an Orbiter Processing Facility. If the landing was made at Edwards, the orbiter was safed on the runway and then towed to the Mate/Demate Device at NASA Dryden. After a detailed inspection of the vehicle, ground crews installed the tailcone and mated the orbiter to the SCA for the ferry flight back to KSC. The tailcone smoothed the aerodynamics around the tail of the 747 and minimized drag on the base area of the orbiter.

ORBITAL MECHANICS

As originally envisioned, space shuttle missions to equatorial orbits would be launched from the Kennedy Space Center (KSC), and those requiring polar orbit (mostly national security and NOAA weather satellites) would be launched from the Vandenberg Launch Site (VLS). Orbital mechanics, the complexities of mission requirements, safety constraints, and the possibility of infringing on foreign air space prohibited polar-orbit launches from KSC. However, following the *Challenger* accident in 1986, NASA and the Air Force canceled plans to activate Vandenberg, and all space shuttle missions were launched from KSC, leaving the United States without a manned polar-flight capability (something, it should be realized, the nation has never possessed).

At the Kennedy Space Center, launch complex 39A (LC-39A) was located at 28 degrees, 36 minutes, 29 seconds north and 80 degrees, 36 minutes, 14 seconds west, and LC-39B was at 28 degrees, 37 minutes, 37 seconds north and 80 degrees, 37 minutes, 15 seconds west. Using a 35-degree launch azimuth from KSC placed the orbiter in an equatorial orbit inclined 57 degrees, meaning the vehicle would never exceed 57 degrees latitude north or south of the equator. A launch azimuth of 90 degrees (due east) placed the vehicle in an equatorial orbit inclined 28.5 degrees. A 120-degree launch azimuth resulted in a 39-degree orbit.

These two launch azimuths, 35 and 120 degrees, represented the upper and lower operational limits from KSC. Angles farther north or south would have traveled over a habitable landmass, potentially allowing the solid rocket boosters or external tank to land on foreign territory. Flights into slightly higher (STS-36 to 62.0 degrees) orbits were accomplished using inefficient dogleg profiles that initially used the approved launch azimuths, then turned toward the desired orbit during second-stage. Ultimately, 53 flights launched into roughly 28.5-degree due-east orbits, 10 to 39 degrees, 47 to 51.6 degrees, and 20 to 57 degrees, with the other four into orbits unique to their missions.

SLC-6 at the Vandenberg Launch Site was at 34 degrees, 34 minutes, 52 seconds north and 120 degrees, 37 minutes, 32 seconds west. Launches from Vandenberg would have had allowable launch limits of 201 and 158 degrees. At a 201-degree launch azimuth, the vehicle would orbit at a 104-degree inclination. Zero degrees would be due north of the launch site, and the orbital trajectory would be within 14 degrees east or west of the north-south pole meridian. With a launch azimuth of 158 degrees, the orbiter would be at a 70-degree inclination, and the trajectory would be within 20 degrees east or west of the polar meridian.

Mission requirements and payload penalties were major factors in selecting a launch site, since the Earth rotates from west to east. The rotational speed differs depending on latitude; at the equator it is 1,038 mph, while it is 911 mph at KSC, 819 mph at Wallops Island, and only 726 mph at the Baikonur Cosmodrome. An easterly launch

After the ET was jettisoned, the crew of many flights maneuvered the orbiter to take photos that were evaluated for foam loss. Note the heat damage to the base region at the back of the tank, caused by the recirculation of the superheated SSME and SRB exhaust gases. (NASA)

After the SRBs were jettisoned, parachutes lowered them into the Atlantic, where they were recovered by two retrieval ships based at Cape Canaveral AFS. The operation included divers inserting a plug in the aft end of the booster, roughly 100 feet under the surface. (NASA)

Three main engines up and running at full power just seconds before Atlantis launches as STS-117. The SSMEs were started 6.5 seconds before SRB ignition to ensure each was operating properly before committing the vehicle to launch. (NASA)

uses this rotation somewhat as a springboard, increasing maximum payload and altitude capabilities. This rotation was also the reason the orbiter was designed with a cross-range capability of 1,100 nm, to provide an abort-once-around capability. Due-east launches provide the maximum payload and altitude capability. Launches other than due east benefit less and less from the rotation as the launch azimuth approaches north or south, and westerly launches decrease performance even further.

The basic payload capability of all orbiters except *Challenger* and *Columbia* for a due-east (28.5-degree orbit) launch from KSC was 56,300 pounds to a 115 nm orbit to support a four-day mission with five crew members. This capability was reduced approximately 120 pounds for each additional nautical mile of altitude. The payload capability for the same mission to a 57-degree orbit was 42,050 pounds. Performance for intermediate inclinations could be estimated by subtracting 500 pounds per degree between 28.5 and 57 degrees. *Challenger* could carry 5,000 pounds less to any given orbit, and *Columbia* could carry about 7,550 pounds less.

From the Vandenberg Launch Site, all orbiters except *Challenger* and *Columbia* could, in theory, carry 29,600 pounds at a 98-degree launch inclination and 110-nm polar orbit. Performance for other inclinations could be estimated by subtracting 660 pounds for each degree between 68 and 98 degrees.

It should be noted that each additional crew member beyond five was chargeable to the payload weight allocation and reduced the payload capability by approximately 500 pounds, including life support and crew escape equipment.

The orbiters were limited to a maximum landing weight of 250,000 pounds for abort landings and 240,000 pounds for nominal end-of-mission landings.

Aborts

A space shuttle launch could be delayed any time prior to the start of ET tanking, then scrubbed or aborted up to SRB ignition. Essentially, it was a matter of semantics: a scrub happened prior to SSME start and an abort happened afterward.

There were nine scrubs that happened after T-31 seconds but before SSME ignition: STS-1, 2, 32/61C, 32/61C (again), 31R, 56, 51, 88, and 93. There were also five prelaunch aborts that occurred after one or more main engines had ignited: STS-14/41D, 26/51F, 55, 51, and 68. Only a single inflight abort was flown during the

The forward reaction control system (FRCS) module contained the thrusters and propellant tanks used to maneuver the orbiter on-orbit, in conjunction with similar hardware in the OMS pods next to the vertical stabilizer. This FRCS is being installed on Discovery for STS-131. (NASA)

flight campaign, the abort-to-orbit of STS-26/51F after SSME-1 shutdown during ascent. Other than that single abort to orbit (ATO), the bulk of the abort software was never executed in flight. Contrary to many reports, STS-93 was not an ATO.

Aborts fell broadly into three categories: performance aborts, systems aborts, and range safety actions.

A performance abort occurred after the loss of one or more main engines. The performance after an engine lost thrust or completely failed was directly related to when the problem occurred. Early degradation or failure while the vehicle was heavy with propellant could preclude achieving a safe orbit, while late engine problems could result in little or no underspeed. By using an abort region determinator computer program, flight controllers in Houston could predict the underspeed resulting from any performance problem and determine whether a safe orbit could be achieved.

A systems abort occurred when one or more orbiter systems failed during ascent. In most cases, the crew had to rely on the insight provided by mission control using vehicle telemetry to fully understand the magnitude of any systems problem during ascent, since only limited data was displayed in the cockpit.

The third category has always been a touchy subject for the manned spaceflight programs. Despite all of the planning, development, and testing, there was always the possibility a problem could result in a vehicle flying out of control toward a populated area. At both KSC and Vandenberg, the Air Force was responsible for public safety in the areas surrounding the eastern and western ranges. To protect populated areas, impact limit lines were drawn around them, and no potentially lethal piece of a vehicle was allowed to land beyond those boundaries. Since any flight termination needed to occur well before an impact limit line was reached, destruct lines were drawn inside the impact limit lines. Any flight passing outside the destruct line was subject to termination.

During first-stage ascent, trajectory deviations could have led to a violation of a destruct line by a vehicle that was still under control, and it could have been possible to return the vehicle toward its nominal trajectory or to safely execute an abort. Therefore, the Flight Director (FD) and flight dynamics officer (FDO) in mission control were in voice communication with the Air Force Mission Flight Control Officer (MFCO, formerly called the Range Safety Officer) on the range during ascent. If the MFCO detected a potential violation, he would have immediately informed the FDO and FD. At that point, the FD would have determined whether the vehicle was controllable or uncontrollable. As long as the FD declared the vehicle controllable, the MFCO would not terminate the flight for trajectory deviations alone. It should be noted a destruct command would not be sent until after the flight crew attempted to separate the orbiter from the rest of the stack, a tactic unlikely to succeed.

If it had ever been necessary to destroy the vehicle, the MFCO would have issued a command to detonate explosive charges that ran the length of the solid rocket boosters. Early in the program the external tank also carried a range safety system but it was deleted after STS-78. The orbiter did not carry a range safety system, since it was not considered a propulsive vehicle.

Although NASA and the Air Force had agreements concerning the destruction of a manned vehicle, in the end the laws and regulations of the United States empowered sole responsibility to the MFCO, whose duty was to protect the public and who reported exclusively to the commander of the appropriate range. The space shuttle operational flight rules warned, "Flight termination action will be taken by the MFCO when, based on assessment of all available data, the secondary ILL [impact limit line] flight termination criteria are violated, regardless of the controllability status." It was an accepted fact that was not open to debate.

Performance and systems aborts were further divided into intact aborts and contingency aborts. Intact aborts were designed to provide the safe return of the orbiter to a planned landing site. Contingency aborts resulted in the flight crew bailing out (or ejecting, for the first four flights) or ditching the orbiter in the ocean, a tactic that was unlikely to be successful.

The types of intact aborts evolved over the course of the flight campaign but generally consisted of four basic scenarios: abort to orbit (ATO), abort once around (AOA), transoceanic abort landing (TAL), and return to launch site (RTLS). Later in the program there was a fifth intact-abort scenario (ECAL) that was a variation of TAL and was generally not considered separately. The software used for STS-1 supported only the RTLS and ATO modes.

Abort to orbit allowed the vehicle to achieve a lower-than-nominal orbit. This mode required less performance and allowed time to evaluate problems and then choose either an early deorbit or an OMS thrusting to raise the orbit and continue the mission. ATO could be selected for either a performance shortfall or certain systems failures. This abort was used by STS-26/51F and the mission was completed successfully.

Abort once around allowed the vehicle to fly once around the Earth and make a normal entry and landing 105 minutes after launch. The AOA mode was designed for cases where vehicle performance was lost to such an extent that it was impossible to achieve a viable orbit, or not enough OMS propellants remained for orbit circularization and/or deorbit. In addition, an AOA could be used when a major system failure made it desirable to land quickly.

Transoceanic abort landing allowed an intact landing, usually on the other side of the Atlantic, approximately 40-45 minutes after launch. These were initially call transatlantic landings (hence, TAL). This mode resulted in a ballistic trajectory that did not require an OMS maneuver for deorbit. The TAL mode was developed to improve the options available when an SSME failed after the last RTLS opportunity but before an AOA could be accomplished, or when a major system failure (e.g., cabin leak) made an AOA undesirable. In a TAL abort, the vehicle continued on a ballistic trajectory across the Atlantic (or mid-Pacific for SLC-6 launches) to land at a predetermined location approximately 45 minutes after launch. The landing site was selected near the nominal ascent ground track to make the most efficient use of SSME propellants.

The return-to-launch-site abort involved flying downrange and then turning around under power to return directly to a landing at, or near, the launch site approximately 25 minutes after liftoff. After SRB separation, the GPCs initiated a powered pitch-around maneuver at 10 degrees per second to orient the stack to a heads-up attitude, pointing toward the launch site. At this time the vehicle was still moving away from the launch site, but the main engines were thrusting to nullify the downrange velocity. Visualize a space shuttle stack traveling backward at hypersonic velocities, at least for a while. Timing the shutdown of the SSMEs was critical, since the vehicle had to have sufficient speed and altitude to glide to the runway but not so much as to end up in the Gulf of Mexico. External-tank-heating considerations meant an RTLS could not be selected earlier than 150 seconds after liftoff, although it would have been impossible prior to T+123 seconds (SRB separation) in any case. Ideally, the vehicle would reach the desired main engine cut off point with less than 2 percent excess propellant remaining in the external tank. After MECO, the ET separation sequence would begin, including an RCS translation that ensured the orbiter did not recontact the ET and that the orbiter had achieved the necessary pitch attitude to begin the glide phase of the RTLS. After the RCS translation maneuver was completed, the glide phase began. From this point through landing, the RTLS was handled similarly to a nominal entry.

The option for an East Coast abort landing (ECAL) was added during the early 1990s for high-inclination launches, although it was available for only a small period during ascent. NASA considered this a variation of TAL, but it is broken out here for clarity. The ECAL capability replaced certain bailout regions with landing capability for two-engine-out scenarios prior to a single-engine TAL capability. Small ECAL capability regions also existed for two-engine-out scenarios prior to RTLS-powered pitch around and for some three-engine-out trajectories. The theory was that it was always better to try and fly toward a land mass, even if you can't make a landing site, rather than fly wings level straight ahead and bail out (which is what regular contingency-abort logic did).

The type of failure and when it occurred during ascent largely determined which type of abort would have been selected. However, there was a definite order of preference for the various abort modes. In cases where performance loss was the only factor, the preferred modes would have been ATO, AOA, TAL, and RTLS, in that order. The mode chosen was the highest one that could be completed with the remaining vehicle performance. In the case of some systems failures, such as crew module air leaks, the preferred mode might be the one that ended the mission most quickly. In these cases, TAL or RTLS might have been preferable to AOA or ATO.

The priorities in choosing an abort mode were not endangering the general public, saving the flight crew, saving the vehicle, and completing the mission, even partially, in that order. A contingency abort would never have been selected if an intact-abort option existed.

Mission control was responsible for calling these aborts because it had more precise knowledge of the vehicle position and performance than could be obtained from onboard systems. If ground communications were lost, the flight crew had onboard methods to determine the abort region, although not as accurately as the ground. It should be noted that the commander—and only the CDR—could initiate an abort by positioning the ABORT MODES switch to the desired mode and depressing the ABORT push button. Aborts could not be initiated by the pilot or mission control.

The crane operators at the Kennedy Space Center were among the best in the world. They had to lift SRB segments, the ET, and the orbiter and then position them within a few tenths of an inch so they could be mated together. This is Atlantis as STS-132. (NASA)

CREW

The flight crew consisted of the commander (CDR), pilot (PLT), mission specialists (MS), and payload specialists (PS). The size of the crew varied depending on the mission objectives. The orbital flight tests had the smallest crews, only a commander and pilot, while STS-30/61A had the largest, with eight crew members (the same number was carried down during STS-71). For most of the flight campaign, each mission typically carried a crew of five (early) or seven (late). The commander, pilot, and two mission specialists sat on the flight deck with the remaining crew members on the middeck. Typically, the mission specialist seated behind the center console on the flight deck acted as a flight engineer to assist the pilot and commander with the myriad switches and controls.

A mission specialist was a career NASA astronaut trained in orbiter systems and payload operations. The mission specialist participated in planning the mission and was responsible for overall coordination between the payload and orbiter. Mission specialists had prime responsibility for experiments for which no payload specialist was assigned and assisted payload specialists as needed.

The payload commander, if designated, was an experienced mission specialist designated who represented the Flight Crew Operations Directorate and the astronaut office (mail code CB) on a Spacelab or complex payload flight. This individual worked with the payload mission managers to identify and resolve issues associated with payload integration, training, crew member qualification, and operational constraints.

A payload specialist was a non-NASA employee whose presence was required to operate a specific payload. In general, a payload specialist worked for the entity that sponsored or manufactured the payload and paid NASA for their flight. Ultimately, only four truly commercial payload specialists flew: Charles Walker on STS-14/41D, STS-23/51D, and STS-31/61B; Sultan bin Salman bin Abdulaziz Al Saud on STS-25/51G; Rodolfo Neri-Vela on STS-31/61B; and Gregory Jarvis on STS-33/51L. Over the years, NASA designated other payload specialists, but the term had largely gone away by the end of the flight campaign and all non-pilots were simply called mission specialists.

Prior to the *Challenger* accident, a space flight participant was an individual the NASA administrator determined to be in the national interest or one that would contribute to other approved NASA objectives. This category included several politicians as well as the Teacher in Space and Journalist in Space projects.

The Air Force and National Reconnaissance Office named thirty-two manned spaceflight engineers, but only two ultimately flew. This is Gary Payton on STS-20/51C. Note the "USAF" logo on his shirt in addition to the traditional NASA worm, and the posters behind him. (NASA)

In 1979, the Air Force, operating as a cover for the then-secret National Reconnaissance Office (NRO), selected 13 military astronauts that they called manned spaceflight engineers (MSE). These were, essentially, military payload specialists. A further 14 were selected in 1982 followed by 5 more in 1985. Only two flew before the program was disbanded in 1988: Gary Payton on STS-20/51C and Bill Pailes on STS-28/51J.

More people flew on space shuttle than all other launch systems, worldwide, combined. Over the course of 135 missions, the orbiters transported 355 individuals for a total of 852 fliers. These 306 men and 49 women represented 16 different countries. Of these, 60 flew on *Challenger*, 160 on *Columbia*, 252 on *Discovery*, 173 on *Atlantis*, and 207 on *Endeavour*. Jerry Ross and Franklin Chang-Diaz flew most often, seven times each, while Story Musgrave was the only person to fly on all five orbiters. The largest crew was eight on STS-30/61A (and the return of STS-71) while the smallest was two, on each of the four orbital flight tests. The youngest person to travel on space shuttle was Sultan Salman Al Saud (28 years, 11 months, 21 days) on *Discovery* during STS-25/51G while the oldest was John Glenn, (77 years, 3 months, 11 days) on *Discovery* during STS-95. Bob Crippen was the first rookie to fly on space shuttle (STS-1) and James Dutton, Dorothy Metcalf-Lindenburger, and Naoko Yamazaki were the last (all on STS-131).

WAKEUP CALLS

Beginning during Project Gemini, mission control played music as it woke orbiting astronauts to begin their days. As Kay Hire explained when she was CAPCOM during STS-79, "the wakeup music is selected by the astronauts working as CAPCOM for the mission. Traditionally, the music relates to mission objectives or to specific crew members … within the highly structured environment of a shuttle flight, the morning wakeup presents an opportunity for levity and a bit of shared camaraderie. It tends to stand out as a human element in an otherwise complex technical enterprise."

MISSION DESIGNATIONS

For the first three years of the flight program, space shuttle missions were designated STS-x, where x was a sequential number based on the order the flight was manifested, not necessarily the order they were launched. The non-sequential launch order bothered NASA public affairs officials, and reportedly NASA administrator James Beggs was superstitious (triskaidekaphobia) about the upcoming STS-13 mission. Therefore, beginning in 1983, each mission was also assigned a designation where the first digit indicated the fiscal year of the scheduled launch (4 for 1984)), the second digit indicated the launch site (1 was Kennedy Space Center and 2 was Vandenberg), and a letter indicated the scheduling sequence. As with the sequential numbers, these codes were assigned when the launches were initially manifested and were not changed as missions were delayed or rescheduled.

Although the public face of NASA religiously used the new alpha-numeric mission designators, many within NASA continued to use the original sequential manifest numbers, and that is what is used in this book since they are what most engineering data reflects. After the *Challenger* accident, everybody returned to the sequential numbering scheme, but it was reset to "26" since *Challenger* had been the 25th mission launched. Unfortunately, under the original numbering scheme, the *Challenger* mission had been STS-33, so restarting with 26 created some duplication. Therefore, when flights resumed in 1988, NASA restarted with STS-26R, the "return-to-flight" suffix used to disambiguate from prior missions. This continued through STS-33R (including STS-29R, despite there not having been an STS-29 actually launched).

After the *Columbia* accident, the program always had a rescue mission planned (and often, standing-by). These launch-on-need (LON) missions carried designations in the STS-3xx series for ISS flights and STS-4xx for the STS-125 Hubble Servicing Mission 4.

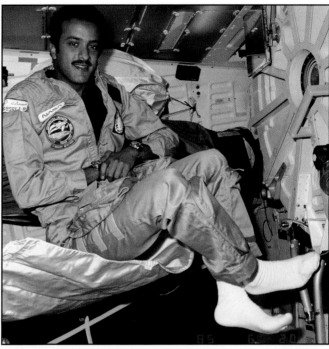

One of only four true commercial payload specialists to fly, Charlie Walker is in front of the McDonnell Douglas electrophoresis operations in space (EOS) experiment where the middeck accommodations rack (MAR) would later fly. (NASA)

Sultan bin Salman bin Abdulaziz Al Saud represented the Arab Satellite Communications Organization (ARABSAT) during the deployment of Arabsat-A on STS-25/51G. Al Saud was also the youngest person to fly on space shuttle, at 28 years, 11 months, 21 days. (NASA)

The largest space shuttle crew was STS-30/61A, which carried eight crew members to operate Spacelab D1. Here, commander Hank Hartsfield shakes hands with George Abbey, the director of flight crew operations at JSC. Other crew members, from left to right, are Guy Bluford, Jim Buchli, Steve Nagel, Bonnie Dunbar, Wubbo Ockels, Ernst Messerschmid, and Reinhard Furrer. The crew is wearing the light blue launch-entry coveralls that were customary before the Challenger accident. (NASA)

Atlantis as STS-125 on LC-39A (foreground) and Endeavour as the STS-400 launch-on-need rescue vehicle (subsequently flown as STS-127) on LC-39B in late April 2009. Some significant changes had been made to LC-39B for the scheduled mid-summer Ares I-X demonstration flight for the ill-fated Constellation Program. The most obvious are the three new lightning towers surrounding the pad. The Space Shuttle Program used a single lightning tower on top of the fixed service structure (FSS). The original hammerhead cranes had long since been removed from the FSS of both pads. If it would have been necessary to launch STS-400, the vehicle would have moved to LC-39A. (NASA)

In August 1972, NASA expected to fly the first of six orbital tests flights in February 1978. By May 1974, a combination of funding and technical issues had delayed the flight to November 1978. However, these were internal working dates and were not formally advertised by the program. Robert Thompson, the first space shuttle program manager, soon established the first official date as March 1979, a date that remained constant for almost four years.

In July 1977, an independent blue-ribbon panel suggested a more reasonable date was July 1979, although NASA declined to change the schedule and instead compressed the processing timeline at KSC by three months. It was a bit laughable since most engineers expected to miss the date by at least a year. In November, an assessment team commissioned by John Yardley, the associate administrator for space transportation systems, concluded all program elements could be ready for the first flight by 29 September 1979. As a result, the program moved the internal planning date to September 1979, but did not change the official date.

An external tank review in April 1978 determined that ET-1 could not be delivered before July 1979, finally forcing NASA to officially change the official March 1979 date. Two months later, NASA administrator Robert Frosch commissioned a detailed assessment of the schedule that recommended changing the date to December 1979. It was a case of sneaking up on the truth.

This date did not last long. As the result of a main engine failure on 27 December 1978, Bob Thompson moved the target to 9 November 1979, the first time a specific date had been selected (instead of just a month). In March 1979, NASA deputy administrator Alan Lovelace conducted an alternate planning study that found a more realistic date was June 1980. Three months later, based on a KSC assessment of the likely processing schedule for the first set of flight hardware, Bob Thompson moved the internal planning date to March 1980, but did not change the official schedule.

In November 1979, the Space Shuttle Program Office determined the most likely date for the first flight was December 1980, although by February 1980, Bob Thompson had conceded a more realistic date was mid 1981. A year later, November 1980, space shuttle had still not flown and Robert Frosch again conducted a review of the program. As a result, Frosch reported the plan was to roll *Columbia* out of the OPF on 23 November 1980 with the intent "to launch by the end of March 1981, although we recognize that this is a tight schedule." He expected the six orbital flight tests to consume about 18 months, leading to an initial operational capability in September 1982. It was a schedule that could not, politically, slip again. Fortunately, everything finally came together and the schedule (mostly) held.

The first major flight hardware milestone at KSC was a hot-fire of the SRB thrust vector control system on 25 May 1979. The first live solid rocket motor segment arrived on 10 September and was stacked onto the waiting aft skirts on 27 September. The aft skirts and their attached segments were moved onto the MLP in High Bay 3 on 26 November. This was the beginning of the STS-1 stacking operation, some 17 months before its eventual launch.

The first flight external tank, ET-1, arrived from the Michoud Assembly Facility on 6 July 1979 and was moved into High Bay 2 for storage. After the first two flight tanks were completed, engineers determined they needed to add insulation to external lines on the ET to prevent ice formation. To maintain the schedule for the first manned orbital flight, NASA elected to modify the tanks at the Kennedy Space Center. However, the changes increased the diameter of the lines, causing interference with the flip-down platforms in High Bay 4, so the tank had to wait in High Bay 2 until the facility modifications were complete. On 3 November 1980, the tank was hoisted from High Bay 4 and mated to the waiting solid rocket boosters in High Bay 3.

The first external tank arrived at the Kennedy Space Center on 6 July 1979. The first two flight tanks included a coat of white fire-retardant latex paint to protect the insulating foam. Engineers subsequently decided this was unnecessary, saving 595 pounds on later tanks. (NASA)

The first solid rocket motor segment arrived on 10 September 1979 and was unloaded in the Vehicle Assembly Building since dedicated SRB facilities would not exist for several years. The non-motor parts of the booster had been at KSC for most of the previous year. (NASA)

The first manned orbital flight had already slipped one year and there looked to be another year of hard work prior to flight. The delays were taking their toll on everybody. During June 1980 interview with the *New York Times*, John Young commented, "By the time we get to fly, I'll be 130 percent trained for the mission." The delays were reportedly costing $1.3 million per week.

Finally, after spending 613 days in the OPF, *Columbia* was rolled-over to the VAB and mated to the stack on 24 November 1980. The entire stack was powered up for the first time on 4 December. Sitting atop MLP-1, the first complete flight-rated space shuttle stack rolled-out to LC-39A on 29 December 1980, making the trip in slightly more than 10 hours.

Engineers conducted an LH_2 tanking test on 22 January 1981 and an LO_2 tanking test on 24 January to verify the ground and flight systems would play nicely and allow firing room personnel to gain some real-life experience. However, during the tanking tests several large pieces of spray-on foam insulation debonded from ET-1. After investigating the problem, Martin Marietta elected not to attempt repairs until after the scheduled flight readiness firing (FRF), so technicians draped a cargo net over the affected area to keep debris from falling and impacting the orbiter.

NASA held a countdown demonstration test (CDDT) from 16 to 20 February 1981 to demonstrate the ability of the launch processing system and provide additional experience for the launch team. Other objectives were to exercise the propellant loading procedures, verify the SSME shutdown sequence, and demonstrate the launch pad and ground-support equipment would survive the SSME firing and shutdown environment. The CDDT culminated in a 20-second flight readiness firing at 08:45 on 20 February 1981.

There were two compelling reasons to conduct the FRF. The first was as an all-up test of the mated space shuttle stack. This was the first opportunity to flow propellants into a flight ET through a real orbiter, although tests at the National Space Transportation Laboratory had already verified the essential design. The FRF also included main engine gimbal tests at 94 and 100 percent power, another capability already demonstrated at the NSTL. The other reason was less obvious from a vehicle perspective. Unlike most previous programs, there had been no dedicated facility verification vehicle, so this was also the first time the major ground system components had flowed cryogenic propellants in any quantity.

A month later, between 13 and 16 March, KSC conducted a launch readiness verification test. With the flight crew aboard *Columbia* and the all subsystems powered up, the launch processing system sequenced the vehicle systems through simulated launch, ascent, return to launch site abort, and entry profiles.

After the FRF, technicians from Martin Marietta repaired the foam on ET-1, a process that took most of a month given the difficulty accessing the areas on the launch pad. Unfortunately, tragedy struck on the morning 19 March when five Rockwell technicians entered the nitrogen-purged aft fuselage: John Bjornstad died at the scene, Forrest Cole died 13 days later, and Nicholas Mullon succumbed after several years from complications resulting from nitrogen exposure.

More tanking tests on 25 and 27 March verified the repairs to the ET and allowed engineers to verify minor tweaks to the launch processing system. Subsequently, the launch readiness review set the date for the first orbital flight test: 10 April 1981. A total of 24 months, 19 days elapsed between when *Columbia* arrived at KSC on 24 March 1979 and the launch of STS-1 on 12 April 1981.

The Goodyear blimp makes a low pass near LC-39A with the STS-1 stack. At this point much of the fixed service structure was still in the red paint it had worn when the pieces had been part of a Saturn launch umbilical tower; slowly NASA would paint the entire structure the bland gray seen during later missions. Otherwise, the configuration of the launch pads remained remarkably stable during the thirty-year flight campaign, with the largest visual difference being the removal of the hammerhead crane from the top of the fixed service structure during the 1990s. (NASA)

STS-1 (OFT-1)

Mission:	1	NSSDC ID:		1981-034A
Vehicle:	OV-102 (1)	ET-1 (SWT)		SRB: BI001
Launch:	LC-39A	12 Apr 1981		12:00 UTC
Altitude:	172 nm	Inclination:		40.30 degrees
Landing:	EDW-23	14 Apr 1981		18:22 UTC
Landing Rev:	37	Mission Duration:		54 hrs 21 mins

Commander:	John W. Young (5)
Pilot:	Robert L. "Crip" Crippen (1)

Payloads:	Up: 10,823 lbs	Down: 0 lbs
	DFI Pallet (9,290 lbs)	
	IECM (816 lbs)	
	ACIP	
	No SRMS	

Notes: Initial planning called the mission "SS-1"
                  Also called first manned orbital flight (FMOF)
                  Flight readiness firing on 20 Feb 1981
                  Scrubbed 10 Apr 1981 (GPC timing issue)
                  First flight of Columbia (OV-102)
                  First use of mobile launch platform one (MLP-1)
                  First time a maiden flight had carried a crew
                  First orbital flight by a winged spacecraft
                  Launch was three years later than the original plan
                  21 TPS tiles missing and 530 damaged
                  Orbiter returned to KSC on 28 Apr 1981 (N905NA)
                  Backup crew was
                     Joe H. Engle and Richard H. Truly

Wakeup Calls:

     13 Apr   "Blast-Off Columbia"
     14 Apr   "Reveille"

Designed largely by John Young and penned by space artist Robert McCall, the crew patch showed an orbiter lifting off and on-orbit around the Earth. The crew member surnames and orbiter name are displayed in the flame from the main engines. (NASA)

In most previous programs, integrated tests were conducted at test sites prior to the vehicles arriving at the launch site. However, on space shuttle, the first time all of the flight systems came together was at KSC. Each element had gone through rigorous testing—there had been 726 tests on 24 SSMEs totaling 110,252 seconds and 7 full-scale SRB firings prior to STS-1—but this was the first time a full set of flight-rated hardware had been assembled. Because of this, the program conducted a flight readiness firing (FRF). Everything was controlled from the launch control center at KSC and the crew was not in *Columbia* for the test. Essentially, the FRF was identical to a launch, except the SRBs were not ignited and the hold-down bolts were not blown. At the end of the 20-second firing on 20 February 1981, the launch team sequentially shut down the three SSMEs to simulate a pre-launch abort. Afterward, engineers combed through the data and carefully inspected the vehicle before declaring it ready for launch.

Engineers scrubbed a launch attempt on 10 April at T-20 minutes due to a timing slew when the backup flight system (BFS) failed to synchronize with the primary avionics software system (PASS). IBM diagnosed the problem and installed a software patch.

The mission was launched at 12:00 UTC on 12 April 1981 as a two-day demonstration the ability of the vehicle to ascend to orbit, conduct on-orbit operations, and return safely. Due to a late engineering reassessment of the "twang" motion after main engine start, STS-1 lifted off 4 milliseconds after T-0, a condition that was corrected for subsequent flights. The primary payload was a developmental flight instrumentation (DFI) pallet that recorded temperatures, pressures, and acceleration levels at various locations on the vehicle. STS-1 marked the first use of solid rockets on a manned vehicle, and the first time a crew had been aboard a new type of spacecraft on its maiden flight.

NASA planned all of the orbital flight test missions to land at Edwards AFB, and *Columbia* settled onto lakebed Runway 23 at 18:22 UTC on 14 April. The primary thermal protection system damage was to the OMS pods, although a few tiles around the nose and main landing-gear doors and the body flap were also a concern. Post-flight inspection showed the orbiter had lost 21 tiles and suffered minor damage to 530 others. In addition, the FRSI in the elevon coves and OMS pods were damaged. Much of the damage was attributed to an overpressure caused at SRB ignition. Approximately 300 tiles were replaced before the next flight, with the minor dings being repaired. Otherwise, *Columbia* was in good condition for a "used" spaceship.

John Young (left) and Bob Crippen give a thumbs-up on the flight deck of Columbia in an Orbiter Processing Facility at KSC. For the four orbital flight tests, Columbia was equipped with Lockheed SR-71 ejection seats and the crew wore David Clark Company full-pressure suits. (NASA)

Looking through an aft flight deck window into the payload bay shows the missing tiles on the OMS pods. The aero and thermal loads on the OMS pods had been severely underestimated during development and the thermal protection would evolve considerably for later flights. The developmental flight instrumentation pallet is in the foreground. (NASA)

The maiden launch of space shuttle. Only the first two missions used painted, white external tanks, and they can be differentiated by the black lightning-protection stripe around the top of the STS-1 tank. The stripe was originally installed on the STS-2 tank, but was removed sometime during processing in the Vehicle Assembly Building. (NASA)

This is what the STS-1 external tank looked like as it fell back to Earth after separation. Note the significant scorching on the back of the tank where hot gases from the SSME and SRB exhaust circulate, and the scorched ogive from ascent aerodynamic heating. (NASA)

The flight readiness firing (FRF) on 20 February 1981 was the first all-up test of the space shuttle stack. The test was essentially identical to a launch except the SRBs were not ignited and the crew was not in the orbiter. Compare the clean ET to the photo at left. (NASA)

STS-2 (OFT-2)

Mission:	2	NSSDC ID:	1981-111A
Vehicle:	OV-102 (2)	ET-2 (SWT)	SRB: BI002
Launch:	LC-39A	12 Nov 1981	15:09 UTC
Altitude:	137 nm	Inclination:	38.00 degrees
Landing:	EDW-23	14 Nov 1981	21:24 UTC
Landing Rev:	37	Mission Duration:	54 hrs 13 mins

Commander:	Joe H. Engle (1 + X-15 flights)
Pilot:	Richard H. "Dick" Truly (1)

Payloads:	Up: 18,778 lbs	Down: 0 lbs
	DFI Pallet (9,290 lbs)	
	IECM (816 lbs)	
	OSTA-1 (SIR-A) Pallet (5,395 lbs)	
	OFT Pallet	
	ACIP	
	SRMS s/n 201	

Notes:
Delayed 09 Oct 1981
Scrubbed 04 Nov 1981 (APU issues)
Only all-rookie space shuttle crew
First re-use of a manned spacecraft
First use of shuttle remote manipulator system
Flight shortened due to fuel cell failure
Shortest mission of the flight campaign
447 TPS tiles damaged
Orbiter returned to KSC on 25 Nov 1981 (N905NA)
Backup crew was
    Jack R. Lousma and C. Gordon Fullerton

Wakeup Calls:

13 Nov	"Pigs in Space" comedy routine #1
14 Nov	"Pigs in Space" comedy routine #2

The second orbital flight test was scheduled seven months after the first to give engineers time to evaluate the data collected on the first flight, and to make minor changes to the vehicle and launch complex. The original launch date of 9 October 1981 was rescheduled when a nitrogen tetroxide spill occurred during the loading of the FRCS module, necessitating repairing 379 tiles. A second attempt on 4 November was delayed by a low reading on a fuel cell oxygen tank pressure transducer, and then aborted at T-31 seconds when clogged filters in an auxiliary power unit caused it to overheat.

Columbia was launched on 12 November after a 2-hour, 40-minute delay to replace a multiplexer/demultiplexer. It was the first manned flight of a "used" spacecraft and Dick Truly became the first person launched on his birthday. This was the first all-rookie crew carried by space shuttle and the last American all-rookie crew. A post-launch inspection revealed significant erosion of the primary o-ring in the right SRB aft field joint. This was the first indication of a serious issue with the joints, and the erosion was the deepest experienced until the loss of the *Challenger* during STS-33/51L.

The orbiter carried a payload of Earth survey instruments from the NASA Office of Space and Terrestrial Applications (OSTA-1), as well as experiments mounted on a prototype Spacelab pallet. The orbiter also carried the same developmental flight instrumentation pallet as STS-1. This flight marked the first test of the shuttle remote manipulator system (SRMS), which was exercised using several different control modes.

Initially, all three fuel cells operated normally on-orbit. However, about two and a half hours later the performance of the fuel cell began to rapidly decrease, so the crew took it offline and shut it down. The other two fuel cells continued to operate normally, but flight controllers elected to shorten the planned five-day flight by three days. Nevertheless, the crew still accomplished more than 90 percent of the flight test objectives, including a test where Joe Engle manually flew significant portions of the entry. This was the shortest mission of the flight campaign.

Modifications to the launch pad sound suppression water system successfully absorbed the SRB overpressure and only 12 significantly damaged tiles were noted when *Columbia* returned to the Edwards lakebed. In addition, 97 tiles had minor damage, mostly caused by debris during ascent. Prior to STS-2, engineers had replaced the original FRSI in the elevon coves with new AFRSI blankets, which held up much better. The elevon/elevon ablator strips performed similarly to STS-1, but engineers were already working toward replacing these with HRSI tiles in the future.

Columbia is depicted along with the crew members surnames and an eagle merged with the American flag. The two stars represent the number of crew members, the second flight of Columbia, and the mission's STS designation. (NASA)

Dick Truly (left) and Joe Engle continuing a comedy they had started during the Approach and Landing Tests by wearing old aviator clothing for a gag crew portrait. The pair wore the same headgear as they egressed Columbia after landing at Edwards. (NASA)

The payload bay was more crowded on STS-2, and the crew tested the SRMS for the first time. The large object at left is the antenna for the OSTA-1 payload, also called SIR-A (shuttle imaging radar). Much improved radar would fly on several subsequent missions. (NASA)

STS-2 is difficult to distinguish from STS-1 since each used a white ET. The most recognizable difference is the lack of a black lightning-protection ring around the ogive of the ET used on STS-2. Note the large hammerhead crane on top of the fixed service structure. (NASA)

Although the reusable tiles got all the publicity, the orbiters initially used ablators in some small areas such as these ablative strips at the elevon/elevon interface between the two control surfaces. Engineers soon replaced the ablative strips with black HRSI tiles. (NASA)

Columbia touches down on Rogers Dry Lake at Edwards AFB. Much to the disappointment of the crew, a malfunctioning fuel cell cut the mission significantly short. Fortunately, each man would get the chance to fly again, Engle on STS-27/51I and Truly on STS-8. (NASA)

STS-3 (OFT-3)

Mission:	3	NSSDC ID:	1982-022A
Vehicle:	OV-102 (3)	ET-3 (SWT)	SRB: BI003
Launch: Altitude:	LC-39A 130 nm	22 Mar 1982 Inclination:	16:00 UTC 38.00 degrees
Landing: Landing Rev:	NOR-17 130	30 Mar 1982 Mission Duration:	16:06 UTC 192 hrs 5 mins
Commander: Pilot:	Jack R. Lousma (2) C. Gordon "Gordo" Fullerton (1)		

Payloads:	Up: 22,710 lbs	Down: 0 lbs
	DFI Pallet (11,048 lbs)	
	IECM (816 lbs)	
	OSS-1 Pallet (8,740 lbs)	
	PDP (344 lbs)	
	ACIP	
	GAS (verification canister)	
	SRMS s/n 201	

Notes:	First unpainted external tank
	Third orbital test flight (OFT-3)
	Edwards lakebed runways flooded
	Only landing at White Sands
	37 TPS tiles missing and 77 damaged
	Orbiter returned to KSC on 6 Apr 1982 (N905NA)
	Backup crew was
	Thomas K. Mattingly II and
	Henry W. Hartsfield, Jr.
	Last mission with a full backup crew

Wakeup Calls:

23 Mar	"On the Road Again"
24 Mar	The Marine Corps Hymn – "Halls of Montezuma"
25 Mar	The Air Force Hymn – "Wild Blue Yonder"
26 Mar	"Sail Away"
27 Mar	"Those Magnificent Men and Their Flying Machines"
28 Mar	"Six Days on the Road"
29 Mar	"This is My Country

Columbia *is depicted in the middle of the blue sphere against the background of the Sun. The three prominent rays represented the STS designation. The payload bay doors are open and the SRMS arm is extended. The patch was designed by Robert McCall. (NASA)*

Experience was lowering the processing time between flights, and *Columbia* spent only 69 days in the Orbiter Processing Facility after STS-2 being prepared for the third orbital flight test, compared to 103 days for between the first and second flights. Likewise, the time in the Vehicle Assembly Building was reduced from 53 days for STS-1 to 17 days for STS-3. The number of days at the pad was reduced from 104 days for STS-1 to 73 days for STS-2 to 34 days for STS-3. For the most part, the processing time would continue to be reduced as the ground operations team gained experience, reaching their best just before the *Challenger* accident.

The launch was delayed for an hour by a failure of a heater on a ground-based nitrogen gas purge line. *Columbia* stayed on-orbit for eight days, making this the longest of the OFT missions. Activities included an SRMS test that removed a package of instruments from the payload bay, but did not release it. The flight included experiments in materials processing, and thermal testing of the orbiter. The latter was accomplished by exposing the tail, nose, and top of the orbiter to the Sun for varying periods of time, rolling it in between tests to stabilize temperatures over the entire vehicle.

For the first time, experiments were carried on the middeck, including CFES (continuous flow electrophoresis system) to study separation of biological components and MLR (monodisperse latex reactor) to produce uniform micron-sized latex particles. The first Shuttle Student Involvement Project (SSIP)—the study of insect motion—also was carried in a middeck locker.

This marked the first flight of an unpainted external tank, saving about 595 pounds. During the flight, both crew members experienced some space adaptation syndrome, the potty malfunctioned, one auxiliary power unit overheated (but worked properly during entry), and three communications links were lost on 26 March. Heavy rains at Edwards caused the landing site to be changed to Northrup Strip at White Sands, New Mexico, although high winds at this site still caused the mission to be extended one day. Some brake damage occurred during landing and an unexpected dust storm caused extensive contamination of the orbiter with the white gypsum dust that is common to that area of the New Mexico desert.

Since Northrup Strip was not equipped with a Mate/Demate Device to lift the orbiter onto the 747 shuttle carrier aircraft, a stiff-leg derrick was moved from MSFC and two large cranes were rented locally. Initially, Air Force was going to move the equipment, but NASA ultimately moved the equipment using two dedicated trains.

Columbia returned to KSC on 6 April 1982 with 36 tiles missing and another 77 damaged with dings and gouges.

Jack Lousma (left) and Gordo Fullerton wearing their David Clark Company S1030A full-pressure suits. The straps connected to their neck rings were used to pull the helmets down since the pressure in the suits tended to push the helmets upward. (NASA)

Heavy rains at Edwards forced NASA to divert landing to another lakebed runway since the program was not ready to use the concrete runway at KSC. STS-3 was the only mission to end at White Sands, and white gypsum dust penetrated everywhere in the orbiter. (NASA)

An unusual view of the payload bay with only one door open. Each payload bay door had a highly polished silver radiator on it. Note the stowed SRMS arm at right and the OSS-1 pallet in the foreground. This was the first flight to move an object using the SRMS. (NASA)

Thirty-six tiles were missing, all either on the top of the body flap under the main engines (shown) or on top of the forward reaction control system module in front of the windscreen. Fortunately, neither area was exposed to extreme temperatures during entry. (NASA)

Northrup Strip was not equipped with a Mate/Demate Device to load the orbiter onto the 747, so a stiff-leg derrick (large device in back) and two cranes were used instead. The same derrick had been used when Enterprise was delivered to MSFC for vibration testing. (NASA)

This was the first mission to use an unpainted external tank, and the stack looked considerably different than it had with the white tanks. Martin Marietta had already painted the nose cone and cable trays so they remained white for the first few unpainted tanks. (NASA)

STS-4 (OFT-4)

Mission:	4	NSSDC ID:	1982-065A
Vehicle:	OV-102 (4)	ET-4 (SWT)	SRB: BI004
Launch:	LC-39A	27 Jun 1982	15:00 UTC
Altitude:	175 nm	Inclination:	28.52 degrees
Landing:	EDW-22	04 Jul 1982	16:11 UTC
Landing Rev:	113	Mission Duration:	169 hrs 10 mins

Commander:	Thomas K. "TK" Mattingly II (2)
Pilot:	Henry W. "Hank" Hartsfield, Jr. (1)

Payloads:	Up: 24,492 lbs	Down: 0 lbs
	DFI Pallet (9,900 lbs)	
	IECM (846 lbs)	
	DoD 82-1 (CIRRIS and UHS)	
	GAS (1)	
	SRMS s/n 201	

Notes: First DoD mission (non-dedicated)
Fourth, and last, orbital test flight (OFT-4)
SRBs not recovered due to parachute failure
First landing on a concrete runway
Orbiter returned to KSC on 15 Jul 1982 (N905NA)

Wakeup Calls:

28 Jun	"Up, Up and Away"
29 Jun	"Hold That Tiger"
30 Jun	(taped greetings to Hartsfield for anniversary)
01 Jul	(none)
02 Jul	"Chariots of Fire Theme"
03 Jul	College fraternity songs
04 Jul	"This is My Country"

This was the first space shuttle mission to be launched on schedule. This was also the first flight that communicated with a classified Air Force Mission Control Center in Sunnyvale, California, mostly to make certain the links worked in preparation for future classified DoD flights from KSC and Vandenberg. Nevertheless, for the most part, the normal NASA Mission Control Center in Houston controlled the mission. These classified missions resulted in some additional hardware in the orbiters, such as secure communications equipment (also used on NASA flights to ensure privacy when needed) and a safe in the crew module to store classified documents.

The payload consisted of the first get-away special (GAS) canister that included nine experiments provided by students from Utah State University. A classified payload, known at the time only as DoD 82-1, was later identified as CIRRIS (cryogenic infrared radiance instrumentation for shuttle) and UHS (ultraviolet horizon scanner), two sensors intended for the "Star Wars" Strategic Defense Initiative for detecting missiles from space. A cover failed to open, so neither worked. On the middeck, the CFES (continuous flow electrophoresis system) and MLR (monodisperse latex reactor) flew for the second time. The crew conducted a lightning survey with handheld cameras, performed medical experiments on themselves for two student projects, and operated the SRMS with the induced environment contamination monitor (IECM) mounted on its end. The IECM was designed to obtain information on emissions from the orbiter in flight. As it had on the two previous flights, the SRMS experienced minor failures: on STS-2 the shoulder joint failed in the backup mode, on STS-3 the wrist camera failed, and on STS-4 the end effector reported incorrect status due to a broken wire. All the issues were easily correctable and to be expected in such a complicated mechanism that could not be adequately tested on the ground.

The only major problem on this flight was the loss of the two solid rocket boosters when the main parachutes failed to deploy properly and they impacted the water at a high velocity and sank. They were later found and examined by remote camera, but were not recovered. Other than the set destroyed during the *Challenger* accident, these were the only boosters that were not recovered (although several others were damaged enough to preclude reuse).

With the landing of STS-4 on the concrete runway at Edwards, the orbital flight test program came to an end with 95 percent of its original objectives completed. Since more than the two pilots would be carried on future flights, NASA disabled the ejection seats on *Columbia* following this flight.

Hank Hartsfield (left) and TK Mattingly pose for the official portrait in their S1030A suits. This was the last flight to use full-pressure suits until the S1035 was phased-in beginning with STS-64 (the S1032 used on STS-26R through STS-88 was a partial-pressure suit). (NASA)

The crew patch showed Columbia trailing the national colors in the shape of her flight number, representing the fourth and final flight of the orbital flight test series. Columbia then streaks into the operational era scheduled to begin with STS-5. (NASA)

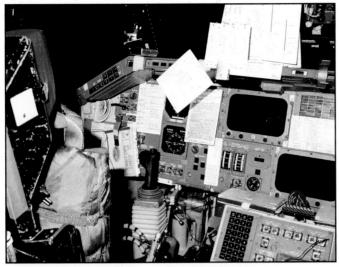

Despite the heavily reliance on computers and automation, most crews plastered the inside of the orbiter with checklists and instructions while they were on-orbit. As the flight campaign continued, NASA permanently affixed velcro to many surfaces to attach things to. (NASA)

Columbia gliding toward her first landing on a concrete runway. The effects of aerodynamic heating during entry had left marks along the side of the fuselage, especially at the transition between the white LRSI tiles and the blankets. Note the T-38 chase plane. (NASA)

Both SRBs were lost after their recovery parachutes failed. NASA sent unmanned submersible vehicles 3,500 feet under the Atlantic to photograph the boosters. Without the parachutes, the SRBs hit the water hard enough to break the tool-steel casings. (National Archives)

Firing Room 1 (FR-1) at KSC was one of two launch control rooms used for most of the flight campaign (the other was Firing Room 3, used for all DoD launches). After the STS-107 accident, NASA modernized Firing Room 4 and used it for the final 29 launches. (National Archives)

Workers at KSC watch as a crane operator lowers the external tank between the already stacked SRBs. Note the feedlines and attach hardware are still painted white since they had been built before the decision to delete the paint on the external tank. (NASA)

The landing of STS-4 marked the end of the orbital flight test effort, having accomplished some 95 percent of its original objectives. Perhaps more importantly, the interval between flights had been trimmed from seven months to four, then three. NASA declared space shuttle "operational," a term that quickly encountered criticism from many quarters because it erroneously suggested the vehicle had attained an airline-like degree of routine operation. More criticism would follow 20 years later. In any event, NASA regarded all flights after STS-4 operational in the sense that payload requirements took precedence over spacecraft testing.

Becoming operational had profound effects on the Space Shuttle Program and its contractors. At the beginning of the space age, there were only a few contractors involved in developing and launching missiles. Principally out of necessity, when a vehicle finally arrived at the launch site, its development contractor came with all the resources necessary to test and fly the vehicle, sharing little information or resources with other programs. Not a particularly efficient way to conduct business, but at the height of the space race, nobody really cared. This began to change as Apollo matured. In January 1963, secretary of defense Robert McNamara and NASA administrator James Webb signed an agreement that consolidated many resources at the Cape Canaveral Air Force Station and Kennedy Space Center. In reality, the practice did not work all that well except for a few laboratories and in non-technical areas such as facility maintenance and housekeeping.

When space shuttle began, it carried over much of the previous practice. Each of the element contractors, Rockwell, USBI, and Martin Marietta (called the RUM contractors, for obvious reason) created their own empire and seldom shared resources with one another. This was partly the result of hard-won experience on previous programs, partly the result of arcane government contracting rules, and partly a desire to be in control of their own destiny. At

the end of the orbital flight tests, there were 14 major contractors supporting space shuttle at JSC, 18 at KSC, and 4 at MSFC. These included the five major hardware contractors (Martin Marietta, Morton Thiokol, Rocketdyne, Rockwell, and USBI) and a myriad of other development (particularly IBM) and support contractors.

In 1983, the NASA Advisory Council issued a report recommending that operating the Space Shuttle Program be "fenced" off from the rest of NASA to ensure an operational program did not decimate other efforts and that operational costs were tracked separately from research and development expenses. In response, NASA made sweeping contractual changes for the space shuttle operational ("ops") era. The government organization was streamlined, with NASA Headquarters giving up much of its control to the Space Shuttle Program Office at JSC.

Although none of these changes directly affected the five hardware contractors, during 1983–84 NASA streamlined all of the support contracts, resulting in three contractors at JSC, three at KSC, and four at MSFC. In particular, Rockwell took over all space shuttle operations at JSC as part of the space transportation system operations contract (STSOC) and Lockheed did the same at KSC and Vandenberg as part of the shuttle processing contract (SPC). These two contracts would continue until the government further combined the efforts into the space flight operations contract (SFOC) awarded to United Space Alliance in 1996.

HARDWARE CHANGES

Between June and October 1982 technicians made numerous small changes to *Columbia* at KSC to prepare her for the initial operational missions. These included installing the payload signal processor and payload data interleaver necessary to accommodate the PAM-D upper stage being used by the STS-5 payloads, installing a mission

At KSC, the conversion of Apollo-era facilities and equipment continued. Here, a section of an Apollo launch umbilical tower is being moved to LC-39B to become part of the space shuttle fixed service structure. Another Apollo mobile launcher and LUT is in the background. (NASA)

The LC-39 area in 1983. The two Orbiter Processing Facilities are complete at the top left and the Vehicle Assembly Building is in the middle top. The area at the bottom would later become the solid rocket booster complex. The launch pads are out of sight to the right. (NASA)

specialist seat on the flight deck behind the center console and the left middeck seat, strengthening the middeck floor, and removing parts of the developmental flight instrumentation pallet. The ejection seats were disabled after STS-4 and removed at Palmdale following STS-9. Around the same time, technicians replaced the ablator strips between the elevons with HRSI tiles after researchers determined the temperatures were not as severe as expected.

In addition, crew comfort items like the potty, galley, and sleepstations began to find their way onto the orbiters. There were also changes to the ground systems at JSC and KSC, with additional flight control and firing rooms coming online, a second launch pad being readied, and other infrastructure being converted from its Apollo configuration to one that supported space shuttle. NASA and the contractors were putting additional processes and procedures in place that reflected the move from development to operations, or at least what passed for operations within the manned spaceflight community. It would be an exciting three and a half years.

This rush to declare the space shuttle operational would come under heavy criticism during the investigation that followed the *Columbia* accident in 2003. The Columbia Accident Investigation Board asserted the program should never have been considered operational and, while not intrinsically unsafe, it was in fact a developmental vehicle. Echoing the sentiments from two decades earlier, the CAIB explained that civilian and military aircraft were evaluated over hundreds or thousands of flights before being declared operational, while the space shuttle had, in 1983, completed just four flights. In fact, twenty years later, in 2003, the program had completed just 113 flights, including two fatal accidents.

MANIFEST MANIA

On 8 June 1979, the program had released a manifest that showed 37 missions between the first orbital test flight in February 1980 and September 1983, approximately one mission per month. The manifest released on 31 March 1980 showed 39 missions from KSC and 4 from Vandenberg through the end of September 1984. These were very conservative manifests.

By the end of the orbital flight tests, the manifest had matured considerably. There were 62 missions scheduled from KSC between November 1982 and September 1987 and 8 from Vandenberg between October 1985 and September 1987. This was an average of 12 flights annually; something considered well within the capability of the launch sites and hardware production lines.

An assessment by the National Research Council during 1983 concluded a four-orbiter fleet could sustain an annual 18-mission manifest, while five orbiters could support 24 missions and six orbiters could manage 30 flights annually. The review found that the existing SSME production and maintenance capability could support the 18-mission manifest but was marginal for 24 flights and could not support anything greater. The solid rocket motor capability was marginal to support even 18 missions annually, although proposed (and later implemented) expansions in Utah would allow up to 24 missions. The solid rocket booster effort could sustain 18 missions and be expanded to 30 relatively easily. The equipment installed at the Michoud Assembly Facility could produce 24 external tanks annually, but the assessment determined sufficient room existed in the factory to manufacture up to 60 tanks per year if needed.

However, the Space Shuttle Program knew it would never receive the funds to create the infrastructure to exceed what the existing facilities could support. Therefore, throughout 1984 and 1985, most manifests showed ramping up to 24 missions annually, but all thoughts of anything greater had already disappeared.

For instance, a manifest published on 13 August 1984 showed 11 missions in FY85, 15 in FY86, 22 in FY87, and 24 each in FY88 and FY89. One published a year later, on 7 June 1985, gave slight concessions to the processing issues being encountered at KSC and showed 14 in FY86, 17 in FY87, 19 in FY88, and 24 each in FY89 and FY90. This was the launch rate (24) that became the Holy Grail, at least until everything changed on 28 January 1986.

The Air Force built a space shuttle launch complex at Vandenberg AFB, California, to support launching into high-inclination polar orbits. Surprisingly, most of the missions scheduled to fly from SLC-6 were not military, but were civilian Earth resources satellites and science missions, This is Enterprise, with ET-23 and an inert set of steel SRBs during the facility verification tests. (Dennis R. Jenkins)

Vandenberg, as completed in 1986. The Payload Preparation Room (PPR) is at the left while the Shuttle Assembly Building (SAB) and Mobile Service Tower (MST) are docked together over the launch mount where the stacked vehicle would have sat. (U.S. Air Force)

STS-5

Mission:	5	NSSDC ID:	1982-110A
Vehicle:	OV-102 (5)	ET-5 (SWT)	SRB: BI005
Launch:	LC-39A	11 Nov 1982	12:20 UTC
Altitude:	162 nm	Inclination:	28.48 degrees
Landing:	EDW-22	16 Nov 1982	14:34 UTC
Landing Rev:	82	Mission Duration:	122 hrs 14 mins

Commander:	Vance D. Brand (2)
Pilot:	Robert F. "Bob" Overmyer (1)
MS1:	Joseph P. "Joe" Allen IV (1)
MS2:	William B. "Bill" Lenoir (1)

Payloads:	Up: 32,080 lbs	Down: 0 lbs
	SBS-C/PAM-D (7,211 lbs)	
	Anik C3/PAM-D (7,374 lbs)	
	GAS (1)	
	No SRMS	

Notes:
First operational flight
First space shuttle with more than two crew
Largest American crew flown to-date
First American crew not to wear pressure suits
Planned EVA not performed (Lenoir and Allen)
Orbiter returned to KSC on 22 Nov 1982 (N905NA)

Wakeup Calls:

12 Nov	"76 Trombones"
13 Nov	"Cotton-eyed Joe"
14 Nov	The Marine Corps Hymn – "Halls of Montezuma"
15 Nov	"The Stroll"
16 Nov	"Take Me Home, Country Roads"

A five-pointed star represented STS-5, the first operational mission. The crew member surnames along with the orbiter name were along the border of the crew patch, with two satellites representing the first operational payloads being delivered on-orbit. (NASA)

This was the second on-time launch and the first "operational" space shuttle mission. This was the first mission where the payloads were installed vertically using the payload changeout room at the launch pad rather than horizontally in the Orbiter Processing Facility. Unlike the previous mission, the SRB parachutes worked and the boosters were successfully recovered. This was the first launch with astronauts taking advantage of the "shirt-sleeve" environment in the crew module, marking the first American crew not to wear pressure suits during ascent, similar to Soviet Voskhod and Soyuz flights prior to the ill-fated Soyuz 11 mission. The crew wore NASA-blue flight suits and Gentex Corp. launch-entry helmets that provided supplemental oxygen in case of smoke in the cabin and also provided limited bump protection for the head. This was the largest American space crew yet, featuring four astronauts that advertised themselves as the "We Deliver" team.

Columbia deployed two Hughes HS-376 communications satellites with attached McDonnell Douglas payload assist modules (PAM-D). The upper stages, using Thiokol Corporation Star 30 solid propellant motors, placed the satellites into near-synchronous orbits, and each satellite then drifted to its final orbital location. Traditionally, many operators assigned satellites a letter prior to them arriving in their final orbital position, then switched to a corresponding number. Satellite Business Systems (a consortium of IBM, Aetna, and Comsat) owned SBS-C (later, SBS-3) and Telesat Canada owned Anik C3. The SBS payload was deployed on the first flight day on-orbit, and the Canadian payload on the second flight day.

In addition, the flight carried a West German-sponsored microgravity experiment in a get-away special (GAS) canister in the payload bay. This flight was supposed to mark the debut of the extravehicular mobility unit (EMU) spacesuits, but a planned extravehicular activity (EVA) by Joe Allen and Bill Lenoir was canceled when a pressure regulator in one suit and a ventilation fan in the other malfunctioned.

During landing the crew performed a maximum braking test, and the left main inboard wheel locked up during the last 50 feet of the rollout because of a brake failure.

When *Columbia* returned to KSC on 22 November 1982, it entered OPF-2 for the modifications needed to carry the Spacelab module for its next flight, STS-9. Other modifications included adding more seats to the crew module, strengthening the mid-fuselage structure, adding crew sleepstations, landing-gear and brake modifications, removing the developmental flight instrumentation, and various upgrades to the thermal protection system.

The STS-5 crew show off their soon-to-become traditional light blue launch-entry coveralls and the new plain-white launch-entry helmets; the helmets would become more colorful on later missions. From left are Joe Allen, Vance Brand, Bob Overmyer, and Bill Lenoir. (NASA)

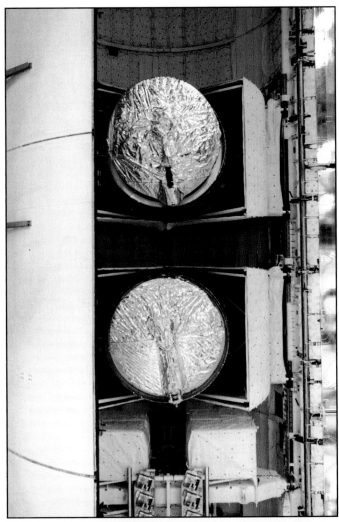

The ANIK-C3 (top) and SBS-C communication satellites are shown on 6 November 1982 as the payload bay doors close. The sunshields will close to keep the satellites cool until they are deployed. Each satellite and its PAM-D upper stage weighed more than 7,000 pounds. (NASA)

Columbia, with her distinctive black chines, sitting on LC-39A awaiting launch. Note the original wing markings, with an American flag on the left wing and "USA" on the right. Compare this to later markings that included the NASA logo and orbiter name. (NASA)

STS-5 was the first flight to use black HRSI tiles in the elevon/elevon interface instead of the ablative strips (see photo on page 49) used on the first four missions. Although difficult to initially install, the tiles proved more durable and required less maintenance. (NASA)

Columbia being towed from the Edwards runway to the NASA Dryden Flight Research Center where she would be safed and lifted onto the Shuttle Carrier Aircraft for the ferry flight back to KSC. Note the convoy trucks that supplied cooling and purge air to the orbiter. (NASA)

STS-6

Mission:	6	NSSDC ID:	1983-026A
Vehicle:	OV-099 (1)	ET-8 (LWT)	SRB: BI006
Launch: Altitude:	LC-39A 155 nm	04 Apr 1983 Inclination:	18:30 UTC 28.48 degrees
Landing: Landing Rev:	EDW-22 82	09 Apr 1983 Mission Duration:	28:53 UTC 120 hrs 24 mins

Commander:	Paul J. Weitz (2)
Pilot:	Karol J. "Bo" Bobko (1)
MS1:	F. Story Musgrave (1)
MS2:	Donald H. "Don" Peterson (1)

Payloads:	Up: 46,971 lbs	Down: 0 lbs
	TDRS-A/IUS (37,546 lbs)	
	GAS (3)	
	No SRMS	

Notes:	First flight of *Challenger* (OV-099)
	Flight readiness firing 18 Dec 1982
	Delayed 20 Jan 1983 (SSME leaks)
	First flight of a Phase I space shuttle main engine
	Flight readiness firing 25 Jan 1983
	First use of mobile launch platform two (MLP-2)
	First use of full-power-level SSMEs
	First use of lightweight SRB cases
	First use of lightweight external tank (LWT)
	First space shuttle EVA
	EVA1 07 Apr 1983 (Musgrave and Peterson)
	Shortest *Challenger* mission
	Orbiter returned to KSC on 16 Apr 1983 (N905NA)

Wakeup Calls:

05 Apr	"Cadets on Parade" / The Air Force Hymn
06 Apr	"Teach Me Tiger"
07 Apr	"Theme from F-Troop"
08 Apr	"The Poor Co-pilot"
09 Apr	"Ode to the Lions"

The sixth space shuttle flight was represented by the hexagonal shape of the crew patch and the six stars in the constellation Virgo. The sign Virgo is also symbolic of the first flight of Challenger. The TDRS/IUS payload was shown above the orbiter, heading for its final orbit. (NASA)

Like *Columbia* before them, each of the orbiters would go through a flight readiness firing (FRF) before its first launch to verify the complex main propulsion system plumbing and electronics worked as expected. In addition to the normal goals, engineers also wanted to determine the "twang" response of the new lightweight external tank and lightweight solid rocket booster cases. In the case of *Challenger*, it would take two tries.

The first launch of STS-6 was originally set for 20 January 1983, but the 20-second FRF on 18 December 1982 resulted in a cracked nozzle on one main engine and a leaky fuel line on another. A second FRF on 25 January 1983 revealed leaks in all three SSMEs. Technicians removed all three engines from the orbiter on the launch pad, and the cracked fuel lines were repaired. Two main engines were then reinstalled following extensive failure analysis and testing, while SSME-1 was replaced with a spare engine after analysis revealed the possibility of further problems.

Meanwhile, as technicians replaced the engines, a severe storm contaminated the TDRS-A payload while it was in the payload changeout room on the rotating service structure at LC-39A. The satellite had to be taken back to its checkout facility where it was cleaned and rechecked, and the PCR and orbiter payload bay were also cleaned before reinstalling TDRS.

The first flight of *Challenger* was finally launched on 4 April 1983, with no unexpected holds during the countdown. This was the first use of full-power-level SSMEs, lightweight ET, and lightweight SRB cases. The first space shuttle EVA, with Story Musgrave and Don Peterson spending 4 hours and 17 minutes outside the ship, highlighted the mission. This time the extravehicular mobility units (EMU) worked well.

Although the TDRS was deployed successfully, the inertial upper stage shut down early and stranded the satellite in a low elliptical orbit. Fortunately, the satellite carried a lot of propellant and nearly two months of short bursts of its attitude control thrusters eventually nudged it into the correct orbit, saving the $100-million communications satellite.

During the post-flight inspection at Edwards, technicians noted the AFRSI blankets on the leading edge of both OMS pods were damaged, causing structural damage. This led to a series of major ground and combined environment tests at the Air Force Arnold Engineering Development Center in Tullahoma, Tennessee. These tests resulted in the development of a new coating for the AFRSI blankets and the addition of black HRSI tile "eyeballs" on the leading edges of the OMS pods.

The official crew portrait shows the then-standard blue flight jackets and trousers. From the left are Don Peterson, Paul Weitz, Story Musgrave, and Karol Bobko. Note the model depicts Columbia with her black chines, despite this being a Challenger flight. (NASA)

Story Musgrave translating down the payload bay. A tether connects him to a safety slidewire running the length of the payload bay sill. Note the extensive use of AFRSI blankets on the leading edge of the OMS pod, unlike the white LRSI tiles used on Columbia. All of the OMS pods would later revert back to using tiles on the leading edge. (NASA)

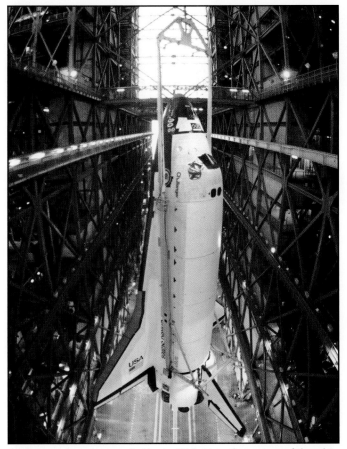

Challenger shown suspended in the VAB. Note the pattern of the white LRSI tiles on the outer portions of the wings and the FRSI blankets used on the inner portions. Challenger was the first orbiter to have her name painted on the forward fuselage instead of the payload bay door. (NASA)

TDRS-A and its IUS being ejected from the payload bay. The solid-propellant upper stage would be ignited after the orbiter had moved a safe distance away. The large cradle that supported the TDRS stack would be returned to Earth. Note the three GAS canisters at left. (NASA)

On 1 March 1983 workers erected platforms at LC-39A and removed the main engines from Challenger to repair the cracked fuel lines. Later in the program, work like this would have required rolling the vehicle back to the VAB and destacking. (National Archives)

STS-7

Mission:	7	NSSDC ID:	1983-059A
Vehicle:	OV-099 (2)	ET-6 (SWT)	SRB: BI007
Launch:	LC-39A	18 Jun 1983	11:33 UTC
Altitude:	162 nm	Inclination:	28.48 degrees
Landing:	EDW-15	24 Jun 1983	13:58 UTC
Landing Rev:	98	Mission Duration:	146 hrs 24 mins

Commander:	Robert L. "Crip" Crippen (2)
Pilot:	Frederick H. "Rick" Hauck (1)
MS1:	John M. Fabian (1)
MS2:	Sally K. Ride (1)
MS3:	Norman E. "Norm" Thagard (1)

Payloads:	Up: 31,893 lbs	Down: 16,944 lbs
	Anik C2/PAM-D (7,374 lbs)	
	Palapa B1/PAM-D (7,575 lbs)	
	SPAS-01 (3,192 lbs)	
	OSTA-2 (3,192 lbs)	
	GAS (7)	
	SRMS s/n 201	

Notes:
First American woman in space
Largest American crew flown to-date
First space shuttle rendezvous (with SPAS-01)
Last flight of a standard-weight external tank
Planned EVA not performed (Thagard and Fabian)
Orbiter returned to KSC on 29 Jun 1983 (N905NA)

Wakeup Calls:

19 Jun	University of Texas fight song
20 Jun	Texas A&M Aggie War Hymn
21 Jun	"Reveille" / "When You're Smiling"
22 Jun	Washington State University fight song
23 Jun	Stanford Hymn
24 Jun	Florida State University fight song

The seven white stars in the black field of the crew patch, as well as the arm extending from the orbiter in the shape of a 7, reveal the STS designation. The five-armed symbol on the right side illustrates the four male/one female crew. (NASA)

This was the third flight launched with no countdown delays. However, this mission suffered the first known bipod ramp foam loss. Images revealed that a 12x20-inch piece of the ET left bipod ramp was missing; a similar event would later cause the loss of STS-107 and its crew. In addition, the thermal protection system foam on the ET showed 65 shallow divots.

According to the majority of the media coverage of this flight, its primary purpose was to carry the first American woman astronaut, Sally Ride, despite its real mission to deploy a pair of communication satellites. The flight also carried the largest crew so far, five persons. Norm Thagard flew ahead of the other members of his astronaut class because NASA needed a physician to study space adaptation syndrome, which had severely affected members of the STS-5 crew. In response, Thagard conducted medical tests of the nausea and sickness frequently experienced by astronauts during space flight.

Two communications satellites, Anik C2 for Telesat Canada and Palapa B1 for Indonesia Telkom, were successfully deployed during the first two days of the mission. Their attached PAM-D upper stages later boosted them into their appropriate orbits. *Challenger* also carried the first Shuttle Pallet Satellite (SPAS-01) built by Messerschmitt-Bolkow-Blohm in West Germany. This spacecraft could operate in the payload bay or be deployed by the SRMS as a free-flyer. It carried ten experiments to study formation of metal alloys in microgravity, the operation of heat pipes, instruments for remote sensing observations, and a mass spectrometer to identify various gases in the payload bay. It was deployed by the SRMS and flew alongside and over *Challenger* for several hours while an American-supplied camera on SPAS-01 took pictures of the orbiter performing various maneuvers. The crew later used the SRMS to grapple the pallet and return it to the payload bay.

Challenger also carried CFES (continuous flow electrophoresis system), MLR (monodisperse latex reactor), and one Shuttle Student Involvement Project (SSIP) experiment on the middeck. Seven GAS canisters in the payload bay held a variety of experiments, including one studying affects of space on social behavior of an ant colony in microgravity. In addition, the crew made the first communication tests using the TDRS-1 satellite deployed during STS-6.

Challenger was supposed to make the first landing on the Shuttle Landing Facility at KSC, but flight controllers waved-off the first KSC landing opportunity due to bad weather. Weather conditions did not improve enough to allow a landing at KSC, so NASA delayed the entry until revolution 98 and finally landed on the lakebed Runway 15 at Edwards.

The crew looking as if they are having a good time on-orbit. In the front, from the left, are Sally Ride and Norm Thagard, while in the back are Bob Crippen, Rick Hauck, and John Fabian. Crippen, commonly called "Crip," was the first person to fly twice on space shuttle. (NASA)

Although STS-6 had used a lightweight tank, STS-7 reverted back to a standard-weight tank since it was not carrying a particularly heavy payload. This was the last standard-weight tank flown. (National Archives)

The first American woman in space on the middeck where the galley would later go. Sally Ride also had the unfortunate distinction of being the only person to serve as a member of both space shuttle accident boards. (NASA)

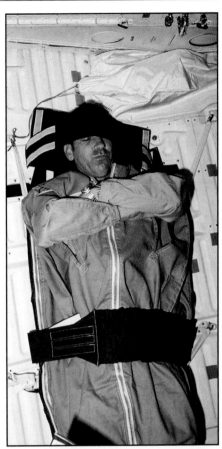

John Fabian demonstrates how to sleep in a zip-up blue sleep restraint device in the locker area of the middeck of Challenger. The head restraint could also be pulled over the eyes as a sleep mask. (NASA)

Challenger as seen by a 70 mm camera onboard SPAS-01. Visible in the cargo bay are the protective cradles for the Anik C2 and Palapa B1 communications satellites, the OSTA-2 pallet, and the SRMS configured to look like a "7" in honor of the mission number. (NASA)

Anik C2 is ejected from its launch cradle on 18 June; note the open sunshield that had kept the satellite from overheating while it was in the cradle. Also visible in the 70 mm exposure are SPAS-01, OSTA-2, some GAS canisters on the left, and the stowed SRMS arm at right. (NASA)

STS-8

Mission:	8	NSSDC ID:	1983-089A
Vehicle:	OV-099 (3)	ET-9 (LWT)	SRB: BI008
Launch:	LC-39A	30 Aug 1983	06:32 UTC
Altitude:	161 nm	Inclination:	28.48 degrees
Landing:	EDW-22	05 Sep 1983	07:42 UTC
Landing Rev:	98	Mission Duration:	145 hrs 09 mins

Commander:	Richard H. "Dick" Truly (2)
Pilot:	Daniel C. "Dan" Brandenstein (1)
MS1:	Guion S. "Guy" Bluford, Jr. (1)
MS2:	Dale A. Gardner (1)
MS3:	William E. Thornton (1)

Payloads:	Up: 30,076 lbs	Down: 0 lbs
	INSAT-1B/PAM-D (7,445 lbs)	
	PFTA (7,350 lbs)	
	DFI Pallet	
	GAS (4 + 8)	
	SRMS s/n 201	

Notes: Delayed 04 Aug 1983 (payload reconfiguration)
First night launch
First night landing
First African-American astronaut
Orbiter returned to KSC on 09 Sep 1983 (N905NA)

Wakeup Calls:

31 Aug	Georgia Tech fight song
01 Sep	Illinois fight song
02 Sep	Penn State fight song
03 Sep	University of North Carolina fight song
04 Sep	"Tala Sawari"
05 Sep	"Semper Fidelis"

The eight stars of the constellation Aquila represented the STS designation. The "spooky" patch shows a calm, sleepy-eyed Dick Truly in the left window (the traditional commander seat, on the right of the patch) and the four wide-eyed rookies in the right window. (NASA)

The original STS-8 manifest showed a three-day mission to deploy the INSAT-1B and TDRS-B satellites. However, concerns about the inertial upper stage (IUS) used to deploy TDRS-A during STS-6 caused NASA to remove TDRS from STS-8, replacing it with a payload flight test article (PFTA) to exercise the SRMS. *Challenger* also carried a Spacelab pallet, previously flown during the OFT flights, in the forward payload bay. This time, however, the pallet carried two experiments instead of developmental flight instrumentation. In addition, the mission carried 260,000 U.S. Postal Service covers in two storage boxes attached to the Spacelab pallet and eight get-away special cans (in addition to four GAS cans that carried actual experiments).

Challenger made the first night launch of the flight campaign, providing a spectacular view for tens of thousands of spectators in central Florida. The time of launch was dictated by the tracking requirements of the INSAT-1B primary payload, although liftoff was delayed 17 minutes due to weather. The five-member crew included the first African-American to fly in space, Guy Bluford.

However, during the post-flight inspection of the SRBs, engineers noted severe corrosion to the 3-inch-thick resin lining protecting the nozzles. NASA later determined this anomaly was due to the particular batch of resin used on this set of boosters. Nevertheless, as a precaution the nozzles on STS-9 were changed, resulting in a two-month delay for that mission.

The crew deployed INSAT-1B on the second day, and then maneuvered the orbiter away before the satellite ignited its PAM-D to carry it to its geosynchronous orbit. In preparation for the upcoming Spacelab 1 mission, the crew tested the communications capabilities of the TDRS-1 satellite deployed on STS-6. The testing was only marginally successful throughout the mission because of numerous signal dropouts and some equipment problems. In all, *Challenger* was able to use the satellite for at least part of 65 of its 98 orbits.

The crew reported there was a noticeable amount of floating debris and dust in the cabin, to the point of making them uncomfortable. Nothing in particular could be done other than vacuuming the larger pieces. A minor cabin pressure leak on 2 September was traced to the waste management system (potty). Late the following day, the crew conducted a live press conference, the first since Apollo 17. The mission ended in the first of what would be 26 night landings, conducted mostly to make sure it would work. To NASA, "night" was defined as from 15 minutes after nautical sunset until 15 minutes before nautical sunrise. Post-flight inspection showed that the performance of the thermal protection system was better than any previous flight.

The STS-8 crew responds to a comment made by Ronald Reagan during a post-flight telephone conversation from JSC later on landing day. From the left are Guy Bluford, Dan Brandenstein, Dick Truly, William Thornton, and Dale Gardener. (NASA)

Florida is one of the most lightning-prone locations on the planet, as vividly shown by this photograph taken by a remote camera set up by Sam Walton of UPI. Fortunately, no damage was recorded from the strike, and the rain and light shower stopped prior to launch. (NASA/UPI)

The payload bay as seen from the aft flight deck, showing the top of the DFI pallet in the foreground followed by the PFTA with its American flag. Note there are no black "eyeball" tiles on the OMS pods; many flights did not carry them until after the Challenger accident. (NASA

Guy Bluford checks out dinner in the middeck of Challenger. The food served aboard space shuttle constantly improved over the years, and a wide variety was available by the end of the flight campaign. Some missions carried galleys that offered additional options. (NASA)

The payload flight test article (PFTA) was a 7,350-pound aluminum structure with a 19.5-foot-long central axle. The crew used the PFTA to gain experience using the SRMS arm (at lower right). The article had several grapple fixtures the arm could connect to. (National Archives)

STS-9 (41A)

Mission:	9	NSSDC ID:	1983-116A
Vehicle:	OV-102 (6)	ET-11 (LWT)	SRB: BI009
Launch:	LC-39A	28 Nov 1983	16:00 UTC
Altitude:	136 nm	Inclination:	57.02 degrees
Landing:	EDW-17L	08 Dec 1983	23:47 UTC
Landing Rev:	167	Mission Duration:	247 hrs 47 mins

Commander:	John W. Young (6)
Pilot:	Brewster H. Shaw, Jr. (1)
MS1:	Owen K. Garriott (2)
MS2:	Robert A. R. "Bob" Parker (1)
PS1:	Ulf D. Merbold (1)
PS2:	Byron K. Lichtenberg (1)

Payloads:	Up: 33,264 lbs	Down: 0 lbs
	Spacelab 1 (32,294 lbs)	
	No SRMS	

Notes: Delayed 30 Sep 1983 (SRB nozzles)
First Spacelab verification flight test
First non-American astronaut on space shuttle
Backup payload specialists were
Wubbo J. Ockels and Michael L. Lampton
Orbiter returned to KSC on 15 Dec 1983 (N905NA)

Wakeup Calls:

Since crew worked two shifts around the clock,
mission control did not send any wakeup calls.

Spacelab 1 was depicted in the payload bay of Columbia, and the nine stars and the path of the orbiter represented STS-9. The flight crew names were at the top of the patch, the two mission specialists were in the middle, and the two payload specialists were at the bottom. (NASA)

The ninth flight was originally scheduled for 30 September 1983, but was initially delayed for 28 days due to SRB issues identified after the STS-8 boosters were recovered. Unfortunately, *Columbia* had already been moved to the launch pad and had to be rolled back to the VAB. Since the entire stack had to be disassembled to replace the nozzles, the orbiter was moved back to the OPF. All of this ultimately resulted in a two-month delay

Columbia, in its last flight prior to an extended modification period, carried the first Spacelab on its verification flight test. Jointly sponsored by the European Space Agency (ESA) and NASA, Spacelab consisted of various pressurized modules and unpressurized U-shaped pallets that remained in the payload bay during flight.

The six-person crew included Owen Garriott, who had spent 56 days aboard Skylab 3 in 1973, and the first foreign citizen to fly on space shuttle, Ulf Merbold from West Germany (representing ESA).

The crew divided into two teams, each working 12-hour shifts. Brewster Shaw, Owen Garriott, and Byron Lichtenberg were the blue team while John Young, Bob Parker, and Ulf Merbold formed the red team. Usually, the commander and the pilot were assigned to the flight deck to monitor the orbiter and communication systems, while the mission and payload specialists worked inside the Spacelab. The crew conducted 73 experiments in atmospheric and plasma physics, astronomy, solar physics, material sciences, technology, life sciences, and Earth observations. The effort went so well that mission management added an extra day on-orbit, making it the longest flight to date at ten days. The mission also marked the first operational use of the TDRS system to route large amounts of science data through its ground terminals to the Payload Operations Control Center (POCC). In addition, Garriott made the first amateur radio transmissions from space. This led to many further space flights incorporating amateur radio as an educational and backup communications tool.

About five hours before entry, a general-purpose computer failed when the RCS jets were fired. Six minutes later, a second GPC also failed, but was successfully reinitialized. Post-flight analysis revealed the computers failed when the RCS jet motion knocked a piece of solder loose and shorted a memory circuit. Earlier, an inertial measurement unit had failed, but had not affected the mission. The landing was delayed 7.5 hours to evaluate the GPC problem. Approximately six minutes after the orbiter landed, an auxiliary power unit unexpectedly shut down, followed five minutes later by a detonation in another APU that resulted in an automatic shutdown. The third APU ran normally until the crew shut it down after landing.

This was the largest crew flown to-date. From the left, sitting, are Owen Garriott, Brewster Shaw, John Young, and Bob Parker. Standing are Byron Lichtenberg and Ulf Merbold. Although large for the time, space shuttle would later standardize of seven-member crews. (NASA)

Racks similar to this one lined the sides of the Spacelab module and contained the experiments, instrumentation, and data processing equipment used on the mission. This rack is being moved inside the Operations and Checkout Building at KSC. (National Archives)

The Kennedy Space Center sits in the middle of the 140,000-acre Merritt Island National Wildlife Refuge. NASA acquired the land in the early 1960s as a buffer zone around the space center and turned over its operation to the U.S. Fish and Wildlife Service. (NASA)

On the left, wearing the headband-like data recorder, is Ulf Merbold of the Max Planck Society. He and Byron Lichtenberg wore the devices during most of their waking hours on the ten-day flight. Robert Parker is partially obscured by a deployed instrument of the fluid physics module at the materials sciences double rack. Contrary to popular perception, this was essentially a verification flight test of the Spacelab equipment, not the first operational Spacelab mission. The Spacelab 1 configuration consisted of a tunnel adapter, pressurized long module, and single pallet. (NASA)

STS-11 (41B)

Mission:	10	NSSDC ID:	1984-011A
Vehicle:	OV-099 (4)	ET-10 (LWT)	SRB: BI010
Launch:	LC-39A	03 Feb 1984	13:00 UTC
Altitude:	166 nm	Inclination:	28.48 degrees
Landing:	KSC-15	11 Feb 1984	12:17 UTC
Landing Rev:	128	Mission Duration:	191 hrs 16 mins

Commander:	Vance D. Brand (3)
Pilot:	Robert L. "Hoot" Gibson (1)
MS1:	Bruce McCandless II (1)
MS2:	Ronald E. "Ron" McNair (1)
MS3:	Robert L. "Bob" Stewart (1)

Payloads:	Up: 33,868 lbs	Down: 0 lbs
	Westar VI/PAM-D (7,307 lbs)	
	Palapa B2/PAM-D (7,556 lbs)	
	SPAS-01A (3,192 lbs)	
	IRT (210 lbs)	
	Manned Maneuvering Units (2)	
	GAS (5)	
	SRMS s/n 201	

Notes:	Delayed 20 Jan 1984 (APU replacement)
	First reflight of a refurbished satellite (SPAS)
	First MMU test
	EVA1 07 Feb 1984 (McCandless and Stewart)
	EVA2 09 Feb 1984 (McCandless and Stewart)
	First KSC landing

Wakeup Calls:

04 Feb	(none, technical fault)
05 Feb	"A Train" (by Contraband with Ron McNair)
06 Feb	"Glory, Glory, Colorado"
07 Feb	Armed forces medley
08 Feb	"Southern to the Top"
09 Feb	"The Greatest American Hero Theme"
10 Feb	The Air Force Hymn – "Wild Blue Yonder"
11 Feb	"In the Mood" (by Contraband with Ron McNair)

Challenger *is flanked by a PAM-D satellite deployment and an astronaut with an MMU and the eleven stars symbolize the STS designation. The crew patch was designed by Robert McCall. The smaller patch was for the manned maneuvering unit. (NASA)*

This was the first mission to use the new numbering scheme, so it was known as 41B by public affairs, but was still called STS-11 by many inside the program. NASA had canceled the previous mission, STS-10, due to delays with the DoD payload.

Launch was originally scheduled for 20 January 1984, but was delayed to replace the auxiliary power units as a precaution after tests based on the STS-9 anomalies revealed possible discrepancies. This *Challenger* mission was highlighted by the introduction of the Martin Marietta-built manned maneuvering unit (MMU), an untethered backpack propulsion unit that allowed astronauts to maneuver in space independent of the orbiter.

On flight day 2, the crew lowered the pressure in the crew module to 10.2 psia in preparation for the first of two extravehicular activities. Lowering the pressure reduced the time the crew needed to "pre-breathe" pure oxygen from three hours (on STS-6 at 14.7 psia) to one hour. All orbiter systems functioned normally during the 72 hours of lower pressure.

Bruce McCandless ventured 320 feet from the orbiter without a tether, while Bob Stewart tested a foot restraint at the end of the SRMS. The pair was supposed to release an inflated balloon for use as a target to practice precision flying with the MMUs, but it burst upon inflation. However, it still formed a large enough target and the EVA crew successfully completed rendezvous operations on the fourth day of the mission as practice for the retrieval of the Solar Maximum satellite during STS-13/41C.

The mission launched Westar VI on the first day, but its PAM-D upper stage never fired, leaving the satellite stranded in a 656x162-nm orbit. NASA delayed the deployment of Palapa B2 while engineers evaluated data from the PAM-D failure. Not believing the failure was systemic, engineers and Indonesian officials approved deploying Palapa on the fourth day of the mission, with identical results (although the orbit was only 639x148 nm). The crew of STS-19/51A retrieved both satellites the following November, ironically on the last mission to use the MMUs.

The Shuttle Pallet Satellite originally carried by STS-7 was flown again, marking the first time a satellite had been refurbished and returned to space. This time, however, SPAS-01A remained in the payload bay due to a failure of the SRMS wrist joint. The mission also carried five GAS canisters, six live rats in the middeck, an IMAX Cinema-360 camera, and another flight of the CFES (continuous flow electrophoresis system) and MLR (monodisperse latex reactor) experiments on the middeck.

For the first time, an orbiter landed on the Shuttle Landing Facility at the Kennedy Space Center.

Posed for an on-orbit crew portrait on the middeck of Challenger *are, clockwise from the lower left, Vance Brand, Bob Stewart, Bruce McCandless, Ron McNair, and Hoot Gibson. The crew used a self-timed 35 mm camera to expose the frame. (NASA)*

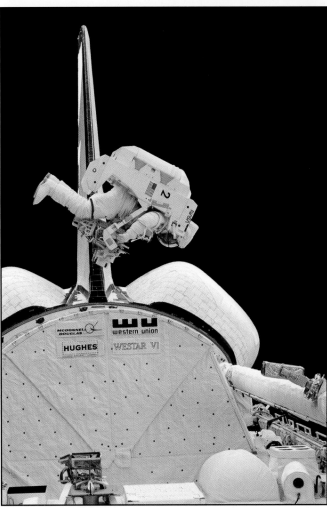

Bruce McCandless (red stripes) tests the foot restraint at the end of the SRMS (top left) as well as the manned maneuvering unit (two lower photos). The top right photo shows the payload bay forward bulkhead with the open airlock hatch. To the left of the hatch, Bob Stewart is checking out his MMU during the second EVA. Each MMU had a dedicated flight support station on the forward payload bay wall, one on each side. This is where the unit was stored during launch and entry. Note the grapple fixture used by the SRMS on SPAS-01A in the two photos at right. (NASA)

STS-13 (41C)

Mission:	11	NSSDC ID:	1984-034A
Vehicle:	OV-099 (5)	ET-12 (LWT)	SRB: BI012
Launch:	LC-39A	06 Apr 1984	13:58 UTC
Altitude:	268 nm	Inclination:	28.45 degrees
Landing:	EDW-17L	13 Apr 1984	13:39 UTC
Landing Rev:	108	Mission Duration:	167 hrs 40 mins

Commander:	Robert L. "Crip" Crippen (3)
Pilot:	Francis R. "Dick" Scobee (1)
MS1:	Terry J. "TJ" Hart (1)
MS2:	James D. A. "Ox" van Hoften (1)
MS3:	George D. "Pinky" Nelson (1)

Payloads:	Up: 38,266 lbs	Down: 0 lbs
	Long-Duration Exposure Facility (21,396 lbs)	
	Solar Max ASE (4,740 lbs)	
	Manned Maneuvering Units (2)	
	SRMS s/n 302	

Notes: First direct ascent trajectory
Both MCC computers failed during ascent
Extended one day due to capture Solar Max
Featured in 1985 IMAX movie *The Dream is Alive*
EVA1 08 Apr 1984 (Nelson and van Hoften)
EVA1 11 Apr 1984 (Nelson and van Hoften)
Orbiter returned to KSC on 18 Apr 1984 (N905NA)

Wakeup Calls:

07 Apr	"A Boy Named Sue"
08 Apr	U.C. Berkeley fight song / Lehigh U. fight song
09 Apr	(unidentified)
10 Apr	"Theme from Rocky"
11 Apr	(none)
12 Apr	(unidentified)
13 Apr	Fight songs from U. of Texas and U. Arizona

The crew patch featured the helmet visor of an astronaut performing an extravehicular activity (EVA). The Sun's rays, Challenger deploying the long duration exposure facility (LDEF), and an astronaut working on Solar Max were reflected in the visor. (NASA)

This was the first flight to use a direct-insertion trajectory, where the main engines carried the orbiter all the way to its initial operational altitude; the orbital maneuvering engines were only used to circularize, and later raise, the orbit as needed.

During ascent, the mission operations computer in Houston failed, as did its backup; a common software problem had stopped processing in both mainframes. For about an hour the flight controllers had no insight into the mission except for radio calls from the crew. Fortunately, the vehicle performed as expected, so there was no particular affect on the mission.

The crew deployed the Long Duration Exposure Facility (LDEF) late on flight day 2. The retrievable, 12-sided LDEF was 14 feet in diameter, 30 feet long, and carried 57 experiments to evaluate how materials survive during long-term exposure to low-earth orbit. NASA had initially intended to retrieve LDEF during STS-23/51D in March 1985, but manifest changes soon moved this to STS-41/61I in September 1986. The *Challenger* accident ultimately postponed the return until January 1990 during STS-32R.

On flight day 3, *Challenger* demonstrated an important capability of space shuttle: the retrieval, repair, and redeployment of the malfunctioning Solar Maximum spacecraft. The Ku-band radar acquired Solar Max at a range of 17 nm, then Bob Crippen and Dick Scobee maneuvered the orbiter within 200 feet of Solar Max.

During the first EVA, Pinky Nelson and Ox van Hoften could not capture the satellite, despite Nelson flying his MMU exactly as planned during three attempts. During all of this, Solar Max began tumbling on multiple axes and Bob Crippen was maneuvering *Challenger* to keep up with Nelson and Solar Max, depleting most of the maneuvering propellant. The crew then attempted to grapple Solar Max using the SRMS, but this was also unsuccessful. Ultimately, ground controllers were able to stop the tumbling using the satellite's magnetic torque bars and put Solar Max into a slow, regular spin. The next morning, TJ Hart successfully grappled the satellite using the SRMS and moved Solar Max into its cradle in the payload bay.

Pinky Nelson and Ox van Hoften successfully repaired the satellite in less than four hours during the second EVA. Because of the shorter-than-planned time required for repair, mission control allowed van Hoften to conduct a performance evaluation of the second MMU. After the EVA, the crew used the SRMS to lift Solar Max out of the payload bay and held the satellite while ground controllers evaluated the repairs before redeploying the satellite.

Mission highlights, including the LDEF deployment and Solar Max repair, appeared in the IMAX movie *The Dream is Alive*.

The official crew portrait. From the left are Bob Crippen, Terry Hart, Ox van Hoften, George Nelson, and Dick Scobee. Although a rookie astronaut, Scobee had a long history with Challenger; he had been one of the 747 pilots on her delivery and STS-6 ferry flights. (NASA)

Ox van Hoften, on the end of the SRMS, changes a faulty attitude control module on the Solar Maximum satellite while it was berthed in the aft end of the payload bay. The empty space in the foreground is where LDEF had ridden to orbit. (NASA)

The repair of Solar Maximum demonstrated one of the key capabilities that had long been advertised for space shuttle. The crew wasted no time proclaiming themselves as the "Ace Satellite Repair Co." in this photo taken on the aft flight deck. (NASA)

Pinky Nelson shows some of the food carried on the flight. Behind him is the original galley with a tray for each crew member arranged on the inside of the doors. Note the packets of wet wipes and condiments (mustard and mayonnaise) at the bottom. (NASA)

The long-duration exposure facility (LDEF) suspended at the end of the SRMS while Challenger passes over southern Florida and The Bahamas. A variety of circumstances would delay the LDEF retrieval for almost five years until STS-32R in January 1990. (NASA)

One of the multiplexer/demultiplexers in Challenger failed before launch but did not impact to the countdown. After an on-time launch, this was the first flight to use a direct-insertion trajectory, with the external tank breaking up near Hawaii during reentry. (NASA)

STS-14 (41D)

Mission:	12	NSSDC ID:	1984-093A
Vehicle:	OV-103 (1)	ET-13 (LWT)	SRB: BI011
Launch:	LC-39A	30 Aug 1984	12:42 UTC
Altitude:	175 nm	Inclination:	28.48 degrees
Landing:	EDW-17L	05 Sep 1984	13:39 UTC
Landing Rev:	97	Mission Duration:	144 hrs 56 mins

Commander:	Henry W. "Hank" Hartsfield, Jr. (2)
Pilot:	Michael L. "Mike" Coats (1)
MS1:	Richard M. "Mike" Mullane (1)
MS2:	Steven A. "Steve" Hawley (1)
MS3:	Judith A. "Judy" Resnik (1)
PS1:	Charles D. "Charlie" Walker (1)

Payloads:	Up: 47,516 lbs	Down: 0 lbs
	SBS-D/PAM-D (7,383 lbs)	
	Syncom IV-2/Orbus-6 (15,306 lbs)	
	Telstar 3C/PAM-D (7,507 lbs)	
	OAST-1 (3,405 lbs)	
	SRMS s/n 301	

Notes:
Originally manifested as STS-12
Flight readiness firing 02 Jun 1984
Delayed 22 Jun 1984 (SSME replacement)
Scrubbed 25 Jun 1984 (GPC failure)
Aborted 26 Jun 1984 (SSME shut down)
First launch abort of the flight campaign
Remanifested with payloads from STS-16/41F
Scrubbed 29 Aug 1984 (GPC software)
First flight of *Discovery* (OV-103)
Orbiter returned to KSC on 10 Sep 1984 (N905NA)

Wakeup Calls:

31 Aug	The Navy Hymn – "Anchors Aweigh"
01 Sep	"Telstar" (by The Ventures)
02 Sep	"Top of the World"
03 Sep	(unidentified)
04 Sep	(unidentified)
05 Sep	(none)

The twelve stars within the blue field indicate that this was the 12th mission of the flight campaign. Discovery's namesake is manifested in a sailing ship, which is linked to the orbiter via a red, white, and blue ribbon, signifying its maiden voyage. (NASA)

Discovery conducted a successful 20-second flight readiness firing on 2 June 1984, although engineers replaced one main engine because a heat shield debonded on the fuel preburner. To accommodate this work, NASA slipped the 22 June launch to 25 June. During that attempt, a general-purpose computer failed at T-32 minutes, so NASA scrubbed the launch. Technicians replaced the offending computer, but a launch attempt the following day was aborted at T-4 seconds when one main engine shut down after 0.22-second. This was the first on-pad abort-after-ignition of the flight campaign and the first since Gemini VI-A on 12 December 1965. After the abort, Steve Hawley commented, "Gee, I thought we'd be a lot higher at MECO [main engine cut-off]!"

Because of the delays, NASA remanifested STS-14/41D to include the most important items from both the original payload and that intended for STS-16/41F, which was then canceled. A 29 August launch attempt failed because of a computer issue, but engineers developed a software patch and scheduled launch for 30 August. The final countdown proceeded smoothly.

The CFES (continuous flow electrophoresis system) experiment was more elaborate than the ones flown previously and Charlie Walker operated it for more than 100 hours during the flight. Walker, an employee of McDonnell Douglas, was the first non-government employee and the first commercial payload specialist to fly in space.

The primary payloads were three communications satellites: SBS-D for Satellite Business Systems, Telstar 3C for Telesat Canada, and Syncom IV-2 (Leasat F2), a Hughes satellite leased to the Navy. The Leasats were the first large communications satellites specifically designed to be carried by space shuttle. The satellites were successfully deployed one per day beginning on flight day 1 and the highlights appeared in the IMAX movie *The Dream is Alive*.

The NASA Office of Aeronautics and Space Technology (OAST-1) payload was a solar array 13 feet wide and 102 feet high that folded into a package only 7 inches deep. It demonstrated the feasibility of large lightweight solar arrays for future applications.

A "pee-cicle" about 12 inches in diameter and 27 inches long formed around the dump nozzle after the crew dumped wastewater from the potty overboard. For the remainder of the mission, the crew used bags to collect urine and there were no further water dumps.

NASA planned the landing for Edwards since this was the first *Discovery* flight. A post-flight inspection revealed the right main gear strut had lost pressure, a condition that caused the orbiter to pull to the right after nose gear touchdown. Otherwise, *Discovery* was in excellent shape.

From the left, seated, are Mike Mullane, Steve Hawley, Hank Hartsfield, and Mike Coats. Standing are Charlie Walker and Judy Resnik. Both the early ocean-going Discovery and the debuting spacecraft are depicted in the background. (NASA)

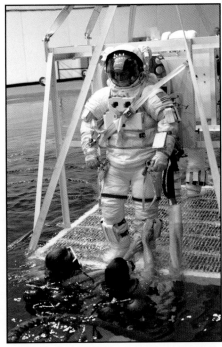

At left, Mike Coats positions himself for a photography lesson during a training session in the 1-g trainer at the Johnson Space Center. In the center, Mike Mullane trains for a contingency EVA in case the crew had to manually close the external tank umbilical doors or the payload bay doors. At right is a view of the launch as seen by John Young in a Shuttle Training Aircraft flying over the Kennedy Space Center. (NASA)

The gold solar array experiment (SAE) panel, part of the OAST-1 payload, with the vertical stabilizer of Discovery silhouetted against the accordion-like array. The frame was exposed through an aft flight deck window with a handheld 70 mm Hasselblad camera. (NASA)

Telstar 3C, the third of three satellites deployed from Discovery, departs the payload bay at 14:35 UTC on 1 September 1984. The crew photographed the release through an aft flight deck window using a handheld 70 mm Hasselblad camera. (NASA)

STS-17 (41G)

Mission:	13	NSSDC ID:	1984-108A
Vehicle:	OV-099 (6)	ET-15 (LWT)	SRB: BI013
Launch: Altitude:	LC-39A 192 nm	05 Oct 1984 Inclination:	11:03 UTC 57.08 degrees
Landing: Landing Rev:	KSC-33 133	13 Oct 1984 Mission Duration:	16:28 UTC 197 hrs 24 mins

Commander:	Robert L. "Crip" Crippen (4)
Pilot:	Jon A. McBride (1)
MS1:	Kathryn D. "Kathy" Sullivan (1)
MS2:	Sally K. Ride (2)
MS3:	David C. "Dave" Leestma (1)
PS1:	Paul D. Scully-Power (1)
PS2:	J. Marc Garneau (1)

Payloads:	Up: 23,465 lbs Down: 0 lbs
	Earth Radiation Budget Satellite (4,949 lbs)
	OSTA-3/SIR-B (4,254 lbs)
	LFC/ORS (4,583 lbs)
	GAS (8)
	SRMS s/n 302

Notes:	First seven person crew
	First crew with two women
	First Canadian astronaut
	EVA1 11 Oct 1984 (Sullivan and Leestma)
	Longest *Challenger* mission
	Backup payload specialists were
	Robert E. Stevenson and Robert B. Thirsk

Wakeup Calls:

06 Oct	*Flashdance*, "What A Feeling"
07 Oct	(unidentified)
08 Oct	(unidentified)
09 Oct	"Theme from Rocky"
10 Oct	"Navy Wings are Made of Gold"
11 Oct	"Take Me Home, Country Roads"
12 Oct	"Theme from Star Wars"
13 Oct	Oklahoma fight song

The crew patch featured the astronaut symbol. This iconic emblem shows a trio of trajectories merging in infinite space, capped by a bright shining star, and encircled by an elliptical orbit. The patch was created by Patrick Rawlings. (NASA)

A perfect countdown resulted in an on-time launch of *Challenger*. This was the third mission to carry an IMAX camera to document the flight and footage from the mission and EVA appeared in the movie *The Dream is Alive*.

The crew used the SRMS to deploy the Earth Radiation Budget Satellite (ERBS) on flight day 1, almost three hours later than planned because the solar array on the satellite did not extend when commanded. The ERBS was the first of three planned satellites designed to measure the amount of energy received from the Sun and reradiated into space. It also studied the seasonal movement of energy from the tropics to the poles.

The major mission activity was operating the shuttle imaging radar (SIR-B) that was part of the OSTA-3 payload. SIR-B was an improved version of a similar device flown as part of the OSTA-1 payload on STS-2. This package included a large format camera (LFC) to photograph Earth, the measurement of air pollution from satellites (MAPS) sensor, and the feature identification and location experiment (FILE) that consisted of two television cameras and two 70 mm still cameras. On flight day 2, the SIR-B antenna would not refold, delaying a planned OMS burn, but eventually the crew used the SRMS to bump the outer antenna leaf into position.

Kathy Sullivan became the first American woman to walk in space when she and Dave Leestma performed an EVA on 11 October. The spacewalkers installed a valve into simulated satellite propulsion plumbing that duplicated that found on Landsat, which was one of the satellites that space shuttle was intended to service. Although engineers designed Landsat, which used the same bus as Solar Max, to be retrieved by a space shuttle, they had not included the plumbing for refueling, hence the need to install the valve. Just before the *Challenger* accident, NASA was developing plans to launch missions from Vandenberg to refuel Landsat-4 and Landsat-5.

Early in the mission, the Ku-band antenna developed a fault that prevented it from being stowed remotely, so Kathy Sullivan and Dave Leestma manually rotated it to the stow position and then to the lock position. Paul Scully-Power, an employee of the U.S. Naval Research Laboratory, performed a series of oceanography observations. In addition, Marc Garneau conducted a series of CANAEX experiments sponsored by the Canadian government that were related to medical, atmospheric, climatic, materials, and robotic science.

Despite persistent conspiracy claims, NASA, the crew and knowledgeable members of the American intelligence community deny a story that the Soviet Union illuminated *Challenger* using the Terra-3 laser complex at Sary-Shagan, Kazakhstan.

From the left in the first row are Marc Garneau, Paul Scully-Power, and Bob Crippen. Behind them are Jon McBride, Dave Leestma, and Sally Ride, with Kathy Sullivan at the top. The crew is pictured in front of one of the 1-g trainers at the Johnson Space Center. (NASA)

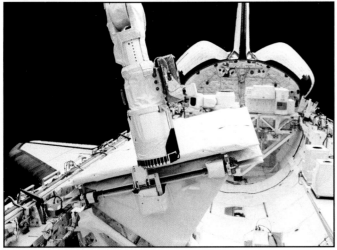

The crew used the SRMS end effector to secure the antenna on the shuttle imaging radar (SIR-B). Behind the radar is the mission peculiar equipment support structure (MPESS) carrying a large format camera and the orbital refueling system demonstration. (NASA)

Bob Crippen in the commander seat during entry at the conclusion of the mission. The superheated air outside the windows is glowing orange. Note the checklists propped against the windscreen; forward visibility was not particularly critical at this point. (NASA)

At the back (bottom) of the payload bay is the LFC/ORS followed by the Earth Radiation Budget Satellite (ERBS); the instruments were wrapped in gold foil and the folded solar panel is at the back. A Spacelab pallet carried the OSTA-3/SIR-B experiment. (NASA)

The crew walks down the ladder at the Shuttle Landing Facility. Crip is shaking hands with George Abbey, the head of flight crew operations at JSC, who was always at the landings. Later crews would be saved the effort of disembarking down the airstairs when the program introduced a modified mobile lounge acquired from the Baltimore/Washington International Thurgood Marshall Airport. (NASA)

Challenger is towed back to the Orbiter Processing Facility for a couple of months of maintenance and updates prior to her next flight in April 1985. Unusual here is the extensive use of black tiles around one of the mid-fuselage vent doors. (NASA)

STS-19 (51A)

Mission:	14	NSSDC ID:	1984-113A
Vehicle:	OV-103 (2)	ET-16 (LWT)	SRB: BI-014
Launch:	LC-39A	08 Nov 1984	12:15 UTC
Altitude:	195 nm	Inclination:	28.48 degrees
Landing:	KSC-15	16 Nov 1984	12:01 UTC
Landing Rev:	127	Mission Duration:	191 hrs 45 mins

Commander:	Frederick H. "Rick" Hauck (2)
Pilot:	David M. "Dave" Walker (1)
MS1:	Joseph P. "Joe" Allen IV (2)
MS2:	Anna L. Fisher (1)
MS3:	Dale A. Gardner (2)

Payloads:	Up: 45,306 lbs	Down: 2,381 lbs
	Anik D2/PAM-D (7,374 lbs)	
	Syncom IV-1/Orbus-6 (15,190 lbs)	
	Westar VI (1,119 lbs, down)	
	Palapa B2 (1,262 lbs, down)	
	SRMS s/n 301	

Notes:	Scrubbed 07 Nov 1984 (KSC weather)
	First satellites retrieved from orbit and returned
	Last use of the manned maneuvering units
	EVA1 12 Nov 1984 (Allen and Gardner)
	EVA2 14 Nov 1984 (Allen and Gardner)

Wakeup Calls:

09 Nov	"Good Morning Starshine"
10 Nov	The Marine Corps Hymn – "Halls of Montezuma"
11 Nov	"Theme from Victory at Sea"
12 Nov	"The Eagle and the Hawk"
13 Nov	(none)
14 Nov	"Theme from For a Few Dollars More"
15 Nov	Medley of several songs by LAV FM97
16 Nov	"Theme from Peter Gunn"

The red and white trailing stripes and the blue background, along with the presence of the Eagle, commemorate 208 years of American history. The two satellites amidst a celestial scene were intended as a universal representation of the versatility of space shuttle. (NASA)

NASA scrubbed the first launch attempt on 7 November 1984 during the T-20-minute hold because of high wind shears in the upper atmosphere. A second attempt the following day proceeded smoothly. This was the only mission during the flight campaign that deployed two communications satellites and retrieved two others. The failure of their upper stages had left Westar VI and Palapa B2 in unusable orbits after being deployed by STS-11/41B. This was the first time an object placed on-orbit by one vehicle had been recovered by a different vehicle and marked the first official downmass returned by space shuttle.

The crew deployed Anik D2 on flight day 1 within one second of the planned time. On flight day 3, Rick Hauck and Dave Walker raised their orbit to rendezvous with Palapa B2. Later that day, but before the rendezvous, the crew deployed Syncom IV-1 (Leasat F1) and confirmed that the omni antenna deployed and the satellite was spinning as expected.

The retrieval of Palapa B2 on flight day 5 involved considerable determination by Dale Gardner and Joe Allen. After *Discovery* rendezvoused with the satellite, Allen inserted a "stinger" into the apogee motor nozzle and used his MMU thrusters to slow the rotation from 12 to 1 rpm. Gardner, standing in a foot restraint on the end of the SRMS operated by Anna Fisher, grappled the satellite and lowered it into the payload bay with Allen and the MMU still attached. Allen cut the omni antenna with a set of shears and attempted to attach clamps to Palapa so he could install the antenna bridge A-frame. Unfortunately, he could not install the clamps because a waveguide on Palapa protruded farther than expected.

The crew changed to a practiced backup procedure where Joe Allen disconnected the MMU from the stinger and doffed MMU-2 in its flight support station. He then got into the portable foot restraint, which had been secured to the starboard side of the Westar pallet and held onto the top of Palapa after the SRMS released it. Allen manhandled Palapa so Gardner could remove the stinger and install the upper stage nozzle cover and adapter to the bottom of the satellite. Together, the pair manually maneuvered Palapa into the payload bay and secured it for return to Earth. The recovery of Westar VI on flight day 7 was not as difficult, although the failure of both floodlights at the back of the payload bay meant some of it was conducted in near darkness. These EVAs marked the last use of the manned maneuvering unit and were the last American untethered spacewalks until STS-64 in 1994.

Ultimately, the insurance companies refurbished the satellites and placed them in service using other launch vehicles in April 1990.

From the left are Dale Gardner, Dave Walker, Anna Fisher, Rick Hauck, and Joe Allen. No mention of the name of the eagle. This was probably the most experienced crew so far, with three of the five astronauts having flown on space shuttle before. (NASA)

Dale Gardner provides a good reference of the size of the middeck lockers. The middeck forward bulkhead was covered with lockers, which held everything the crew needed for the mission; food, clothing, tools, etc. Note the multiple foot restraints on the floor. (NASA)

Dale Gardner, wearing a manned maneuvering unit (MMU), captured the spinning Westar VI over the Bahama Banks. He would soon pass the satellite off to Anna Fisher, operating the shuttle remote manipulator system arm, who would berth it in the payload bay. (NASA)

An unusually good photograph of the payload bay with the doors closed. The extremely reflective radiators made photography difficult. The two empty return cradles are in the foreground and Anik D2 is in its launch cradle with the sunshield open. (NASA)

A beautiful launch against a threatening sky. The vehicle was typical of the paint and markings used prior to the Challenger accident. Note the large black photo-reference blocks on the solid rocket boosters and the various shades of foam on the external tank. (NASA)

Discovery in the OPF after the mission showing Palapa B2 (bottom) and Westar VI in the payload bay, looking forward. The location resulted in a more forward center-of-gravity than was usually allowed. These were the first satellites retrieved from orbit and brought back to Earth. (NASA)

STS-20 (51C)

Mission:	15	NSSDC ID:	1985-010A
Vehicle:	OV-103 (3)	ET-14 (LWT)	SRB: BI015
Launch:	LC-39A	24 Jan 1985	19:50 UTC
Altitude:	185 nm	Inclination:	28.45 degrees
Landing:	KSC-15	27 Jan 1985	21:24 UTC
Landing Rev:	49	Mission Duration:	73 hrs 33 mins

Commander:	Thomas K. "TK" Mattingly II (3)
Pilot:	Loren J. Shriver (1)
MS1:	Ellison S. "El" Onizuka (1)
MS2:	James F. "Jim" Buchli (1)
PS1:	Gary E. Payton (1)

Payloads:	Up: (38,500 lbs)	Down: 0 lbs
	USA-8/IUS (Magnum/ORION)	
	SRMS s/n 301	

Notes:	Crew was originally assigned to STS-10
	Delayed 23 Jan 1985 (KSC weather)
	First dedicated Department of Defense mission
	First flight of an Air Force MSE
	Classified payload
	Significant SRB seal erosion
	Backup payload specialist (MSE) was
	Keith C. Wright
	Shortest *Discovery* mission

Wakeup Calls:

Classified mission. Either nothing played or nothing released.

There were two slight variations in the crew patch, both apparently made by AB Emblem. The logo above and patches worn in the crew photo at right have longer "tabs" on the bottom compared to the patches worn by the crew during the mission. (NASA)

This crew began training in October 1982 for a November 1983 launch of STS-10 using *Challenger*. The failure of the inertial upper stage during STS-6 caused the Air Force to cancel STS-10 until engineers determined what had plagued the IUS. Eventually, the same payload (DoD 84-1, also called USA-8) and crew were assigned to STS-20/51C. Initially the flight was still on *Challenger*, but concerns about her thermal protection system resulted in the mission being switched to *Discovery*. The first launch attempt on 23 January 1985 was postponed one day due to freezing weather in Florida, although engineers also noted two minor technical anomalies. The countdown on 24 January proceeded smoothly.

As they were disassembling the boosters, engineers found a leak path in the forward field joint of the left motor and the center field joint on the right motor. On the right motor, the leak had gone through the primary o-ring and the secondary o-ring showed significant heating effects, but no erosion." There were also large quantities of dark black soot between the primary and secondary o-rings in both field joints, the first time engineers had seen soot in a field-joint and also the first time that more than one field joint displayed damage. Engineers at Thiokol determined this was caused by extremely cold weather in the days preceding launch. A year later, the Rogers Commission reported STS-20/51C experienced the worst SRB blow-by of any mission prior to the *Challenger* accident. Significantly, the temperature at KSC was only 62°F during the launch of STS-20/51C, the coldest launch until *Challenger*.

This was the first dedicated flight for the Department of Defense, although small DoD payloads had flown on nine previous flights. For the first time, NASA did not provide pre-launch commentary until T-9 minutes. The Air Force identified the payload only as USA-8 although many pundits believe the primary payload was a TRW Magnum/ORION electronic signals intelligence (ELINT) satellite. These reportedly weighed between 5,000 and 6,000 pounds and had antennas that extended to 250–300 feet in diameter. The inertial upper stage delivered the satellite to a geosynchronous orbit above the Indian Ocean where it reportedly listened to communications from the Soviet Union, China, and neighboring countries.

This marked the first flight of one of the DoD manned spaceflight engineers and, as late as 2009, Gary Payton would only say that the payload was "still up there and still operating."

By all reports, the orbiter returned on the first landing opportunity at KSC. All the official Air Force statements ever said was that the flight, "... successfully met its mission objectives." This was the shortest mission for *Discovery*.

Kneeling, from the left, are Loren Shriver and Ken Mattingly, while standing are Gary Payton, Ellison Onizuka, and Jim Buchli. Beginning in 1979, the Air Force selected thirty-two military astronauts, including Payton, that it called manned spaceflight engineers. (NASA)

The orbiter used an unusual double-delta planform where the chines were much more severely swept-back than the main wing. Nevertheless, the shape proved remarkable stable throughout the flight regime and provided adequate landing characteristics. (NASA)

The crew walks around Discovery after wheels-stop. A deployed air-data probe may be seen near the nose. Note the orbital maneuvering system pods do not have the characteristic black "eyeballs" that were being added to the fleet before the Challenger accident. (NASA)

The post-flight inspection would show that both SRBs had significant issues with the field joint seals. The Rogers Commission later reported this mission experienced the worst blow-by prior to the Challenger accident. Significantly, the temperature at KSC was only 62°F. (NASA)

STS-23 (51D)

Mission:	16	NSSDC ID:	1985-028A
Vehicle:	OV-103 (4)	ET-18 (LWT)	SRB: BI018
Launch:	LC-39A	12 Apr 1985	13:59 UTC
Altitude:	249 nm	Inclination:	28.51 degrees
Landing:	KSC-33	19 Apr 1985	13:56 UTC
Landing Rev:	110	Mission Duration:	167 hrs 55 mins

Commander:	Karol J. "Bo" Bobko (2)
Pilot:	Donald E. "Don" Williams (1)
MS1:	M. Rhea Seddon (1)
MS2:	S. David "Dave" Griggs (1)
MS3:	Jeffrey A. "Jeff" Hoffman (1)
PS1:	Charles D. "Charlie" Walker (2)
PS2:	Edwin J. "Jake" Garn (1)

Payloads:	Up: 35,794 lbs	Down: 0 lbs
	Anik C1/PAM-D (7,386 lbs)	
	Syncom IV-3/Orbus-6 (15,190 lbs)	
	GAS (2)	
	SRMS s/n 301	

Notes:
Originally scheduled for 19 Mar 1985
Carried part of the canceled STS-22/51E payload
Walker flew instead of Patrick P. R. Baudry
EVA1 16 Apr 1985 (Hoffman and Griggs)
Inboard right main tire blew during landing

Wakeup Calls:

13 Apr	"Top of the World"
14 Apr	"Stargazer"
15 Apr	"Skybird"
16 Apr	"Rescue Aid Society" from The Rescuers
17 Apr	"Ride of the Valkyries"
18 Apr	"Rocket Man"
19 Apr	"America"

The Colonial American flag encircling the crew patch symbolized a continuity of technical achievement and progress since colonial times. The name Discovery preceding the flag represented the spirit of discovery and exploration of new frontiers. (NASA)

This *Discovery* flight was a composite mission, carrying part of its original payload and part of that from the canceled STS-22/51E. NASA had targeted 19 March 1985 for launch, but delayed this to allow the payloads and crews to be shuffled. The crew was entirely from the canceled mission except Charlie Walker substituted for Patrick Baudry whose flight experiments were no longer on the manifest. STS-22/51E was to have been a *Challenger* flight and the crew patch selected for the remanifested STS-23/51D was that of the canceled mission with the substitution of orbiter names. The final countdown on 12 April proceeded smoothly, but was delayed 55 minutes at the T-9-minute hold for weather.

The mission featured the first flight of an elected official, Senator Jake Garn (R-UT), chairman of the Senate committee with oversight of the NASA budget.

The crew deployed Anik C1 on the first day and Syncom IV-3 (Leasat F3) on the second day. However, the crew noted the omni antenna on the satellite did not deploy and the Orbus-6 upper stage did not fire 45 minutes later. An analysis team on the ground determined a switch lever on the satellite that started the event sequencer did not completely trip during the deployment sequence. Mission management extended the mission two days and wanted the crew to rendezvous with the satellite to activate the switch lever. Houston sent the crew instructions for making a "flyswatter" they would attach to the SRMS end-effector during a spacewalk on flight day 5. The crew completed the fabrication and Jeff Hoffman and Dave Griggs conducted the first contingency EVA of the flight campaign.

Discovery rendezvoused with the satellite the following day, but despite Rhea Seddon swatting the switch lever at least twice with the SRMS, the omni antenna did not deploy and the subsequent motor firing did not take place. The crew left the satellite in an 180x250 nm orbit and it was eventually repaired on STS-27/51I.

Mission management delayed landing one revolution due to a rain shower at the Shuttle Landing Facility. The inboard tire on the right main landing-gear burst 33 feet after the brake locked up at 20.6 knots some 113 feet before the orbiter stopped rolling. Afterward, NASA management directed all future landings would be at Edwards until the nose-wheel steering system could be activated and tested since this would relieve stress on the brakes. The investigation into the *Challenger* accident ultimately drove the installation of an improved nose-wheel steering system and more-capable structural-carbon brakes. In addition, the post-flight inspection showed substantial thermal damage to the outboard forward corner of the left outboard elevon.

Front row, from the left, are Bo Bobko, Don Williams, Rhea Seddon, and Jeff Hoffman. In the back are Dave Griggs, Charlie Walker, and Jake Garn (R-UT). Garn was a U.S. Senator and Walker represented the McDonnell Douglas Corporation. (NASA)

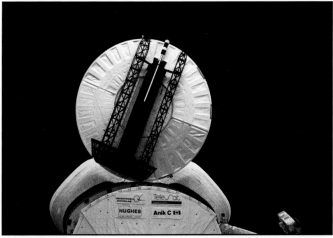

The Hughes HS-381 Syncom IV series of satellites "rolled" out of the payload bay rather then being ejected as with most of the other communications satellites. Note the sunshield for Anik C1, in the foreground, is closed (compare to it being open at right). (NASA)

When the Syncom IV-3 satellite did not activate as planned, the crew fabricated a "flyswatter" and attached it to the end of the SRMS arm. Despite Rhea Seddon swatting the switch at least twice, the satellite did not respond. It was eventually repaired on STS-27/51I. (NASA)

Anik C1 and its attached PAM-D upper stage being deployed from the payload bay. A pair of get-away special cans are on the side of the payload bay at left and the SRMS is stowed on the sill at right. Note that neither OMS pod has a black eyeball yet. (NASA)

As they did for every mission, workers at Hangar AF on Cape Canaveral Air Force Station disassembled and inspected the recovered solid rocket boosters. At a bit more than 12 feet in diameter and 150 feet long, weighing about 150,000 pounds empty, it was a daunting task. After they were disassembled, the empty motor cases were loaded onto rail cars and shipped back to Thiokol in Utah to be refurbished and loaded with new propellant for a future mission. The structural pieces, such as the aft and forward skirts, stayed in Florida for refurbishment. (NASA)

STS-24 (51B)

Mission:	17		NSSDC ID:		1985-034A
Vehicle:	OV-099 (7)		ET-17 (LWT)		SRB: BI016
Launch:	LC-39A		29 Apr 1985		16:02 UTC
Altitude:	192 nm		Inclination:		57.00 degrees
Landing:	EDW-17L		06 May 1985		16:12 UTC
Landing Rev:	111		Mission Duration:		168 hrs 09 mins

Commander: Robert F. "Bob" Overmyer (2)
Pilot: Frederick D. "Fred" Gregory (1)
MS1: Don L. Lind (1)
MS2: Norman E. "Norm" Thagard (2)
MS3: William E. Thornton (2)
PS1: Lodewijk van den Berg (1)
PS2: Taylor G. Wang (1)

Payloads: Up: 31,377 lbs   Down: 0 lbs
Spacelab 3 (30,643 lbs)
GAS (2)
No SRMS

Notes: Orbiter was originally configured for STS-22/51E
First operational Spacelab mission
Significant SRB seal erosion on both boosters
Backup payload specialists were
   Mary H. Johnston and Eugene H. Trinh
Orbiter returned to KSC on 11 May 1985 (N905NA)

Wakeup Calls:

Since crew worked two shifts around the clock,
mission control did not send any wakeup calls.

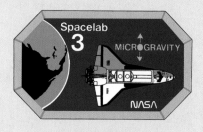

Discovery and its science module payload were featured on the crew patch for the Spacelab 3 mission. The seven stars, representing the crew, of the constellation Pegasus surround the orbiting spaceship above the flag draped Earth. At top is the Spacelab 3 emblem. (NASA)

On 15 February 1985, NASA rolled *Challenger* to LC-39A configured with the TDRS-B and Anik C1 satellites for the STS-22/51E mission. Issues with both payloads had already delayed the launch from 20 February to 28 February, to 3 March, and then to 7 March. The ground team rolled the stack back to the VAB on 5 March after a design issue emerged on TDRS-1 that had been deployed during STS-6. The problem potentially affected TDRS-B, so NASA canceled STS-22/51E less than a week before launch. *Challenger* was reconfigured with the Spacelab payloads planned for her next mission. The final countdown on 29 April proceeded smoothly until T-4 minutes when a failure in the launch processing system caused a short hold.

This was the first flight of an operational Spacelab with a variety of multi-disciplinary experiments. Spacelab 3 carried 15 primary experiments, 14 of which were successful. There were five basic disciplines, including materials sciences, life sciences, fluid mechanics, atmospheric physics, and astronomy, with numerous experiments in each. The gravity gradient attitude of the orbiter proved quite stable, allowing the delicate materials processing and fluid mechanics experiments to proceed successfully. Two monkeys and 24 rodents flew in special cages as part of the life sciences experiments.

Similar to the previous Spacelab mission, the crew divided roughly in half to cover 12-hour shifts, with Bob Overmyer, Don Lind, William Thornton, and Taylor Wang forming the gold team, and Fred Gregory, Norm Thagard, and Lodewijk van den Berg on the silver team. A temporary Payload Operations Control Center (POCC) at JSC supported the crew members. The mission met virtually all of the scientific objectives and several experiments acquired additional data beyond their pre-launch expectations. Researchers estimated more than 250 gigabits of data and 3 million video images were obtained from all instruments combined.

After the crew closed the payload bay doors, the port-aft door latches did not report being secure. However, visual indications showed them in the latched position and entry proceeded as planned.

Although it is not mentioned in the final MSFC propulsion report, there was a major problem with the solid rocket boosters. The primary o-ring on the left nozzle had failed to seal and eroded completely through in three locations and there was a heavy coating of black soot between the primary and secondary seals. Although not as severe, there was also evidence of erosion and blow-by on the nozzle seal of the right motor. Unfortunately, NASA and Thiokol management never truly acknowledged what was happening and the program continued to fly for another eight months.

Sitting, from the left, are Bob Overmyer and Fred Gregory, while standing are Don Lind, Taylor Wang, Norm Thagard, William Thornton, and Lodewijk van den Berg. They are dressed in the standard launch-entry trousers and polo shirts without any mission patches. (NASA)

A shot from Hawaii of the external tank disintegrating during reentry. NASA planned each mission so that any debris from the tanks impacted usually empty areas of the Indian or Pacific Oceans. (NASA)

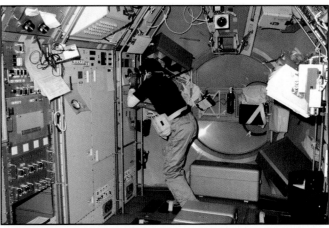

Lodewijk van den Berg observing the growth of mercuric iodide crystal in the vapor crystal growth system (VCGS) in Spacelab. The module is relatively uncluttered, unlike many later missions. (NASA)

A frame of 35 mm film captures the external tank liquid oxygen umbilical a few seconds after the tank was released from Challenger. The silver ball in the center was the physical attach point. (NASA)

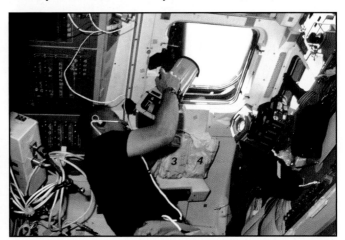

Bob Overmyer aims a Linhof camera through one of the overhead windows on the aft flight deck. Every mission carried an extensive array of camera equipment, mostly to photograph Earth. (NASA)

Challenger on the lakebed at Edwards AFB after the mission. The ground crew is hooking up the cooling trucks to the T-0 umbilicals on the aft fuselage and the crew support team is waiting for the crew to disembark. At this point, the orbiter still made extensive use of white LRSI tiles on the forward fuselage, giving an unusual stair-step pattern where it bordered the AFRSI blankets on the mid-fuselage. Note the black eyeball on the leading edge of the OMS pod and the extensive use of black tiles around one of the mid-fuselage vent doors. (NASA)

STS-25 (51G)

Mission:	18	NSSDC ID:	1985-048A
Vehicle:	OV-103 (5)	ET-20 (LWT)	SRB: BI019
Launch:	LC-39A	17 Jun 85	11:33 UTC
Altitude:	210 nm	Inclination:	24.48 degrees
Landing:	EDW-23	24 Jun 85	13:13 UTC
Landing Rev:	112	Mission Duration:	169 hrs 39 mins

Commander:	Daniel C. "Dan" Brandenstein (2)
Pilot:	John O. Creighton (1)
MS1:	John M. Fabian (2)
MS2:	Steven R. "Steve" Nagel (1)
MS3:	Shannon W. Lucid (1)
PS1:	Patrick Baudry (1)
PS2:	Sultan bin Salman bin Abdulaziz Al Saud (1)

Payloads:	Up: 46,694 lbs	Down: 0 lbs
	Morelos-A/PAM-D (7,591 lbs)	
	Arabsat-A/PAM-D (7,695 lbs)	
	Telstar 3D/PAM-D (7,546 lbs)	
	SPARTAN-101 (2,217 lbs)	
	GAS (6)	
	SRMS s/n 301	

Notes:
First Arab and Muslim in space
Youngest space shuttle flyer
(Al Saud at 28 years, 11 months, 21 days)
Backup payload specialists were
Jean-Loup Chrétien and
Abdulmohsen Hamad Al-Bassam
Orbiter returned to KSC on 28 Jun 1985 (N905NA)

Wakeup Calls:

18 Jun	"Eye in the Sky"
19 Jun	"I Feel the Earth Move"
20 Jun	"Oklahoma!"
21 Jun	"Proud Mary"
22 Jun	"Sailing"
23 Jun	Untitled Saudi music / "Jonathan Livingston Seagull"
24 Jun	"Wedding March" / "Get Me to the Church on Time"

The crew patch illustrated the advances in aviation technology in the United States within the relatively short span of the twentieth century. The surnames of the crew members for the mission appeared near the center edge of the circular design. (NASA)

There were no unplanned holds during the countdown and a French "spatialnaut" and a Saudi prince added an international flavor to the crew. Salman Al Saud was the first Arab, the first Muslim, and the first member of a royal family to fly in space.

As engineers evaluated the SRBs during disassembly, they noted evidence of primary o-ring erosion on both nozzles, although the amount of damage was within the experience base and the primary nozzles did seal, albeit, apparently, later than expected.

The crew successfully deployed Morelos-A within one second of the planned time on flight day 1 and within 1 nm of the desired 190-nm orbit. Morelos-A successfully made it to its final orbit with all systems operating normally. On flight day 2 there was some unplanned excitement prior to the deployment of Arabsat-A for the Arab Satellite Communications Organization: there were indications the solar arrays deployed while the satellite was still in its cradle. However, after opening the sunshield, the crew surveyed the satellite using the SRMS end effector camera and all appeared normal. The crew deployed Arabsat-A on time and watched the PAM-D fire.

The crew deployed Telstar 3D on flight day 3, again within one second of the planned time and used the SRMS wrist camera to observe the PAM-D firing. The AT&T satellite made it to its final orbit with all systems operating normally. The day also included an attempt to operate the high-precision tracking experiment (HPTE) for the "Star Wars" Strategic Defense Initiative, but two errors in the crew activity plan caused the orbiter to be oriented incorrectly. However, the crew successfully operated two French experiments and an automated directional solidification furnace (ADSF).

The crew deployed the free-flying SPARTAN-101 (Shuttle Pointed Autonomous Research Tool for Astronomy) on flight day 4. The following day saw another attempt at operating the HPTE, this time successfully. The major activity of flight day 6 was the rendezvous with SPARTAN, after which the crew grappled the spacecraft and berthed it in the payload bay.

The post-flight thermal protection system inspection revealed approximately 315 debris hits, of which 144 were greater than 1 square inch in area. This was a significantly more than had occurred during any of the previous ten flights and engineers believed the most likely cause was foam from the external tank. A review of ET separation photographs showed a number of large divots in the intertank area. An analysis established a reasonable level of confidence that any foam released during ascent would be limited to minor areas and result in small debris hits on the orbiter. It was part of a larger problem, but it would be another seventeen years before anybody noticed.

Kneeling in front are Dan Brandenstein (left) and John Creighton. Standing, from the left, are Shannon Lucid, Steve Nagel, John Fabian, Salman Al Saud, and Patrick Baudry. Al Saud was the youngest person to fly on space shuttle. (NASA)

Discovery at the moment of liftoff from LC-39A. Note the black photo-reference mark on the upper part of the solid rocket booster. The color of the external tank changed (oxidized) depending on how old it was and how much sunlight it had been exposed to. (NASA)

Arabsat-A is deployed from the payload bay. Despite being somewhat larger than the Hughes satellites, it weighed about the same and used the same PAM-D upper stage. The PAM did not provide quite as much energy as expected, but the satellite made it to its final orbit. (NASA)

Looking up from the back of the payload bay shows (from bottom) Telstar 3D, Arabsat-A, Morelos-A, and SPARTAN-101. Arabsat was built by Aérospatiale, based on the Spacebus 100 satellite bus, while the other two satellites were based on the Hughes HS-376 bus. (NASA)

An unusual view looking down from the aft flight deck window as technicians transfer Morelos-A from the Payload Changeout Room into the payload bay while Discovery was on the launch pad. SPARTAN-101 has not been installed yet. (NASA)

STS-26 (51F)

Mission:	19	NSSDC ID:	1985-063A
Vehicle:	OV-099 (8)	ET-19 (LWT)	SRB: BI017
Launch: Altitude:	LC-39A 174 nm	29 Jul 1985 Inclination:	21:00 UTC 49.49 degrees
Landing: Landing Rev:	EDW-23 127	06 Aug 1985 Mission Duration:	19:46 UTC 277 hrs 45 mins

Commander:	C. Gordon "Gordo" Fullerton (2)
Pilot:	Roy D. Bridges, Jr. (1)
MS1:	Karl G. Henize (1)
MS2:	F. Story Musgrave (2)
MS3:	Anthony W. "Tony" England (1)
PS1:	Loren W. Acton (1)
PS2:	John-David F. Bartoe (1)

Payloads:	Up: 34,400 lbs	Down: 0 lbs
	Spacelab 2 (33,555 lbs)	
	PDP (344 lbs)	
	SRMS s/n 302	

Notes:	Aborted 12 Jul 1985 (SSME)
	Abort-to-orbit 29 Jul 1985 (SSME)
	Only inflight abort of the flight campaign
	Fiftieth NASA manned space flight from KSC
	Verification test flight of this configuration
	Tested "Coca-Cola Space Can"
	Backup payload specialists were
	George W. Simon and Dianne K. Prinz
	Orbiter returned to KSC on 11 Aug 1985 (N905NA)

Wakeup Calls:

Since crew worked two shifts around the clock,
mission control did not send any wakeup calls.

Challenger *is ascending toward the heavens in search of new knowledge with Spacelab 2. The constellations Leo and Orion are in the positions they will be in during the flight. The nineteen stars signify the 19th space shuttle flight. The design was by Houston artist Skip Bradley. (NASA)*

During a 12 July launch attempt, a chamber coolant valve in one main engine was slow to move and caused an on-pad abort at T-3 seconds after all three main engines had ignited. NASA rescheduled launch for 29 July after technicians replaced the valve, actuator, and controller. The final countdown was slightly delayed due to an erroneous command in the backup flight system but the launch itself was nominal.

About two minutes after liftoff, a sensor in SSME-1 began to fail followed by the second, redundant sensor at T+221 seconds. The general-purpose computers shut the engine down at T+343 seconds, and Gordo Fullerton elected to abort-to-orbit (ATO). The remaining two engines throttled up higher than normal and fired for longer than planned. Coupled with a longer than usual OMS burn, the orbiter had sufficient energy to reach orbit. This was the only complete main engine failure and the only in-flight abort during the flight campaign.

This flight carried a Spacelab payload consisting of an igloo and three pallets that contained scientific instruments dedicated to life sciences, plasma physics, astronomy, high-energy astrophysics, solar physics, atmospheric physics and technology research. The infrared telescope (IRT), a 6-inch helium-cooled infrared telescope, experienced some problems, thought to be heat emissions from the orbiter corrupting long-wavelength viewing, but still returned useful data. Another problem was that a piece of mylar insulation broke loose and floated in the line-of-sight of the telescope.

However, the payload that received the most publicity was the carbonated beverage evaluation, which was an experiment in which both Coca-Cola and Pepsi attempted to make their carbonated drinks available to astronauts. The Coke dispenser was, largely, a serious experiment, but the Pepsi effort was mostly a publicity stunt. Post-flight, the crew revealed that they preferred Tang, mostly because it could be mixed with chilled water, whereas there was no refrigeration equipment on board to chill the soda.

The crew deployed the free-flying plasma diagnostics package (PDP) on flight day 3 and retrieved it the following day. During flight day 6 the mission management team extended the mission one day after the crew conducted an inspection of the underside of the vehicle using the SRMS. Mission control ordered the inspection after technicians at KSC found several large pieces of foam insulation that apparently fell off from the ET during launch. The inspection showed a large number of small diameter tile hits, but no tiles were missing and there were no concerns about entry. The post-flight inspection showed 553 debris hits of which 226 were greater than 1 inch. This was the largest number of debris hits on any mission to date.

From the lower left are red team members Loren Acton, Roy Bridges, and Karl Henize. The blue team included John-David Bartoe, Story Musgrave, and Tony England. Remaining neutral, Gordo Fullerton wore a striped blue-red shirt. (NASA)

The solar optical universal polarimeter (SOUP) experiment among the cluster of Spacelab 2 hardware in the payload bay. Various components of the instrument positioning system (IPS) are at the center of the frame and the SRMS arm is stowed at the right. (NASA)

A good view showing Challenger climbing away from the launch pad with the Atlantic Ocean in the background. Note the shock cones in the main engine exhaust and the camera-defeating color (temperature) of the solid rocket booster exhaust. (NASA)

In the days before email, Tony England (left) and Roy Bridges are surrounded by some of the teleprinter copy transmitted from the flight controllers. Eventually, the equivalent of several football fields' length of paper was filled with data from the ground. (NASA)

Challenger about to touch down on the lakebed at Edwards AFB. Like all delta-wing aircraft, the orbiter landed nose-high. However, the relative location and length of the landing-gear caused a hard "slap" when the nose gear finally touched down. (NASA)

STS-27 (51I)

Mission:	20	NSSDC ID:	1985-076A
Vehicle:	OV-103 (6)	ET-21 (LWT)	SRB: BI020
Launch:	LC-39A	27 Aug 1985	10:58 UTC
Altitude:	242 nm	Inclination:	28.54 degrees
Landing:	EDW-23	03 Sep 1985	13:17 UTC
Landing Rev:	112	Mission Duration:	170 hrs 18 mins

Commander:	Joe H. Engle (2)
Pilot:	Richard O. "Dick" Covey (1)
MS1:	James D. A. "Ox" van Hoften (2)
MS2:	John M. "Mike" Lounge (1)
MS3:	William F. "Bill" Fisher (1)

Payloads:	Up: 43,988 lbs	Down: 0 lbs
	ASC-A/PAM-D (7,591 lbs)	
	AUSSAT-1/PAM-D (7,508 lbs)	
	Syncom IV-4/Orbus-6 (15,190 lbs)	
	SRMS s/n 301	

Notes:	Scrubbed 24 Aug 1985 (KSC weather)
	Scrubbed 25 Aug 1985 (GPC failure)
	EVA1 31 Aug 1985 (von Hoften and Fisher)
	EVA2 01 Sep 1985 (von Hoften and Fisher)
	Orbiter returned to KSC on 08 Sep 1985 (N905NA)

Wakeup Calls:

28 Aug	"Waltzing Matilda"
29 Aug	"Over the Rainbow" from The Wizard of Oz
30 Aug	"I Saw the Light"
31 Aug	"I Get Around"
01 Sep	"Lucky Old Sun"
02 Sep	"Stormy Weather"
03 Sep	"Living in the USA"

The crew patch was based on a strong patriotic theme with the basic colors of red, white, and blue and a dominant American bald eagle in aggressive flight. The 19 stars signified the intended mission sequence and the shock wave represented the orbiter during entry. (NASA)

NASA planned this *Discovery* launch for 24 August 1985, but scrubbed it at T-5 minutes due to thunderstorms and lightning. The storm resulted in more than 1 inch of rain and several lightning strikes near at the launch site. During a second attempt on 25 August, a general-purpose computer failed. The launch team scheduled the next attempt for 27 August and the countdown was trouble-free except for a brief delay caused by an unauthorized ship in the SRB recovery area.

As the crew prepared to deploy AUSSAT-1, the left side of its sunshield could not be opened. Engineers determined the structure was binding on a bracket on the top of the satellite, so the crew used the SRMS to open the sunshield. To avoid overheating the satellite in the payload bay, mission control accelerated the AUSSAT deployment from revolution 17 to revolution 5 and delayed the deployment of ASC-A from revolution 7 to revolution 8.

Flight day 2 included two runs of the PVTOS (physical vapor transport organic solid) experiment and preparing to rendezvous with Syncom IV-3 (Leasat F3), which had been left in a useless orbit during STS-23/51D. The crew successfully deployed Syncom IV-4 (Leasat F4) on flight day 3 and the post-ejection sequencer functions occurred on time as visually verified by omni antenna deployment and spacecraft spin-up. The Orbus-6 upper stage successfully placed the satellite in the desired geosynchronous orbit. On-orbit checkout began on 4 September but Hughes later lost all communications with the satellite and declared it a loss.

Discovery rendezvoused with the stranded Syncom IV-3 on flight day 4 and Bill Fisher and Ox van Hoften conducted what was then the longest space shuttle EVA the following day. The effort was slowed when the SRMS elbow joint stopped working in the primary mode and the crew switched to a backup mode for the remainder of the mission. Because of the SRMS issue, flight controllers replanned the retrieval into two EVAs instead of one. The first spacewalk successfully deployed the omni antenna and installed a set of shorting plugs, safe-and-arm pins, and a spin bypass unit on the wayward satellite. During the second EVA, Fisher and van Hoften removed the upper stage motor cover, installed the new one, removed the safe-and-arm pins, and started the firing timers, permitting commands from the ground to activate the spacecraft and eventually send it into its proper geosynchronous orbit. During a subsequent pass over Guam, engineers at Hughes successfully commanded the spacecraft to turn on its encoders and telemetry transmitters.

The crew depressurized the cabin after the EVAs, stowed their gear, and returned to Edwards on the first landing opportunity.

From the left, Joe Engle, Ox van Hoften, Dick Covey, Bill Fisher, and Mike Lounge, all wearing the standard-issue light blue launch-entry coveralls. The American flag was creatively arranged to mimic the prominent flag on the crew patch. (NASA)

The ASC-A and its attached PAM-D upper stage for the American Satellite Company rises from the payload bay on 27 August 1985. This communications satellite used the RCA AS-3000 bus. (NASA)

The day after Discovery launched as STS-27/51I, emergency crews at Edwards AFB used Enterprise to practice rescuing the crew during an emergency landing simulation on the dry lakebed. (NASA)

Joe Engle commands a single-engine OMS burn to adjust the orbit between deploying AUSSAT-1 and ASC-1. The crew deployed the two satellites only four hours apart. (NASA)

The Flight Control Room (FCR-2) at the Mission Control Center in Houston. The two original flight control rooms exclusively supported the first sixty-nine space shuttle missions and partially supported the next seven as NASA brought the new White Flight Control Room online. Many of the consoles and much of the equipment in the original flight control rooms dated from the Apollo era. Contrary to public perception, and many television shows and movies, the control rooms were not especially large or particularly fancy. (NASA)

STS-28 (51J)

Mission:	21	NSSDC ID:	1985-092A
Vehicle:	OV-104 (1)	ET-25 (LWT)	SRB: BI021
Launch:	LC-39A	03 Oct 1985	15:16 UTC
Altitude:	278 nm	Inclination:	28.50 degrees
Landing:	EDW-23	07 Oct 1985	17:01 UTC
Landing Rev:	64	Mission Duration:	97 hrs 45 mins

Commander:	Karol J. "Bo" Bobko (3)
Pilot:	Ronald J. "Ron" Grabe (1)
MS1:	David C. "Dave" Hilmers (1)
MS2:	Robert L. "Bob" Stewart (2)
PS1:	William A. "Bill" Pailes (1)

Payloads:	Up: (classified lbs)		Down: 0 lbs
	USA-11/12 (DSCS-III B4/B5)/IUS (classified lbs)		
	No SRMS		

Notes:
First flight of *Atlantis* (OV-104)
Flight readiness firing on 05 Sep 1985
Second dedicated Department of Defense mission
Backup payload specialist (MSE) was
   Michael W. Booen
Orbiter returned to KSC on 11 Oct 1985 (N905NA)

Wakeup Calls:

Classified mission. Either nothing played or nothing released.

Designed by the crew the patch pays tribute to the Statue of Liberty and the ideas it symbolizes. The historical gateway figure bore additional significance for Bo Bobko and Ron Grabe, who were both born in New York City. (NASA)

Engineers conducted a successful 20-second flight readiness firing on 5 September 1985 to verify *Atlantis* was ready for her maiden voyage. Other than a minor problem with an indication on a combustion chamber valve, the FRF was successful.

During pre-launch checkout on 3 October 1985, there was an intermittent liquid hydrogen prevalve indication. The ground team verified the valve was working satisfactorily and the mission management team approved launch after a short delay. Otherwise the countdown proceeded smoothly.

Bo Bobko became the first astronaut to fly on three different orbiters (*Challenger*, *Discovery*, and *Atlantis*), and the only astronaut to fly on the maiden voyages of two orbiters (*Challenger* on STS-6 being the other). Prior to Bill Pailes being assigned to the flight, rumors had Mike Mullane assigned as MS3.

The post-launch pad inspection and film analysis showed *Atlantis* had lost at least two gap fillers from her thermal protection system. The debris assessment team found one in the pad flame trench, which engineers determined came from the lower surface of the aft-most section of the vertical stabilizer. Photo analysts noted another gap filler separated from the lower inboard side of the right inboard elevon. Gap fillers prevented hot gases from leaking through the small gaps between tiles and impinging the aluminum skin of the orbiter and were important components, but not necessarily critical.

This was the second dedicated DoD mission and, officially, the USA-11/12 payload was classified. However, even before launch, pundits deduced the payload was a pair of third-generation Defense Satellite Communications System (DSCS-III) satellites (B4 and B5) on a single inertial upper stage. The Air Force finally declassified the payload on 3 September 1994, confirming the speculation.

After the crew deployed the payload on flight day 1, the inertial upper stage delivered the satellites to their respective geosynchronous orbits. The DSCS-III satellites had a design life of ten years, although several have far exceeded this. Given its (at the time) classified nature, the Air Force issued a its usual statement calling the mission "... successful ..."

The post-flight thermal protection system inspection showed two areas had suffered damage where the gap fillers had fallen out. In one case, several tiles and two carrier panels were severely damaged. The other area showed only minor heat damage to the filler bar. This was a learning exercise in what type of damage the orbiter could accept during entry. Despite the issues, engineers found the thermal protection system was in remarkably good overall shape.

Seated, from the left, are Bob Stewart, Bo Bobko, and Ron Grabe. Behind them are Dave Hilmers and Bill Pailes. Of the thirty-two military manned spaceflight engineers named, Pailes was only the second, and last, to actually fly. The first was Gary Payton on STS-20/51C. (NASA)

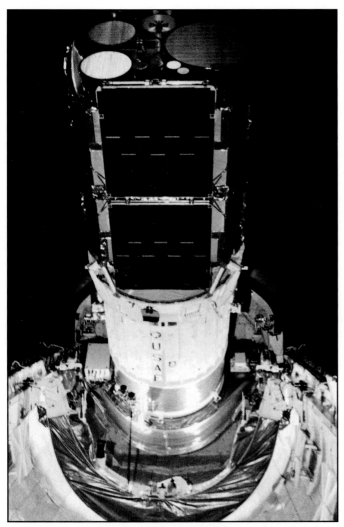

The inertial upper stage carried two DSCS-III (B4 and B5) satellites back-to-back. Like all the IUSs, and the stillborn Shuttle/Centaur, the stage was carried on large aerospace support equipment and tilted out of the payload bay before being ejected. (NASA)

Bo Bobko in the left seat studies a checklist in preparation for the deorbit burn and entry. He is wearing the normal launch-entry coveralls and the colorful version of the launch-entry helmet. (NASA)

Bill Pailes is looking from the flight deck into the middeck through the left passageway in the floor. Note the USAF patch on his right shoulder. He is holding onto the flight ladder that provided access between the levels in the crew compartment when in 1-g. (NASA)

A glamour shot of the primary DSCS-III payload. The Martin Marietta (formerly GE General Electric) Astro Space Division spacecraft was 6 feet high, 6 feet wide, 5 feet long, and weighed 2,580 pounds. Each satellite cost about $100 million. (U.S. Air Force)

As with all the orbiters, Atlantis underwent a 20-second flight readiness firing before her first mission. In this case there was a small anomaly with an engine, but nothing that delayed the launch of the second mission dedicated to the Department of Defense. (NASA)

STS-30 (61A)

Mission:	22	NSSDC ID:		1985-104A
Vehicle:	OV-099 (9)	ET-24 (LWT)		SRB: BI022
Launch:	LC-39A	30 Oct 1985		17:00 UTC
Altitude:	180 nm	Inclination:		56.99 degrees
Landing:	EDW-17L	06 Nov 1985		17:46 UTC
Landing Rev:	112	Mission Duration:		168 hrs 45 mins

Commander:	Henry W. "Hank" Hartsfield, Jr. (3)
Pilot:	Steven R. "Steve" Nagel (2)
MS1:	Bonnie J. Dunbar (1)
MS2:	James F. "Jim" Buchli (2)
MS3:	Guion S. "Guy" Bluford, Jr. (2)
PS1:	Reinhard A. Furrer (1)
PS2:	Ernst W. Messerschmid (1)
PS3:	Wubbo J. Ockels (1)

Payloads:	Up: 31,911 lbs	Down: 0 lbs
	Spacelab D1 (30,369 lbs)	
	GAS (1)	
	SRMS s/n 302	

Notes:	Mission paid for by West Germany
	Spacelab hardware provided by Europe
	Largest crew launched on a single vehicle
	Only mission to carry eight crew members up
	Backup payload specialist was
	Ulf D. Merbold
	Last successful flight of *Challenger*
	Orbiter returned to KSC on 11 Nov 1985 (N905NA)

Wakeup Calls:

Since crew worked two shifts around the clock,
mission control did not send any wakeup calls.

The crew patch included a colorful scene depicting Challenger *carrying a Spacelab long module and an international crew from America and Germany, with the European Space Agency participation marked by the ESA emblem next to Ockels' name. (NASA)*

After a countdown with no delays, *Challenger* was launched 30 October 1985 on the Spacelab D1 mission. The German Test and Research Institute for Aviation and Space Flight (DFVLR) paid NASA $59 million for the flight opportunity. This was the only mission to carry eight crew members and is the largest crew launched on a single vehicle (STS-71 carried eight down-only).

For the first time, someplace other than Houston controlled non-orbiter portions of the mission. Most of the experiments were directed by the German Space Operations Center located in Oberpfaffenhofen in Bavaria.

As was typical for Spacelab missions, the crew divided into two teams that each worked 12-hour shifts. The blue team consisted of Hank Hartsfield, Steve Nagel, Bonnie Dunbar, Reinhard Furrer, and Wubbo Ockels while the red team included Jim Buchli, Guy Bluford, and Ernst Messerschmid.

The crew began activating Spacelab D1 about three hours after launch and completed the task two hours later. All but three of the 76 experiments successfully activated, although the furnace in the MEDEA (experiment modules and apparatus) rack malfunctioned. Some of these experiments had predecessors that had flown on earlier missions, allowing researchers to prepare second-generation experiment regimens. Almost all of the experiments took advantage of the microgravity environment to perform work not possible, or more difficult to do, on Earth.

The crew also successfully deployed the Defense Advanced Research Projects Agency GLOMR (global low orbiting mission relay) satellite that had malfunctioned during STS-24/51B.

One unusual item was the vestibular sled, a European Space Agency contribution consisting of a seat that could be moved backward and forward with precisely controlled accelerations and stops. By taking detailed measurements on a crew member strapped into the seat, scientists gained data on the functional organization of the human vestibular and orientation systems, and the vestibular adaptation processes under microgravity. The acceleration experiments by the sled riders were combined with thermal stimulations of the inner ear and optokinetic stimulations of the eye.

Because of the malfunction of the MEDEA furnace during the early part of the mission, NASA held discussions with the science community on extending the mission one day to obtain additional materials processing data. However, Spacelab power-usage could not be reduced enough to save sufficient consumables, so the mission management team declined the extension.

This marked the last successful mission of *Challenger*.

Crew portrait in front of the crew compartment trainer (CCT) at JSC. Sitting, from the left, are Wubbo Ockels, Hank Hartsfield, and Bonnie Dunbar. Standing are Guy Bluford, Ernst Messerschmid, Steve Nagel, Jim Buchli, Ulf Merbold (backup PS), and Reinhard Furrer. (NASA)

In their light blue launch-entry coveralls, the crew walks from the Operations and Checkout Building led by Hank Hartsfield and Bonnie Dunbar. The crew is followed by John Young, chief of the astronaut office, and George Abbey, director of flight crew operations. (NASA)

The view from the aft flight deck windows showed the top of the tunnel leading from the crew module to Spacelab and the front face of the pressurized module. Note the Spacelab D1 logo on the left, ESA logo on the right, and the mis-matched OMS pod eyeballs. (ESA)

Challenger touching down on Rogers Dry Lake at Edwards during her last landing. Her next flight would end poorly. The orbiter had landed at Kennedy Space Center twice and Edwards seven times. Note the discolored white LRSI tiles on the forward fuselage. (NASA)

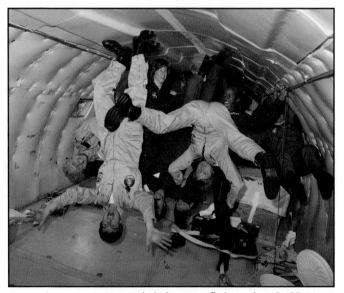

At one time, crew training included a zero-g flight on the KC-135 "Vomit Comet." Here, Bonnie Dunbar, Guy Bluford, and Wubbo Ockels (upside down) are in dark blue NASA flight suits while Reinhard Furrer, Ulf Merbold, and Ernst Messerschmid are in light blue shuttle suits. (NASA)

STS-31 (61B)

Mission:	23	NSSDC ID:	1985-109A
Vehicle:	OV-104 (2)	ET-22 (LWT)	SRB: BI023
Launch: Altitude:	LC-39A 209 nm	27 Nov 1985 Inclination:	00:29 UTC 28.45 degrees
Landing: Landing Rev:	EDW-22 109	03 Dec 1985 Mission Duration:	21:35 UTC 165 hrs 05 mins

Commander:	Brewster H. Shaw, Jr. (2)
Pilot:	Bryan D. "OC" O'Connor (1)
MS1:	Jerry L. Ross (1)
MS2:	Mary L. Cleave (1)
MS3:	Sherwood C. "Woody" Spring (1)
PS1:	Charles D. "Charlie" Walker (3)
PS2:	Rodolfo Neri Vela (1)

Payloads:	Up: 47,509 lbs	Down: 0 lbs
	Morelos-B/PAM-D (7,573 lbs)	
	AUSSAT-2/PAM-D (7,634 lbs)	
	Satcom Ku2/SCOTS (12,258 lbs)	
	EASE/ACCESS (4,685 lbs)	
	GAS (1)	
	SRMS s/n 303	

Notes: EVA1 29 Nov 1985 (Ross and Spring)
EVA2 01 Dec 1985 (Ross and Spring)
Backup payload specialists were
    Robert J. Wood and Ricardo Peralta y Fabi
Orbiter returned to KSC on 07 Dec 1985 (N905NA)

Wakeup Calls:

26 Nov	The Air Force Hymn – "Wild Blue Yonder"
27 Nov	(unidentified)
28 Nov	(unidentified)
29 Nov	"America the Beautiful"
30 Nov	(unidentified)
01 Dec	(unidentified)
02 Dec	Notre Dame Victory March
03 Dec	"Born in the U.S.A."

Whatever symbolism was intended for their patch is lost to history other than it was designed by the crew. The emblem was surrounded by the surnames of the five NASA crew members and a tab beneath showing the names of the two payload specialists. (NASA)

The second night launch of the flight campaign took place on 27 November 1985 with no unplanned delays. This *Atlantis* mission carried the first Mexican astronaut (Vela).

Immediately after reaching orbit the crew activated the SRMS and deployed Morelos-B, the second Hughes HS-376 satellite for the Mexican Secretariat of Communications and Transport (SCT). On flight day 2, the crew deployed AUSSAT-2, the second Hughes HS-376 for AUSSAT Pty Limited in Australia (now SingTel Optus Pty Limited). The crew also began preparing the three extravehicular mobility units (EMU) for two planned spacewalks and deployed Satcom Ku2 on flight day 3. This RCA 4000-series communications satellite was part of a series of satellites owned by RCA American Communications, although General Electric purchased the company the following year and renamed it GE Americom.

On flight day 4, Jerry Ross and Woody Spring conducted the first EVA, focused on human performance while assembling an experimental erectable truss structure. First, the pair assembled the 11-foot ACCESS (assembly concept for construction of erectable space structures) experiment in the payload bay. The pair accomplished the task in 55 minutes, rather than the planned 120 minutes, so they disassembled it and built it again. Then they turned their attention to EASE (experimental assembly of structures through EVA), a task that involved putting 64-pound beams together to create a 12-foot three-sided pyramid to assess the abilities of free-floating astronauts to move large objects. EASE was scheduled to be assembled six times, but Ross and Spring managed eight.

The second EVA came on flight day 6, which was spent assessing the ability of astronauts to handle large structural elements and of the SRMS to support some future space station assembly. Jerry Ross and Woody Spring assembled nine bays of the ACCESS, then placed parts for the tenth bay on the SRMS. Ross stepped into the manipulator foot restraint and Mary Cleave positioned him within reach of the top of the ACCESS girder, where he assembled the tenth bay. Ross and Spring then switched places and changed a beam on the tower to simulate structural repair. After successfully completing these tests, the astronauts disassembled ACCESS and Spring assembled EASE from the SRMS. The astronauts reported the most difficult part of the spacewalks was torquing their own masses while holding the EASE beams. The ACCESS worked well, while EASE required too much free floating.

The mission management team shortened the flight by one revolution due to deteriorating weather conditions at Edwards, where *Atlantis* landed on the concrete runway since the lakebeds were wet.

Kneeling, from the left, are Bryan O'Connor and Brewster Shaw. Standing are Charlie Walker, Jerry Ross, Mary Cleave, Woody Spring, and Rodolfo Vela. Note the Secretariat of Communications and Transportation (SCT) patch worn by Vela. (NASA)

The STS-31/61B payloads in the payload canister (background) ready to be taken to LC-39A and loaded into Atlantis. From the top are EASE/ACCESS, Morelos-B, Satcom Ku2, and AUSSAT-2. In the foreground is TDRS-B waiting to fly on Challenger as STS-33/51L. (NASA)

Until the advent of digital cameras, night launches were notoriously difficult to photograph, mostly because of the intense light generated by the solid rocket boosters. Nevertheless, they were very interesting to watch for those fortunate enough to have seen one. (NASA)

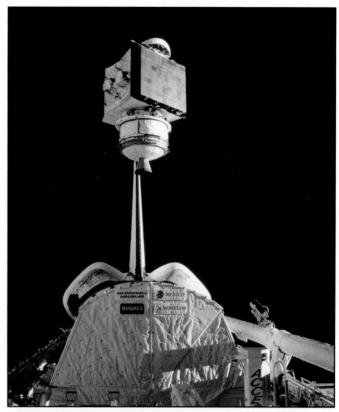

Satcom Ku2 was based on the RCA 4000 bus and was pushed into its geosynchronous orbit by a McDonnell Douglas shuttle compatible orbit transfer stage (SCOTS). The stage was also marketed as the PAM-D2, although it shared little with the original PAM-D. (NASA)

Morelos-B was based on the Hughes HS-376 bus and used a McDonnell Douglas payload assist module delta-class (PAM-D) to get to its desired geosynchronous orbit. Compare the shapes of the PAM-D in this photo to the SCOTS/PAM-D2 in the top photo. (NASA)

STS-32 (61C)

Mission:	24	NSSDC ID:	1986-003A
Vehicle:	OV-102 (7)	ET-30 (LWT)	SRB: BI024
Launch:	LC-39A	12 Jan 1986	11:55 UTC
Altitude:	184 nm	Inclination:	28.45 degrees
Landing:	EDW-22	18 Jan 1986	14:00 UTC
Landing Rev:	98	Mission Duration:	146 hrs 04 mins

Commander:	Robert L. "Hoot" Gibson (2)
Pilot:	Charles F. "Charlie" Bolden, Jr. (1)
MS1:	George D. "Pinky" Nelson (2)
MS2:	Steven A. "Steve" Hawley (2)
MS3:	Franklin R. Chang-Diaz (1)
PS1:	C. William "Bill" Nelson II (1)
PS2:	Robert J. "Bob" Cenker (1)

Payloads:	Up: 32,733 lbs	Down: 0 lbs
	MSL-2 (20,111 lbs)	
	Satcom Ku1/SCOTS (12,351 lbs)	
	GAS (12)	
	Hitchhiker (1)	
	No SRMS	

Notes:	Delayed 18 Dec 1985 (KSC workload)
	Scrubbed 19 Dec 1985 (right SRB HPU)
	Scrubbed 06 Jan 1986 (LO$_2$ system issue)
	Scrubbed 06 Jan 1986 (again, launch window)
	Delayed 07 Jan 1986 (TAL weather)
	Scrubbed 09 Jan 1986 (SSME prevalve)
	Delayed 10 Jan 1986 (KSC weather)
	First flight of orbiter experiment (OEX) package
	Backup payload specialist was
	Gerard E. Magilton
	Orbiter returned to KSC on 23 Jan 1986 (N905NA)

Wakeup Calls:

12 Jan	Theme from *Monty Python's Flying Circus*
13 Jan	(unidentified)
14 Jan	"Heart of Gold"
15 Jan	(unidentified)
16 Jan	(unidentified)
17 Jan	(unidentified)
18 Jan	"Stars and Stripes Forever"

Columbia *is featured inside colorful entry shockwaves in the patch designed by the crew to represent the seven individuals who will fly the vehicle on its seventh mission. Representations of the American flag and the constellation Draco flank the core. (NASA)*

Fresh from modifications in Palmdale, *Columbia* sported a collection of orbiter experiments (OEX) that included the SILTS (shuttle infrared leeside temperature sensing) pod atop the vertical stabilizer. This experiment used an infrared camera to observe entry heating-effects on the left wing and part of the fuselage. The camera was only used for six missions, but the pod remained on *Columbia* until she was lost.

Originally, NASA planners expected to launch STS-32/61C on 18 December 1985, but the mission encountered seven launch delays. The final countdown on 12 January went smoothly, but the delays caused serious workload problems at the various NASA centers and contractor sites. This mission tied with STS-73 for the most delayed launches, each having launched on their seventh attempt.

Significantly, the post-flight inspection showed major issues with both SRBs. Engineers found a gas path and soot on the primary o-ring on the aft field joint and nozzle joint of the left SRM and the nozzle joint of the right SRM. The o-ring on the aft field joint of the left motor suffered heat effects across a 14-inch section. In addition, they found a gas path and soot on the igniter joint of both motors. It bode ill for the next launch, only two weeks in the future.

This mission carried the second Materials Science Laboratory (MSL-2). Initially, STS-32/61C was also going to carry two communications satellites, Satcom Ku1 and Syncom IV-5 (Leasat F5). However, planners were never certain that Syncom would fly on this mission, so they always maintained an option to fly a GAS bridge assembly and Hitchhiker-G1 instead. As it turned out, Syncom was not ready and eventually flew on *Columbia* during STS-32R.

The crew deployed Satcom Ku1 about 9.5 hours after launch and a pair of Boeing EC-135E advanced range instrumentation aircraft (ARIA) verified the shuttle compatible orbit transfer stage (SCOTS, also called PAM-D2) firing that pushed the satellite to geosynchronous orbit. Afterward, the crew activated MSL-2, the infrared imaging experiment, and five of the GAS experiments. The crew activated three additional GAS experiments on flight day 2 and the remainder on flight day 3. A little later, Franklin Chang-Diaz hosted the first Spanish television broadcast from space, consisting mostly of a tour of the vehicle.

NASA shortened the flight by one day to ease workload issues at KSC getting ready for the next mission, STS-33/51L. The mission management team canceled plans for landing at KSC on flight day 6 because of weather. Despite not wanting to return to Edwards because of the schedule impacts associated with landing 2,300 miles from the launch site, *Columbia* headed back to Edwards one revolution later.

Sitting, from the left, are Charlie Bolden and Hoot Gibson. Standing are Bob Cenker, Bill Nelson (D-FL), Steve Hawley, Pinky Nelson, and Franklin Chang-Diaz. The crew did two official portraits: this one and one in their blue flight suits. They also did at least one gag portrait. (NASA)

A nice study of the Columbia stack on LC-39A. A close look will reveal the shuttle infrared leeside temperature sensing (SILTS) pod surrounded by an expanse of black HRSI tiles atop the vertical stabilizer on Columbia. At this point each launch pad a large hammerhead crane on top, although these were later abandoned, then removed. The red parts of the gantry were recycled parts of the Apollo launch umbilical towers and would eventually receive the same boring gray paint as the new parts of the fixed service structure. At right, an interesting view of launch. (NASA)

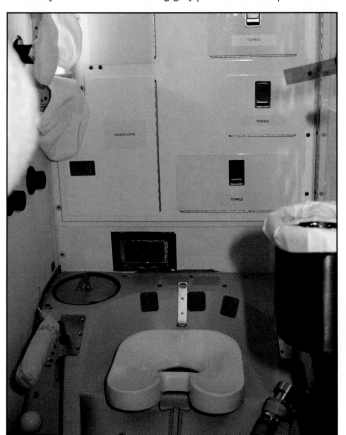

One of the luxuries afforded by space shuttle was a real waste management system. Previously, astronauts had used diapers and fecal bags to collect waste. Although its operation was a bit unusual, the potty was not that much different than a toilet on Earth. (NASA)

Charlie Bolden, Hoot Gibson, and Pinky Nelson get ready for a meal. Each astronaut could choose their own menu from a selection prepared by nutritionists at JSC. The quality and taste of the food evolved (mostly, improved) considerably during the flight campaign. (NASA)

STS-33 (51L)

Mission:	25	NSSDC ID:	n/a		
Vehicle:	OV-099 (10)	ET-26 (LWT)		SRB:	BI026
Launch:	LC-39B	28 Jan 1986		16:38 UTC	
Altitude:	(154) nm	Inclination:		(28.45) degrees	
Landing:	--	--		--	
Landing Rev:	(98)	--		--	

Commander:	Francis R. "Dick" Scobee (2)
Pilot:	Michael J."Mike" Smith (1)
MS1:	Ellison S. "El" Onizuka (2)
MS2:	Judith A. "Judy" Resnik (2)
MS3:	Ronald E. "Ron" McNair (2)
PS1:	Gregory B. "Greg" Jarvis (1)
SFP:	S. Christa McAuliffe (1)

Payloads:	Up: 52,655 lbs	Down: 0 lbs
	TDRS-B/IUS (37,636 lbs)	
	SPARTAN-203 (SPARTAN-Halley) (2,500 lbs)	
	SRMS s/n 302	

Notes: First Teacher in Space flight
Last flight of a Phase I space shuttle main engine
Backup payload specialist was
   L. William Butterworth
Backup space flight participant was
   Barbara R. Morgan

Wakeup Calls:

   Mission did not make orbit.

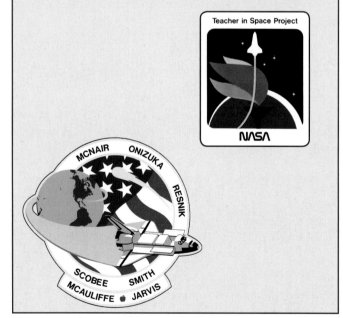

The crew patch continued the trend of using themes based around the American flag, this time with a silver Halley's Comet streaking over it to represent the SPARTAN-Halley payload. The name of the first teacher in space, Christa McAuliffe, is followed by a symbolic apple. (NASA)

NASA added STS-33/51L to the manifest during early 1984 as a *Columbia* mission carrying TDRS-C and the EOS-1 (electrophoresis operations in space) experiment. However, the next 18 months would see ten major changes to the payload and moving the mission to *Challenger*. NASA announced the STS-33/51L crew on 27 January 1985. One, in particular, had a link to *Challenger*. The commander, Dick Scobee, was a former Air Force test pilot with 7,000 hours in 45 types before he became an astronaut in 1978. Coming from a large aircraft background, he frequently piloted the shuttle carrier aircraft and was one of the pilots that flew N905NA when it delivered *Challenger* in July 1982 and when it ferried the orbiter back to KSC following STS-30/61A in November 1985.

The primary payloads were the second Tracking and Data Relay Satellite (TDRS-B) and SPARTAN-203 (Shuttle Pointed Autonomous Research Tool for Astronomy), which would observe Halley's Comet while it was too close to the Sun to be imaged by observatories on Earth. But the highlight of the mission was the first flight of the highly publicized Teacher in Space Project.

NASA held the flight readiness review on 15 January 1986. In theory, this review addressed all aspects of flight processing including any issues or concerns. The orbiter and external tank projects reported being ready. Lawrence Mulloy presented for the solid rocket motor project, indicating there were "no major problems or issues." Oddly, the subject of joints or o-rings, and their poor performance on many previous missions, was never mentioned.

The majority of the KSC presentation addressed a few minor problems encountered during the launch of STS-32/61C a couple of days before. In addition, KSC engineers explained that STS-33/51L would be the first launch from LC-39B. Modifications to convert the facility from Apollo to space shuttle had begun in 1979 and NASA declared the launch pad operational on 2 December 1985. Since the pads were essentially identical, this was not expected to offer any particular problems.

At the end of the flight readiness review, each of the element managers and company executives signed a certificate of flight readiness (CoFR) indicating everything was ready for launch. And so it was for STS-33/51L when the group certified *Challenger* was "… flight ready." Despite serious concerns by some engineers, there was no mention of booster joint issues in the CoFR signed by Thiokol vice president Joseph Kilminster on 9 January 1986.

Challenger was destroyed at T+73.631 seconds on 28 January 1986 while traveling 1,977 mph at 26,000 feet, some eighteen miles east of the Kennedy Space Center. There were no survivors.

The final crew of Challenger. In the first row, from the left, are Mike Smith, Dick Scobee, and Ron McNair. In the back are El Onizuka, Christa McAuliffe, Greg Jarvis, and Judy Resnik. McAuliffe was a "space flight participant" but was better known as a "Teacher in Space." (NASA)

The primary payload was the second Tracking and Data Relay Satellite (TDRS-B) and its attached interim upper stage (IUS). The crew of STS-6 had deployed the first TDRS satellite in 1983 and the space agency was anxious to add to the constellation. (NASA)

For the first few seconds, it looked like a picture-perfect launch on a particularly cold Florida winter morning. As events soon proved, looks could be deceiving. A bit more than seventy-three seconds later, Challenger would be falling from the sky in pieces. (NASA)

The two Teacher in Space finalists, Barbara Morgan (left) and Christa McAuliffe, during a tour of LC-39A prior to STS-30/61A. Morgan would become a professional astronaut and finally get to fly on STS-118 in August 2007, more than twenty years later than expected. (NASA)

The Challenger stack atop a mobile launch platform (MLP) arriving at the newly completed LC-39B on 22 December 1985. STS-33/51L would be first launch from Pad B. For all intents, the two pads were identical, although there were small differences in some details. (NASA)

CHALLENGER ACCIDENT

A significant amount of ice accumulated on LC-39B the night before the 28 January 1986 launch of STS-33/51L, creating considerable concern among engineers. Kennedy Space Center had implemented a freeze protection plan, last used during STS-20/51C, that included adding 1,400 gallons of antifreeze to the sound suppression water troughs. At 06:30 UTC, the ambient air temperature was 29 °F and there was ice up to 3 inches thick on various areas of the pad. By 15:55 the ice on the mobile launch platform had begun to melt and icicles were falling from the upper levels of the fixed service structure. For the most part, this was happening at locations away from the vehicle.

During the early morning, engineers in Utah, Huntsville, and Florida debated the potential effects of the extreme cold weather and developed several suggestions, but nobody was listening.

The crew woke early, ate breakfast, received a weather briefing, and dressed in the tailored blue launch-entry coveralls that were customary at the time. The ambient temperature was 36°F measured at ground level approximately 1,000 feet from the vehicle, some 15°F colder than any previous space shuttle launch. Nevertheless, the flight crew and launch team gave a "go" for launch during the scheduled T-9-minute hold. The final flight of *Challenger* began at 16:38:00.010 UTC on 28 January 1986.

From liftoff until loss of signal, flight controllers in mission control saw no indications of a problem, although post-flight analysis uncovered various anomalies in the telemetry. It should be noted much of the telemetry was recorded for post-flight analysis, not monitored in real-time. However, even if a flight controller had been monitoring the specific data, there was less than six seconds between the first indication of a concern and the destruction of the vehicle; hardly sufficient time to recognize a problem, check the data, and take any meaningful action. Not that there was any meaningful action to take. During the period of flight while the solid rocket boosters were thrusting, there were no survivable abort options. There was nothing the crew or flight controllers could have done to avert the accident.

There were no alarms in the cockpit and the crew apparently had no indication of a problem before the rapid break-up of the vehicle. The first evidence of the accident came from live video coverage and when radar began tracking multiple targets. In mission control, Brian Perry, the flight dynamics officer (FIDO), reported, "RSO [range safety officer] reports vehicle exploded" and 30 seconds later added the Air Force had sent the destruct command to the solid rocket boosters. It was, officially, a very bad day.

Air Force, Coast Guard, NASA, and Navy teams spent the next three months searching the Atlantic Ocean for the remains of *Challenger* and her crew. By 30 January, twelve aircraft and ten ships were scouring a 1,200-square-mile area. The search teams located the main debris field at the edge of the Gulf Stream in water 100 to 1,200 feet deep. High seas complicated the initial efforts, although hundreds of pounds of floating debris, mostly from the external tank, were recovered, brought to Port Canaveral, and unloaded at the Navy Trident Basin.

THE ROGERS COMMISSION

President Ronald Reagan, seeking "a thorough and unbiased investigation of the accident," chartered a presidential commission to investigate the accident. The commission had 120 days to submit its final report to the president and NASA administrator, a remarkably short time to investigate a complex accident, especially when most of the physical evidence was at the bottom of the Atlantic Ocean.

Ronald Reagan named William Rogers, former attorney general under Dwight Eisenhower and secretary of state under Richard Nixon, to chair the commission. Other members included astronauts Neil Armstrong and Sally Ride; David Acheson, former vice president of Communications Satellite Corporation; Eugene Covert, the head of aeronautics and astronautics at MIT; Richard Feynman, 1965 Nobel Laureate in Physics; Robert Hotz, former editor-in-chief of *Aviation Week & Space Technology*; Donald Kutyna (USAF, retired), a former

Although a solid rocket booster was the cause of the accident, both SRBs continued to fly after the rest of the stack was destroyed. Eventually, the Air Force range safety officer destroyed the boosters, although this greatly complicated the accident investigation. (NASA)

Members of the Rogers Commission examine a solid rocket motor field joint in an ordnance storage facility adjacent to the Vehicle Assembly Building on 4 February 1986. The propellant burned from the center hole outward along the entire length of the booster. (NASA)

manager of the DoD space shuttle effort; Robert Rummel, former vice president of Trans World Airlines; Joseph Sutter, executive vice president of Boeing Commercial Airplane Company and arguably the father of the 747; Arthur Walker, Jr. professor of applied physics at Stanford University; Albert Wheelon, executive vice president of Hughes Aircraft Company; and test pilot Charles Yeager.

Ultimately, the commission and its staff interviewed more than 160 individuals and held 35 formal sessions that generated some 12,000 pages of transcript. Nearly 6,300 documents, totaling more than 122,000 pages, and hundreds of photographs were examined and made a part of the permanent record. In addition to the work of the commission and its staff, more than 1,300 NASA civil servants were involved in the investigation, supported by 1,600 people from other government agencies and 3,100 from NASA contractors. All of the materials, except the privileged interviews, were ultimately transferred to the National Archives and Records Administration. The remains of *Challenger* were entombed in former Minuteman III test silos at Complex 31/32 on Cape Canaveral Air Force Station.

A HISTORY OF FAILURE

The Rogers Commission believed the problems with the solid rocket motor began, "with the faulty design of its joint and increased as both NASA and contractor management first failed to recognize it as a problem, then failed to fix it, and finally treated it as an acceptable flight risk." The field joint Thiokol had proposed for space shuttle was elegant, probably fail-safe, and expensive. As the design progressed, largely under pressure from NASA, Thiokol changed the design to lower manufacturing costs and processing time at the launch sites. The final field joint design was a tang and clevis with two rubber-like o-rings as seals.

On 30 September 1977, Thiokol tested the strength of the steel cases by simulating a motor firing. Although the test was successful in that it demonstrated the case met its strength requirements, engineers discovered that, contrary to expectations, the tang and clevis in the field joint bent away from each other instead of toward each other. This reduced, instead of increased, the pressure on the o-rings in the milliseconds after ignition, reducing their ability to seal. Engineers called this phenomenon "joint rotation." To better understand the concept of joint rotation, consider this. Soon after ignition, the combustion pressure, somewhat more than 900 psi, inside the motor caused the sides of the steel casings to bulge outward very slightly. The maximum bulge was halfway between the joints, causing the tang to rotate relative to the clevis and creating a larger gap that had to be sealed by the o-rings. Even with

joint rotation, the primary o-ring normally sealed the gap but the joint rotation could cause the secondary o-ring to lose contact with the tang. If the primary o-ring should fail (e.g., because of erosion) the hot gases could escape through the gap between the tang and secondary o-ring. This could lead to all sorts of bad consequences.

Engineers found varying degrees of o-ring anomalies in the nozzle joints and field joints after 7 test firings and 12 flights prior to the *Challenger* accident. Most of these involved finding soot between the primary and secondary o-rings. However, engineers saw some level of erosion on the field joint primary or secondary o-rings on STS-2, 11/41B, 13/41C, 14/41D, 20/51C, 30/61A, and 32/61C. A more serious problem, the actual blow-by of exhaust gases past the primary o-ring occurred on STS-20/51C and 30/61A.

Roger Boisjoly, a Thiokol engineer who specialized in seals, became the champion of change. Finally, in September 1985 engineers recommended including a capture feature that would minimize joint rotation in the field joints. NASA rejected the recommendation and told Thiokol to resubmit it with a life-cycle cost-benefit analysis. Two Thiokol managers, Allan McDonald and Howard McIntosh, developed a justification that showed the new cases would last longer, offsetting their higher initial costs. Interestingly, McDonald had already instructed the case subcontractor to stop-work on the next batch of cases pending resolution of the issue. These unfinished cases were one reason the program could incorporate the capture feature so quickly after the *Challenger* accident.

THE FINDINGS

William Rogers delivered the final report to Ronald Reagan on 6 June 1986. "The commission concluded that the cause of the *Challenger* accident was the failure of the pressure seal in the aft field joint of the right solid rocket motor. The failure was due to a faulty design unacceptably sensitive to a number of factors. These factors were the effects of temperature, physical dimensions, the character of materials, the effects of reusability, processing, and the reaction of the joint to dynamic loading." Contrary to popular perception, it was not cold weather or the o-rings that were at fault; it was a bad joint design. Cold weather, and its effects on the o-rings, did not help, but they were not the root cause. The report listed numerous other technical and organizational failures and provided NASA with nine major recommendations to help ensure a safe return to flight.

The reader is encouraged to read the Rogers Commission report, which is available online, as well as the various personal accounts of the events surrounding the accident, particularly *Truth, Lies, and O-Rings* by Allan McDonald.

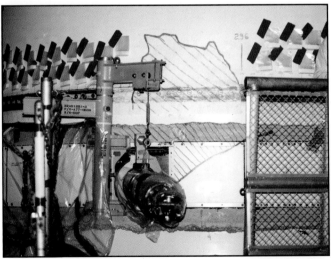

The size and shape of the breach in the right solid rocket motor case marked out on the STS-36/61G (BI028) booster on 5 May 1986. Note how the breach encompasses the lower external tank attach strut and part of the external tank attach ring. (NASA)

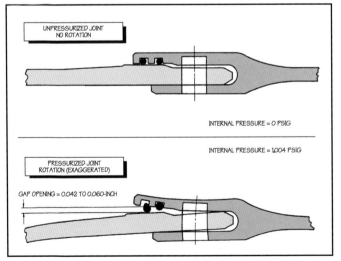

An exaggerated sketch of the joint rotation phenomenon showing how the tang and clevis could separate by as much as 0.060-inch during the first 600 milliseconds after ignition. The capture feature added after the accident largely mitigated this issue. (Rogers Commission)

Return-to-Flight I

After the *Challenger* accident, everything changed: the vehicle, the processes, the organization, and the culture. But it was not easy and, unfortunately, did not last. James Fletcher, the NASA administrator, put Dick Truly, a Navy adMiral and former astronaut, in charge of the return-to-flight efforts. Truly understood what he had to do, but he also realized he could never make the space shuttle, or any other spacecraft, as safe as everybody wanted. He opined, "Flying in space is a bold business. We cannot print enough money to make it totally risk-free. But we are certainly going to correct any mistakes we may have made in the past and we are going to get it going again." As long as we continue to use chemical rockets to escape the gravity well, there will be risk. If we want to continue exploring space, we need to understand and accept this fact.

Her designers had envisioned space shuttle as a satellite launcher and the first 24 missions showed the vehicle was quite adept at the role. In anticipation of these missions, NASA invested considerable treasure in processing facilities, aerospace support equipment, ground support equipment, and other infrastructure. Many of the initial unknowns, such as how maneuverable the vehicle would be on-orbit or how agile the SRMS would be, were quickly answered in positive ways. At the time of the accident, NASA had an impressive list of payloads waiting to fly on space shuttle. Since the beginning of the program, the agency had accepted reservations for 174 deployable payloads and had flown 33 of them, so 141 were still on the manifest. Of the 98 attached payloads accepted, 20 had flown, resulting in 78 waiting. There were 150 accepted middeck payloads, of which 59 had flown. An incredible 533 GAS experiments had been accepted with 479 waiting. It was an enviable backlog, most of which would eventually disappear.

Nobody had ever believed the extremely high flight rates portrayed in many of the early mission models but, in the months before the *Challenger* accident, NASA was still expecting to ramp up to 24 missions per year. After the accident, the agency began lowering expectations for the space shuttle and the government restarted the production lines for the expendable launch vehicles (Atlas, Delta, and Titan) that space shuttle had been designed to replace. Going forward, only missions that demonstrated they needed a man in the loop would be flown on space shuttle; satellite launches returned to the expendable boosters.

There was another casualty of the accident. As part of a public relations campaign, NASA had instituted the space flight participant program to fly non-astronaut crew members. The first two campaigns were for a Teacher in Space and a Journalist in Space. Christa McAuliffe was the first to fly; on STS-33/51L. After the accident, NASA quietly canceled the entire program.

HARDWARE CHANGES

Once it became evident a failure of a solid rocket motor field joint caused the accident, Dick Truly directed MSFC to establish a team that included participation from JSC, KSC, Langley, and the astronaut office to oversee the modifications to the joint. The agency also appointed a 12-person advisory panel that included six experts from outside NASA to further evaluate the modifications.

In the field joint used on STS-33/51L, the lower end of the upper segment contained the tang and upper end of the lower motor segment the clevis. The seal between the two segments was provided by two o-rings installed in grooves in the inner arm of the clevis,

compressed by a flat sealing surface on the tang. When the propellant ignited, going from atmospheric pressure to full thrust (~912 psia) in only 600 milliseconds, the case motor expanded outward. Because the joint area was stiffer than the case wall on either side, it expanded less than the rest of the case. This non-uniform expansion was the primary cause of the joint rotation. This motion could cause the o-rings to become unseated and lose their ability to seal. Cold weather aggravated the situation by making the rubber o-rings less pliant.

Ironically, the solution was already at hand. Engineers had developed a "lock-seal" for an improved joint destined for the lightweight composite solid rocket motors that were going to be used at the Vandenberg Launch Site to compensate for the less-efficient launch profiles that had to be flown from the West Coast base. Therefore, the lock-seal conceived for the filament-wound cases formed the basis of the capture feature used in the redesigned solid rocket motor (RSRM). Thiokol took advantage of the change to add a wiper o-ring in the capture feature that sealed against the inner surface of the clevis. Once engineers settled on adding the capture feature, things happened quickly. NASA and Thiokol held the engineering review on 10 July 1986 and the preliminary requirements review on 19 September. NASA assessed the design developed from these requirements at the preliminary design review on 10 October and approved the final design at the critical design review on 4 February 1988. Using the raw case sections that Allan McDonald had put on hold in July 1985, workers began finishing the RSRM test hardware and first flight cases.

Congress and the White House authorized NASA to assemble a set of structural spare parts into a new orbiter, OV-105, to replace *Challenger*. The new orbiter included all of the improvements and modifications incorporated on the other orbiters since the beginning of the flight campaign. In addition, a few new items were aboard, including a 40-foot drag chute, improved auxiliary power units, upgraded AP-101S general-purpose computers, improved inertial measurement units, new tactical air navigation systems, an enhanced master events controller, enhanced multiplexer-demultiplexers, and solid-state star trackers. Over the next few years, NASA incorporated these improvements into the rest of the fleet. As listed on the paperwork that transferred *Endeavour* to the California Science Center, the vehicle cost $1,980,674, 785. This did not include the $389 million structural spares or the parts from the logistics system that were used to outfit the orbiter during construction.

By September 1988, NASA had completed 76 major orbiter modifications including adding a rudimentary crew escape system, installing new structural-carbon brakes, redesigning the nose-wheel steering, and recertifying the 17-inch propellant disconnects. Engineers had also made 185 modifications to the ground systems.

In addition to the hardware changes, NASA made, or attempted to make, several organizational and cultural changes based on findings from the Rogers Commission and other sources. Whether these changes were successful was the subject of great debate after a second accident board found many of the same failings 17 years later.

Recovering from the *Challenger* accident ultimately cost more than $12,000 million including building *Endeavour*, restarting the various expendable launch vehicle production lines, developing the complementary expendable launch vehicle that became Titan IV (something that had started before the accident), modifying a variety of commercial and government satellites to be launched on ELVs, and procuring Atlas, Delta, and Titan launch vehicles.

GETTING READY

NASA selected *Discovery* for the first return-to-flight mission. Preparations for the return-to-flight began in earnest on 30 October 1986 when workers moved the orbiter from storage in the Vehicle Assembly Building to the Orbiter Processing Facility where many major components were removed and sent to their vendors for servicing. *Discovery* was powered up on 3 August 1987 and remained in OPF-1 while technicians implemented the various modifications and outfitted the orbiter to carry the TDRS-C/IUS payload that would be deployed on STS-26R. Flight processing began in mid-September when technicians began reinstalling the major components as they returned from the vendors. Technicians installed OASIS (observation and analysis of smectic islands in space) on 19 April while *Discovery* was still in the OPF, but the primary TDRS/IUS payload would be installed on the launch pad.

The first redesigned solid rocket motor segments arrived at KSC on 1 March 1988. Technicians began stacking the left booster on MLP-2 in VAB High Bay 3 on 28 March and continued with the right booster on 5 May. The forward assemblies and nose cones were attached 27–28 May and Lockheed mated the external tank to the completed solid rocket boosters on 10 June.

Discovery was rolled-over to the VAB on 21 June 1988 and mated to the rest of the stack on 24 June. The vehicle was rolled-out to LC-39B on 4 July for the final launch preparations. On 15 July, workers found a minor hydrazine leak in an orbital maneuvering system pod, but this did not affect a wet countdown demonstration test on 1 August. However, the test revealed several hydrogen leaks.

On 4 August, NASA attempted a 20-second flight readiness firing (FRF), but sluggish gaseous oxygen flow control valves in two main engines aborted the test prior to SSME start. Technicians soon repaired the valves and the flight readiness firing took place on 10 August with no major issues. Nine days later technicians repaired the OMS and hydrogen leaks and installed the TDRS/IUS payload on 29 August. A full-up terminal countdown demonstration test (TCDT) with the flight crew in the orbiter took place on 8 September.

Thirty-two months after the *Challenger* accident, all seemed ready for the return to flight.

Discovery on LC-39B awaiting launch as STS-26R. After the Challenger accident, NASA returned to the original STS numbering scheme, but it was reset to "26" since Challenger had been the 25th mission. Unfortunately, under the original scheme, Challenger had been STS-33, so restarting with 26 created some duplication. When flights resumed in 1988, NASA restarted with STS-26R, the "return-to-flight" suffix used to disambiguate from prior missions. This continued through STS-33R, including STS-29R, despite there not having been an STS-29 actually launched. (NASA)

STS-26R

Mission:	26	NSSDC ID:	1988-091A
Vehicle:	OV-103 (7)	ET-28 (LWT)	SRB: BI029
Launch:	LC-39B	29 Sep 1988	15:37 UTC
Altitude:	177 nm	Inclination:	28.40 degrees
Landing:	EDW-17L	03 Oct 1988	16:38 UTC
Landing Rev:	64	Mission Duration:	97 hrs 00 mins

Commander:	Frederick H. "Rick" Hauck (3)
Pilot:	Richard O. "Dick" Covey (2)
MS1:	John M. "Mike" Lounge (2)
MS2:	David C. "Dave" Hilmers (2)
MS3:	George D. "Pinky" Nelson (3)

Payloads:	Up: 46,448 lbs	Down: 0 lbs
	TDRS-C/IUS (37,514 lbs)	
	OASIS-1 (1,233 lbs)	
	No SRMS	

Notes:
Flight readiness firing scrubbed 04 Aug 1988
Flight readiness firing 10 Aug 1988
First flight after *Challenger* accident
First flight of a Phase II space shuttle main engine
First all-veteran crew since Apollo 11
First flight since STS-4 to use pressure suits (LES)
First American crew to use partial-pressure suits
Orbiter returned to KSC on 08 Oct 1988 (N905NA)

Wakeup Calls:

30 Sep	"Goooooood Morning Discovery!!!" (Robin Williams)
01 Oct	Parody of "I Get Around"
02 Oct	"Harvey Mudd" College Fight Song
03 Oct	"Fun, Fun, Fun"

The crew patch represented a new beginning (sunrise), a safe mission (stylized launch and plume), the building upon the traditional strengths of NASA (the red vector), and a remembrance of seven colleagues who died aboard Challenger *(the seven-starred Big Dipper). (NASA)*

The crew for STS-26R included most of the original crew for STS-35/61F since they were the furthest along in training at the time of the *Challenger* accident. NASA had hoped to return to flight much sooner than how it actually played out. Rick Hauck, Mike Lounge, and Dave Hilmers had been assigned to STS-35/61F along with Roy Bridges as pilot. Bridges never flew again after the *Challenger* accident, but served as director of the Kennedy Space Center. Dick Covey replaced him on this flight.

The return-to-flight of *Discovery* was delayed almost two hours because the winds aloft were significantly different than anticipated and pre-launch analysis indicated the orbiter might exceed its structural limits. In addition, technicians had to repair minor issues in two of the new S1032 launch-entry suits. Additional analysis at T-70 minutes determined sufficient safety margins still existed and the countdown resumed after technicians repaired the suits. Post-launch inspection showed the solid rocket boosters had a variety of small problems, but overall behaved as expected with no o-ring erosion.

Once on-orbit, the deployment and downlink operations of the Ku-band system appeared normal, but the antenna failed to respond to commanded angles. Flight controllers told the crew to stow the antenna, but initial attempts failed. Engineers developed a revised procedure and the crew stowed the antenna on the first attempt. The alternative would have been to jettison the antenna to allow the payload bay doors to be closed. The crew deployed TDRS-C and its attached inertial upper stage slightly more than six hours after launch. On 2 October, the day before the mission ended, the five-man crew paid tribute to the seven crew members lost in the *Challenger* accident.

The landing at Edwards occurred on the first opportunity. The automatic flight control system performed the entry and terminal area energy management acquisition phases, although Rick Hauck took control just prior to the left heading alignment circle turn and continued through the touchdown and rollout phases. CAPCOM Blaine Hammond welcomed the crew back to Earth, saying the mission was "a great ending to a new beginning."

Post-flight inspection revealed a large damage area on the lower right wing. Fortunately, the boundary layer transitioned from laminar to turbulent flow later than expected, probably due to increased smoothness (shaving) of tiles on the forward part of the orbiter. This late transition minimized the heating on the damaged area. There were 411 impact sites, with 55 of them larger than 1-inch, but inspection showed there was no structural damage. Engineers determined the damage was caused by a 12-inch-long piece of cork from the forward field joint on the right SRB.

From the left: Dave Hilmers, Dick Covey, Pinky Nelson, Rick Hauck, and Mike Lounge. This was the first all-veteran crew since Apollo 11, with every crew member having flown at least one prior mission. Interestingly, Covey had been the CAPCOM during the STS-33/51L launch. (NASA)

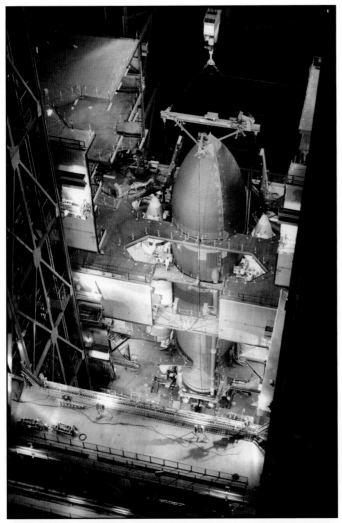

The external tank (ET-28) used by STS-26R had been delivered to KSC on 4 November 1985. Technicians performed 16 minor modifications to it after the accident and then mated it to the waiting solid rocket boosters on 10 June 1988. (NASA)

Discovery lifts-off as STS-26R on 29 September 1988. This was the first flight to use LC-39B after the ill-fated STS-33/51L launch of Challenger. Despite a couple of minor glitches during the count, this time everything worked well and the return-to-flight went according to plan. (NASA)

The improved Phase II main engines began arriving at KSC in January 1988 and technicians installed the first engine in Discovery on 10 January, with the others following on 21 and 24 January. The Phase II changes focused on improving durability and safety. (NASA)

Liberty Star sails past the jetty as she enters Port Canaveral towing the right solid rocket booster. Engineers were anxious to examine the boosters to ensure the modifications made to the field joints after the Challenger accident had worked. (NASA)

STS-27R

Mission:	27	NSSDC ID:	1988-106A
Vehicle:	OV-104 (3)	ET-23 (LWT)	SRB: BI030
Launch:	LC-39B	02 Dec 1988	14:30 UTC
Altitude:	244 nm	Inclination:	57.00 degrees
Landing:	EDW-17L	06 Dec 1988	23:37 UTC
Landing Rev:	68	Mission Duration:	105 hrs 06 mins

Commander:	Robert L. "Hoot" Gibson (3)
Pilot:	Guy S. Gardner (1)
MS1:	Richard M. "Mike" Mullane (2)
MS2:	Jerry L. Ross (2)
MS3:	William M. "Shep" Shepherd (1)

Payloads:	Up: 40,000 lbs	Down: 0 lbs
	USA-34 (Lacrosse) (38,000 lbs)	
	OASIS-1 (1,233 lbs)	
	SRMS s/n 201	

Notes:	Scrubbed 1 Dec 1988 (KSC winds)
	Third dedicated Department of Defense mission
	Payload was classified at the time
	Extensive damage to thermal protection system
	Orbiter returned to KSC on 13 Dec 1988 (N905NA)

Wakeup Calls:

03 Dec	Army Fight Song and Navy Fight Song
04 Dec	Theme from *Rawhide*
05 Dec	Darth Vader parody with "Theme from Star Wars"

The crew patch depicted a space shuttle lifting off against the multi-colored backdrop of a rainbow, symbolizing the successful return-to-flight of the Space Shuttle Program. The seven stars were in memory of the crew of Challenger lost on STS-33/51L. (NASA)

The launch team scrubbed the 1 December 1988 attempt because of winds aloft. The following day, the winds aloft were higher than limits, but balloon data showed they were receding and would be at 102-percent of the limit at T-0. The mission management team approved a waiver and the count continued.

This was the third mission dedicated to the Department of Defense and was termed "... successful ..." by the Air Force. Although classified at the time, NASA now confirms the USA-34 payload was a Lockheed Martin Lacrosse (Onyx) surveillance satellite for the National Reconnaissance Office. Speculation was the Lacrosse weighed nearly 40,000 pounds and did not use an upper stage.

In the days before email, mission control sent complicated instructions to the crew using what was, in effect, a graphics-capable fax machine. A single-page photograph could take more than five minutes to transmit from the ground to the orbiter; plain text, naturally, took less time. Shortly after reaching orbit, the crew reported the text and graphics system (TAGS) printer had a paper jam that rendered it inoperable for the remainder of the mission. Only 46 pages were received prior to the failure. This forced the crew to rely on the older teleprinter, which was slow at any time and even more so when using the secure communications link demanded by the classified mission.

Atlantis reportedly returned to Edwards on the first landing opportunity and communications were maintained during entry using TDRS-West. Hoot Gibson and Guy Gardner successfully conducted braking and nose wheel steering tests during the rollout.

The post-flight inspections revealed significant tile damage with 298 damage sites greater than 1 square inch and a total of 707 damage sites on the lower surface. The majority of the damage was concentrated outboard of a line leading to the LO_2 umbilical. One tile was missing on the right side slightly forward of the L-band antenna. Fortunately, the damage was confined to the cavity around the antenna door and the structure acted as a heat sink to prevent more damage. Engineers found a piece of ablator from the right SRB nose cap embedded in a blanket on the right OMS pod. In addition, an AFRSI-covered fiberglass carrier panel was lost from the right OMS pod, but the door underneath the panel was not damaged since it was in a relatively low-heat area. Excepting *Columbia* on STS-107, this was the most serious thermal protection system damage sustained during the flight campaign.

Although NASA implemented many of the recommendations that came out of the subsequent investigation, somehow the process fell apart when it came time to use it for STS-107.

From the left are Guy Gardner, Bill Shepard, Hoot Gibson, Mike Mullane, and Jerry Ross. They are wearing the David Clark Company S1032 launch-entry suits that became standard after the Challenger accident. Note the name tags use only first names except Gibson. (NASA)

Guy Gardner converses with Hoot Gibson (almost completely out of view at left) from the pilot seat of Atlantis on-orbit. This gives a good view of the forward instrument panel, heads-up display, and the back of the original heavyweight crew seats. (NASA)

From the left, Guy Gardner, Jerry Ross, and Mike Mullane take a break from moving equipment on the middeck. This shows how small the middeck was, especially with the internal airlock (the hatch is barely visible at far right) installed. (NASA)

Atlantis being lifted from the transfer aisle to its waiting ET-SRB stack in High Bay 3 of the Vehicle Assembly Building. Note the articulated sling that supported the orbiter and helped translate it from its normal horizontal orientation to vertical for stacking. (NASA)

Guy Gardner, Mike Mullane, and Bill Shepard examine the damaged thermal protection system after landing. This is the area where one tile was completely missing, fortunately over a heavy antenna structure that acted as a heat sink and prevented extensive damage. (NASA)

ET-23 was the first of four external tanks that had initially been sent to Vandenberg to be launched from KSC after the west coast facility was closed. By the time it arrived at the turn basin (shown here), it was a very well-traveled tank. (NASA)

STS-29R

Mission:	28	NSSDC ID:		1989-021A
Vehicle:	OV-103 (8)	ET-36 (LWT)		SRB: BI031
Launch:	LC-39B	13 Mar 1989		14:57 UTC
Altitude:	178 nm	Inclination:		28.45 degrees
Landing:	EDW-22	18 Mar 1989		14:37 UTC
Landing Rev:	80	Mission Duration:		119 hrs 39 mins

Commander: Michael L. "Mike" Coats (2)
Pilot: John E. Blaha (1)
MS1: Robert C. "Bob" Springer (1)
MS2: James F. "Jim" Buchli (3)
MS3: James P. "Jim" Bagian (1)

Payloads: Up: 47,276 lbs Down: 0 lbs
TDRS-D/IUS (37,546 lbs)
OASIS-1 (1,223 lbs)
No SRMS

Notes: Originally scheduled for Jan 1989
Delayed 18 Feb 1989 (SSME replacements)
Delayed 11 Mar 1989 (MEC failure)
First downlink via TDRS during entry
Orbiter returned to KSC on 24 Mar 1989 (N905NA)

Wakeup Calls:

14 Mar "I Got You (I Feel Good)"
15 Mar The Marine Corps Hymn – "Halls of Montezuma"
16 Mar "Theme from Star Trek"
17 Mar "Heigh-Ho, Heigh-Ho, It's Off To Work We Go"
18 Mar Astronauts' children shouting greetings

The folded ribbon border gave a sense of three-dimensional depth to the crew patch that used the colors of the American flag. The stylistic OMS burn symbolizes the forward momentum of the flight campaign. The seven stars were a tribute to the crew of Challenger. (NASA)

During late January 1989, NASA reassessed the expected launch date of 18 February to ensure there was sufficient time to replace suspected faulty LO_2 turbopumps on all three *Discovery* main engines. At the time, mission management felt a late-February was possible, but difficulties with a master events controller postponed the launch an additional two weeks. The launch on 13 March was delayed almost two hours by morning ground fog and upper winds. High surface winds at Edwards caused NASA to move the abort-once-around (AOA) landing site to Northrup Strip at White Sands, New Mexico.

The crew deployed TDRS-D approximately six hours after launch and the inertial upper stage delivered the satellite to its geosynchronous orbit at 41 degrees west longitude, east of Brazil.

Discovery also carried eight secondary payloads, including two Shuttle Student Involvement Project (SSIP) experiments. One of these, using four live rats with tiny pieces of bone removed from their bodies, tested whether the environmental effects of space flight inhibited bone healing. The other student experiment flew 32 eggs to determine the effects of space on fertilized chicken embryos.

All of the secondary payloads operated successfully throughout the mission except SHARE (space station heat pipe advanced radiator elements), which experienced vapor bubbles in the liquid channels after less than 30 minutes of operation. SHARE was an evaluation of a potential cooling system for Space Station Freedom. CHROMEX (chromosome and plant cell division in space experiment) showed the effects of microgravity on root development. Good crystals were obtained from the PCG (protein crystal growth) experiment. A ground-based experiment used the orbiter as a calibration target for the Air Force Maui Optical Site (AMOS) in Hawaii, something that would be common throughout the flight campaign.

The crew also used a 70 mm IMAX camera to film a variety of scenes for *Blue Planet*, including the effects of floods, hurricanes, wildfires, and volcanic eruptions.

While the crew was closing the payload bay doors, the left aft limit switch failed to indicate closed. They terminated the automated closure sequence and used a manual sequence to close and latch the doors. This was the first mission where the orbiter continued to downlink data throughout entry using TDRS.

Discovery landed at Edwards on the first opportunity. Post-flight inspection showed two long gouges on the right main landing-gear door and engineers believed these were caused by tire pressure instrumentation they found on the runway after landing.

From the left are Jim Bagian, John Blaha, Bob Springer, Mike Coats, and Jim Buchli in their S1032 partial-pressure suits. The five person crew meant one person was alone on the middeck during ascent and entry: Bagian on the way up and Springer on the way down. (NASA)

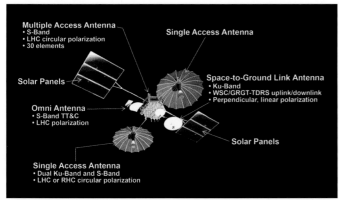

A depiction of how the first-generation TDRS satellites looked during operations. Comparing this drawing to the pre-deployment photo at right shows how engineers folded everything to allow it to fit in the payload bay, or on top of an expendable launch vehicle. (NASA)

Jim Bagian snapped this photo of Jim Buchli sitting in the flight engineer seat during entry. The mission specialist in this seat assisted the pilots by operating controls on the center console and overhead panels. Mike Coats is in the left seat up front. Although difficult to discern, the color in the forward windows is the superheated air the orbiter passed through during entry. A personal egress air pack (PEAP) is visible on pilots seat back with the orange umbilical coming out of it. (NASA)

The nose cone of the lightweight tanks carried a 14-inch aluminum lightning rod. The early nose cones consisted of multiple pieces of sheet metal held together with more than 1,100 fasteners. These were replaced by a single-piece composite nose cone beginning on ET-81. (NASA)

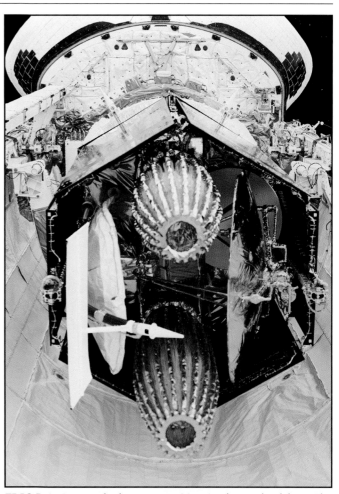

TDRS-D in its pre-deployment position in the payload bay. This head-on view shows the stowed solar panels on each of the six sides, the Ku-band and S-band dish antennas on each side, and the folded gold signal-access antennas (bullet shaped objects) in the middle. The satellite was supported by a forward cradle and aft-frame tilt actuator that were part of the aerospace support equipment (ASE). Note the eyeballs on both OMS pods in the background. (NASA)

Bob Springer on the middeck of the crew compartment trainer (CCT) at JSC. This is a great shot of how the middeck looked when the internal airlock was installed. Springer is wearing a blue S1032 launch-entry suit. Although identical to the orange suits, the blue ones never flew. (NASA)

STS-30R

Mission:	29	NSSDC ID:	1989-033A
Vehicle:	OV-104 (4)	ET-29 (LWT)	SRB: BI027
Launch:	LC-39B	04 May 1989	18:47 UTC
Altitude:	176 nm	Inclination:	28.87 degrees
Landing:	EDW-22	08 May 1989	19:44 UTC
Landing Rev:	65	Mission Duration:	96 hrs 56 mins

Commander: David M. "Dave" Walker (2)
Pilot: Ronald J. "Ron" Grabe (2)
MS1: Mark C. Lee (1)
MS2: Norman E. "Norm" Thagard (3)
MS3: Mary L. Cleave (2)

Payloads: Up: 47,783 lbs   Down: 0 lbs
Magellan/IUS (40,118 lbs)
No SRMS

Notes: Scrubbed 28 Apr 1989 (SSME-1 failure)
First orbital replacement of GPC
Deployed Magellan probe to Venus
First American planetary mission in 11 years
Shortest *Atlantis* mission
Orbiter returned to KSC on 15 May 1989 (N905NA)

Wakeup Calls:

05 May	"Theme from Superman"
06 May	Medley from crew alma maters
07 May	"Theme from Rocky"
08 May	"A Hard Day's Night"

Designed by the crew, the patch depicted the joining of the manned and unmanned space programs. The Sun and planets are shown with an orbital curve connecting Earth and Venus. A Spanish caravel similar to the Magellan logo commemorated the 16th-century journey. (NASA)

Atlantis spent three months in OPF-2 as technicians repaired the damage from the previous flight and conducted the normal post- and pre-mission processing. NASA originally scheduled the launch for 28 April 1989, the first day of a 31-day window when the Earth and Venus were properly aligned for the Magellan mission. However, engineers scrubbed the launch at T-31 seconds because of a problem with the liquid hydrogen recirculation pump on and main engine and a vapor leak in the LH_2 recirculation line. Discussions with the crew after the scrub revealed the shoulder harness for Lee could not be tightened. Technicians replaced the seat and harness, and NASA rescheduled the launch for 4 May.

The launch director held that attempt at T-9 minutes because of unacceptable weather at the Shuttle Landing Facility in case of an RTLS abort. The weather was slowly clearing, and *Atlantis* launched during the final five minutes of the 64-minute window. The crew deployed the long-delayed Magellan Venus mapping probe a little more than six hours after launch, the first American planetary probe in 11 years. Magellan deployed its solar panels and the first burn of the inertial upper stage took place 60 minutes after deployment. Just over five minutes later the IUS fired again, sending Magellan on a 15-month journey to Venus. Magellan arrived at Venus in August 1990 and began a 243-day mission of mapping the planet surface with radar.

Secondary payloads included AMOS (Air Force Maui Optical Site calibration), FEA (fluids experiment apparatus), and MLE (mesoscale lightning experiment).

The only major anomaly during the flight was when one of the four general-purpose computers failed, but the crew installed an onboard spare. After the crew performed the initial program load (the IBM term for boot), they placed the new GPC into the redundant set and it operated satisfactorily for the remainder of the mission. It was the first time a computer had been replaced while on-orbit. In addition, on flight day 2 the water dispensing system in the galley malfunctioned, complicating meal preparation for the remainder of the mission.

Just before landing, NASA switched the Edwards runway from 17L to 22 due to high crosswinds, so this marked the first use of a concrete runway since the return-to-flight.

Unlike the most recent *Atlantis* mission, the post-flight inspection showed that STS-30R had minimal damage to its thermal protection system. In fact, the mission report noted, "The flight had the least amount of ascent damage of any flight." This was the shortest mission for *Atlantis*.

From the left are Ron Grabe, Dave Walker, Norm Thagard, Mary Cleave, and Mark Lee. Interestingly, neither Walker nor Grabe is wearing military astronaut wings, although this was their second flight into space (the military awarded the wings after the first flight). (NASA)

Atlantis with the landing-gear down and locked. After the orbital flight tests, it was unusual to have a chase plane during landing except for the Shuttle Training Aircraft looking for rain, so there are not many good shots of the orbiter on final approach. (NASA)

The launch was delayed for 43 minutes because of bad weather at the Shuttle Landing Facility (in case of an RTLS abort), but it cleared enough to allow a good launch of Atlantis five minutes before the 64-minute Venus transit window closed. (NASA)

Magellan and its attached inertial upper stage leaving the payload bay at the start of its eleven-year transit to Venus. The probe arrive at Venus on 7 August 1990 and collected science data until it entered the Venusian atmosphere on 13 October 1994. (NASA)

STS-28R

Mission:	30		NSSDC ID:		1989-061A
Vehicle:	OV-102 (8)		ET-31 (LWT)		SRB: BI028
Launch:	LC-39B		08 Aug 1989		12:37 UTC
Altitude:	166 nm		Inclination:		57.00 degrees
Landing:	EDW-17L		13 Aug 1989		13:38 UTC
Landing Rev:	81		Mission Duration:		121 hrs 00 mins

Commander: Brewster H. Shaw, Jr. (3)
Pilot: Richard N. "Dick" Richards (1)
MS1: James C. "Jim" Adamson (1)
MS2: David C. "Dave" Leestma (2)
MS3: Mark N. Brown (1)

Payloads: Up: 20,953 lbs Down: 0 lbs
USA-40 (SDS-B1)/Orbus-6 (20,953 lbs)
GAS (2)
No SRMS

Notes: Fourth dedicated Department of Defense mission
Orbiter returned to KSC on 21 Aug 1989 (N905NA)

Wakeup Calls:

Classified mission. Either nothing played or nothing released.

Designed by the crew to portray the pride the American people have in their manned space program, the patch depicts America (the eagle) guiding the space program safely home from a mission. The contrails created by the eagle and orbiter represent the national flag. (NASA)

The four-hour launch window for this classified mission began at 11:30 UTC on 8 August 1989 and *Columbia* was launched a little over an hour into the window on the fourth dedicated DoD mission. During ascent, Dick Richards experienced a seat failure that resulted in him sliding backward and being unable to reach the switches on the instrument panel. When he could no longer reach the switches, he moved the seat forward about 3 inches, although it immediately began moving aft to the stops. Once on-orbit, the crew repaired the seat and it functioned normally during entry.

During post-flight debriefings, the crew reported they felt a large thump/thud during ascent that shook the entire vehicle. This was coincident with the aerosurfaces moving from the droop position to the null position so engineers believed it was a normal event.

Columbia entered a 57-degree orbit and deployed its classified payload. Early reports speculated the primary payload was an Advanced KH-11 Kennen photoreconnaissance satellite, but subsequent observations by amateur space watchers indicated the satellite was spin-stabilized (it "flashed" as it spun), something not consistent with an imaging reconnaissance satellite.

Later reports suggested the payload was a Satellite Data System communication relay satellite that was later transferred to a highly elliptical orbit. The SDS-B satellites were used to relay data from reconnaissance satellites to a ground station at Fort Belvoir, Virginia. Reportedly these were modified versions of the Hughes HS-381 (Leasat) or HS-389 (Intelsat VI) commercial communications satellite bus designated HS-386. These satellites were designed to be launched by space shuttle, although the fourth and last in the constellation was launched on a Titan IVA.

The mission landed at Edwards on the first opportunity. During entry, the boundary-layer transition from laminar to turbulent flow was unusual in that it occurred as early as 900 seconds after entry interface in areas toward the aft end of the orbiter and at 1,200 seconds in areas toward the forward fuselage. This did not result in any TPS or structural temperature limits being violated and engineers later identified a protruding gap filler as the likely cause of the early transition. This event formed part of the rationale for removing a protruding gap filler during an EVA on STS-114 in 2005.

For unexplained reasons, the 156-knot landing speed was approximately 30 knots lower than expected and was, by a small margin, the slowest of the program. The low nose-gear-touchdown velocity resulted in a slightly higher than usual pitch rate at nose gear contact, but within limits. None of it resulted in anything untoward. The Air Force again termed the mission "… successful …"

From the left are Dick Richards, Mark Brown, Brewster Shaw, Jim Adamson, and Dave Leestma. Note the two Naval Aviators have gold name tags, while the two Air Force and single Army (Adamson) pilots have silver, reflecting the tradition of each service. (NASA)

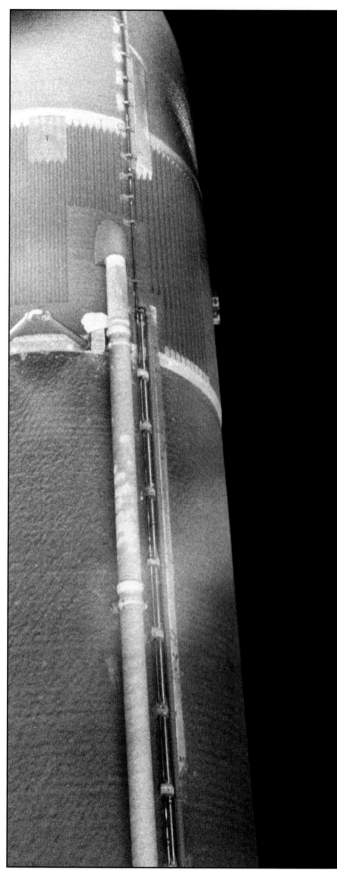

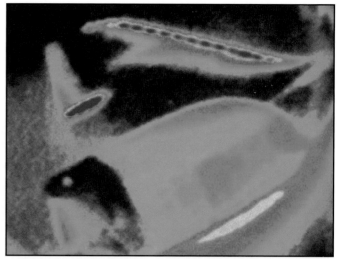

A photo from the SILTS (shuttle infrared leeside temperature sensing) sensor during entry. The left OMS pod is at the bottom of the photo. The wing leading edge is shown red-hot as is the elevon cove. (NASA)

Mark Brown pulls some food containers out of a middeck locker. The tray he is working with is a half-height tray, meaning two fit into the locker stacked on top of each other. This is an original blue fiberglass tray; the program later used lightweight black carbon-fiber trays. (NASA)

A post-separation photo of ET-31 shows a large divot of missing foam just in front of the right bipod ramp. A similar loss would be devastating for Columbia on STS-107. Note the scorching near the nose of the tank from the SRB forward separation motors. (NASA)

The crew walking toward the Astrovan that carried them between the Operations and Checkout Building in the KSC industrial area and the launch pad. From the left are Mark Brown, Jim Adamson, Dave Leestma, Dick Richards, and Brewster Shaw. (NASA)

STS-34

Mission:	31	NSSDC ID:	1989-084A
Vehicle:	OV-104 (5)	ET-27 (LWT)	SRB: BI-032
Launch: Altitude:	LC-39B 177 nm	18 Oct 1989 Inclination:	16:54 UTC 34.33 degrees
Landing: Landing Rev:	EDW-23L 80	23 Oct 1989 Mission Duration:	16:34 UTC 119 hrs 39 mins

Commander:	Donald E. "Don" Williams (2)
Pilot:	Michael J. "Mike" McCulley (1)
MS1:	Shannon W. Lucid (2)
MS2:	Franklin R. Chang-Diaz (2)
MS3:	Ellen S. Baker (1)

Payloads:	Up: 47,865 lbs	Down: 0 lbs
	Galileo/IUS (43,980 lbs)	
	SSBUV (1,215 lbs)	
	No SRMS	

Notes:	Delayed 12 Oct 1989 (SSME controller)
	Scrubbed 17 Oct 1989 (KSC weather)
	Was to use Shuttle/Centaur upper stage
	First flight since *Challenger* not to carry an "R"
	Orbiter returned to KSC on 29 Oct 1989 (N905NA)

Wakeup Calls:

19 Oct	Medley of "Hail Purdue," "Reveille," "Anchors Aweigh"
20 Oct	Medley of university fight songs
21 Oct	"Bohemian Rhapsody"
22 Oct	"Centerfield"
23 Oct	"Fly Like An Eagle"

The Galileo spacecraft overlaying the orbiter symbolized the joining together of both manned and unmanned space programs. The sunrise depicted the expansion of our knowledge of the solar system and other worlds while Jupiter awaited the arrival of Galileo. (NASA)

The first flight since the *Challenger* accident not to carry an "R" suffix was originally scheduled for 12 October 1989, the first day of a 41-day launch window during which the planets were properly aligned for a direct flight to Jupiter. Launch operations were not affected by a small group of demonstrators near KSC protesting the use of a radioisotope thermoelectric generator (RTG) on the Galileo spacecraft. However, the mission management team postponed the launch until 17 October because of a faulty main engine controller and scrubbed that attempt due to weather at the launch site. The countdown on 18 October was held at T-5 minutes to update the general-purpose computers to change the primary TAL site to Zaragoza, Spain, because of rain at Ben Guerir, Morocco.

This mission had been designed to use a Convair Shuttle/ Centaur G-Prime upper stage that was canceled for safety reasons after the *Challenger* accident. Because the substitute inertial upper stage was not as powerful as Centaur, Galileo embarked on a Venus-Earth-Earth gravity assist (VEEGA) trajectory that swung around Venus, the Sun, and Earth on its six-year journey to Jupiter. The crew successfully deployed Galileo about 6.5 hours after launch. Galileo became the first spacecraft to orbit an outer planet and to penetrate the atmosphere of an outer planet. The spacecraft arrived at Jupiter on 7 December 1995 and operated until 21 September 2003 when flight controllers purposefully sent it into the atmosphere to burn up, preventing it from contaminating the planet with Earth organisms.

Besides Galileo, the payload bay held two canisters containing the SSBUV (shuttle solar backscatter ultraviolet) experiment that provided data to help researchers calibrate the ozone sounders on free-flying satellites and to verify the accuracy of atmospheric ozone and solar irradiance data and their associated models.

The crew operated an IMAX camera, last flown on STS-29R, and Werner Herzog used the footage in his 2005 film *The Wild Blue Yonder*. Franklin Chang-Diaz and Ellen Baker conducted an experiment that photographed and videotaped the veins and arteries in the retinal wall to provide detailed measurements that might provide a possible relationship between cranial pressure and motion sickness. Baker, a medical doctor, also tested the effectiveness of motion sickness medication to combat space adaptation syndrome.

A bad weather forecast at Edwards forced mission control to advance landing one revolution, but the crew was also instructed to conserve consumables in case the mission had to be extended up to three days because of unsatisfactory weather conditions at the landing sites. On the last day, mission management team moved the landing up an additional revolution, again due to weather.

The crew taking a break from some final training at the Shuttle Landing Facility; note the Mate/Demate Device in the background. From the left are Mike McCulley, Franklin Chang-Diaz, Ellen Baker, Shannon Lucid, and Don Williams, all in their S1032 launch-entry suits. (NASA)

A good study of IUS-19 with Galileo at the top. This particular inertial upper stage had been assigned to a classified DoD payload but was repurposed for Galileo after NASA canceled the liquid-propellant Shuttle/Centaur stage following the Challenger accident. (NASA)

Galileo and its inertial upper stage begin their six-year journey to Jupiter as they drift out of the payload bay. The switch to the less-powerful IUS forced flight planners to use a Venus-Earth-Earth gravity assist (VEEGA) trajectory that swung around Venus, the Sun, and Earth. (NASA)

It was soon enough after the return-to-flight that the news media still covered each launch. The payload generated even more interest. KSC had not yet established a permanent press facility, so all of the television stations and networks brought their own vans. (NASA)

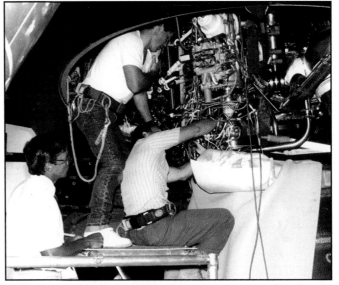

Technicians replace the space shuttle main engine controller after the 17 October scrub. The launch pad was setup to repair or replace engines while the vehicle was vertical. In fact, many of the technicians felt it was easier to work on in that orientation. (NASA)

STS-33R

Mission:	32	NSSDC ID:	1989-090A
Vehicle:	OV-103 (9)	ET-38 (LWT)	SRB: BI034
Launch:	LC-39B	23 Nov 1989	00:23 UTC
Altitude:	302 nm	Inclination:	28.45 degrees
Landing:	EDW-04	28 Nov 1989	00:31 UTC
Landing Rev:	79	Mission Duration:	120 hrs 07 mins

Commander:	Frederick D. "Fred" Gregory (2)
Pilot:	John E. Blaha (2)
MS1:	Manley L. "Sonny" Carter, Jr. (1)
MS2:	F. Story Musgrave (3)
MS3:	Kathryn C. "Kathy" Thornton (1)

Payloads:	Up: (38,500 lbs)	Down: 0 lbs
	USA-48/IUS (Magnum/Orion)	
	No SRMS	

Notes:	Delayed 20 November (SRB issue)
	Fifth dedicated Department of Defense mission
	Orbiter returned to KSC on 04 Dec 1989 (N905NA)

Wakeup Calls:

Classified mission. Either nothing played or nothing released.

Designed by the five crew members, the crew patch featured a stylized falcon soaring into space to represent the American commitment to manned space flight. The single gold star on a field of blue honored the late S. David Griggs, originally assigned to this crew. (NASA)

NASA delayed the launch of *Discovery* on 20 November 1989 to allow the replacement of suspect integrated electronics assemblies on both SRBs. The 23 November countdown proceeded normally until T-5 minutes when ground controllers called a short hold because of a minor ground purge-flow problem. The rest of the launch proceeded smoothly.

The original pilot for this mission was Dave Griggs, a veteran of STS-23/51D. Unfortunately, Griggs was killed when his North American AT-6D Texan crashed while practicing for an airshow near Earle, Arkansas on 17 June 1989. John Blaha replaced Griggs and the crew honored his memory with a single gold star on the blue field of the crew patch. This was first American crew member change since Ken Mattingly was exposed to rubella three days before Apollo 13 and was replaced by Jack Swigert.

This was the fifth dedicated Department of Defense mission. *Aviation Week* claimed *Discovery* initially entered an 110x280 nm orbit, and then executed three OMS burns, the last on its fourth orbit, to circularize the orbit at 280 nm, although official documentation indicates a final 302 nm orbit. The number of OMS burns agrees with NASA reports, which say there were five burns (these three plus single ascent and deorbit burns). The crew deployed the classified primary payload on revolution 7 and the inertial upper stage successfully placed it in a geosynchronous transfer orbit. This was the eighth IUS launched from space shuttle and the seventh successfully deployed.

Most speculation indicated the payload was the second TRW Magnum/Orion electronic signals intelligence (ELINT) satellite. The first Orion was reportedly launched by STS-20/51C. According to Jim Slade of ABC News, the satellite was intended to eavesdrop on military and diplomatic communications from the Soviet Union and China. This satellite reportedly replaced the one launched by STS-20, which was running out of the stationkeeping propellant.

Acknowledged secondary payloads included CRUX-B (cosmic ray upset experiment) in the payload bay and AMOS (Air Force Maui Optical Site calibration), APE-B (auroral photography experiment), CLOUDS (cloud logic to optimize use of defense systems), RME-III (radiation monitoring experiment), and VFT-1 (vision function tester) on the middeck.

The landing was scheduled for 26 November, but was postponed because of strong winds at Edwards. On the following day, the winds were still too strong during the first opportunity, but subsided sufficiently for the second opportunity. Typically, the Air Force simply made their standard "… successful …" announcement.

A (mostly) military crew poses in their uniforms. From the left, are Kathy Thornton, Sonny Carter, Fred Gregory, John Blaha, and Story Musgrave. Although not military, Thornton had been a physicist at the United States Army Foreign Science and Technology Center. (NASA)

As much as STS-33R would be cloaked in secrecy, so its crew was touched by tragedy. In June 1989, John Blaha (left) was assigned as pilot to replace the late Dave Griggs. Less than 18 months later, Sonny Carter (second from the left) was killed in a commercial air crash. (NASA)

Discovery touches down on the concrete runway at Edwards AFB at the end of the mission. Mission management postponed the landing one day because of high winds at Edwards, a condition that initially repeated itself the following day but ultimately subsided. (NASA)

The rules specified the exterior views of the orbiter and stack were not classified unless the payload bay doors were open. Firing Rooms 3/4 in the Launch Control Center at KSC were used for secure processing, as was Fight Control Room 2 in the Mission Control Center at JSC. (NASA)

The stack moved to LC-39B just before Halloween and the crew obviously enjoyed the opportunity to celebrate. Here, Story Musgrave holds a plastic jack-o-lantern, fitting since the orange launch-entry suits were often referred to as "pumpkin suits." (NASA)

Discovery is lifted off the orbiter transporter system in the transfer aisle of the VAB at KSC. The OTS had been built to move orbiters around Vandenberg AFB, California, but found its way to KSC after the West Coast site was abandoned after the Challenger accident. (NASA)

STS-32R

Mission:	33		NSSDC ID:		1990-002A
Vehicle:	OV-102 (9)		ET-32 (LWT)		SRB: BI035
Launch:	LC-39A		09 Jan 1990		12:35 UTC
Altitude:	193 nm		Inclination:		28.50 degrees
Landing:	EDW-22		20 Jan 1990		09:37 UTC
Landing Rev:	172		Mission Duration:		261 hrs 01 mins

Commander:	Daniel C. "Dan" Brandenstein (3)
Pilot:	James D. "WxB" Wetherbee (1)
MS1:	Bonnie J. Dunbar (2)
MS2:	Marsha S. Ivins (1)
MS3:	G. David Low (1)

Payloads:	Up: 26,458 lbs	Down: 21,396 lbs
	Syncom IV-5/Orbus-6 (15,190 lbs)	
	LDEF (retrieved) (21,396 lbs)	
	SRMS s/n 201	

Notes:	Delayed 18 Dec 1989 (pad modifications)
	Scrubbed 08 Jan 1990 (KSC weather)
	First launch from LC-39A since STS-32/61C
	First use of mobile launch platform three (MLP-3)
	Orbiter returned to KSC on 26 Jan 1990 (N905NA)

Wakeup Calls:

09 Jan	"What's More American"
10 Jan	Parody for Low based on "The Banana Boat" song
11 Jan	Parody for crew of "Let it Snow"
12 Jan	Parody of "Hello Dolly" ("Hello LDEF")
13 Jan	Notre Dame Victory march (for Wetherbee)
13 Jan	(from crew) "Attack of the Killer Tomatoes"
14 Jan	"Bow Down to Washington" (fight song)
15 Jan	"Glory, Glory, Colorado" (fight song)
16 Jan	"Danny Boy" (Brandenstein birthday)
17 Jan	"Washington and Lee" (fight song)
18 Jan	"Born to be Wild"
19 Jan	The Navy Hymn – "Anchors Aweigh"

The crew patch depicted the orbiter rendezvousing with LDEF from above. The Syncom satellite is successfully deployed and on its way to geosynchronous orbit. The stars form the STS designation and the seven major rays of the Sun remember the Challenger crew. (NASA)

A delay in completing modifications to LC-39A canceled the planned 18 December 1989 *Columbia* launch. The space center typically shut down for facility maintenance over the Christmas holidays, postponing the launch until after the first of the year. A launch attempt on 8 January 1990 continued smoothly until T-9-minutes when the launch director extended the hold because of weather at the launch site. Despite continuing bad weather, he resumed the count in an attempt to launch during the 58-minute window. However, the weather did not improve and he scrubbed the launch. The attempt on 9 January 1990 proceeded smoothly.

A post-flight review of ET separation imagery revealed a section of missing foam from the left forward bipod strut. This included four divots between 18 and 24 inches in diameter and another divot about 6 inches in diameter. Nobody would truly appreciate the meaning of this anomaly for another 13 years.

This mission deployed Syncom IV-5 (Leasat F5) during flight day 2. The Orbus-6 upper stage fired about 45 minutes after deployment boosted the Navy communications satellite to its geosynchronous orbit at 182 degrees west.

The crew of STS-13/41C had deployed the Long Duration Exposure Facility (LDEF) on 7 April 1984. At the time, NASA expected to retrieve the spacecraft in March 1985. After the return-to-flight, retrieving LDEF became critical because a high solar flux had accelerated the rate of orbital decay, with researchers believing the spacecraft would be too low to retrieve by February 1990. Engineers calculated the exact liftoff time for STS-32R about 12 hours before launch using the latest tracking data on LDEF to ensure an efficient rendezvous. On flight day 4, Dunbar grappled LDEF using the SRMS and berthed the spacecraft in the payload bay. Originally, researchers at NASA Langley had expected to fly LDEF several times, but ultimately this was the end of its only mission.

The crew used the SRMS to perform a survey of the orbiter thermal protection system since technicians at KSC had found a piece of RTV (room-temperature-vulcanizing silicone) used for tile repair during the post-launch beach walk down. Ultimately, engineers identified the RTV as coming from the tip of the right outboard elevon. The crew maneuvered the SRMS elbow camera to view the elevon and mission control concluded the missing RTV did not present a problem for entry, not that there was anything that could have been done about it in any case.

The deorbit maneuver had a 51-degree out-of-plane component and was the longest OMS deorbit burn of the flight campaign (tied with STS-67 at 299.2 seconds).

An on-orbit portrait has Marsha Ivins, Bonnie Dunbar, and David Low in front with Dan Brandenstein and James Wetherbee behind them. The photo was taken by a 35 mm camera with a self-timer and was used at their 30 January 1990 post-flight press conference at JSC. (NASA)

One of the primary missions objectives was deploying the Syncom IV-5 communications satellite. The satellite enjoyed a long career, finally being retired in late 2015 after serving customers in North America and Australia for more than twenty-five years. (NASA)

The Long Duration Exposure Facility (LDEF) was a large satellite and took up most of the available volume in the payload bay. Here it is shown prior to be removed from Columbia after the mission. Note the SRMS arm on the port payload bay sill. (NASA)

Launch as seen from one of the Shuttle Training Aircraft. At least one of the STAs, modified Gulfstream II business jets, usually flew weather reconnaissance missions around the launch site to confirm what meteorologists and radar forecast. (NASA)

Bonnie Dunbar used the SRMS arm to grapple LDEF, about four and a half years later than NASA had originally expected. A series of higher priority payloads, and then the Challenger accident, delayed retrieving the satellite until orbital decay became a concern. (NASA)

Mission:	34	NSSDC ID:	1990-019A
Vehicle:	OV-104 (6)	ET-33 (LWT)	SRB: BI036
Launch: Altitude:	LC-39A 132 nm	28 Feb 1990 Inclination:	07:50 UTC 62.00 degrees
Landing: Landing Rev:	EDW-23L 72	04 Mar 1990 Mission Duration:	18:10 UTC 106 hrs 18 mins

Commander:	John O. Creighton (2)
Pilot:	John H. Casper (1)
MS1:	Pierre J. Thuot (1)
MS2:	David C. "Dave" Hilmers (3)
MS3:	Richard M. "Mike" Mullane (3)

Payloads:	Up: (28,500 lbs)	Down: 0 lbs
	USA-53 (AFP-731/Misty)	
	No SRMS	

Notes:	Delayed 22 Feb 1990 (crew illness)
	Delayed 23 Feb 1990 (crew illness)
	Delayed 24 Feb 1990 (crew illness)
	Scrubbed 25 Feb 1990 (range safety failure)
	Scrubbed 26 Feb 1990 (KSC weather)
	Sixth dedicated Department of Defense mission
	Exceeded 57-degree flight rule limit
	Flew "dog-leg" trajectory to reach 62-degree orbit
	Orbiter returned to KSC on 13 Mar 1990 (N905NA)

Wakeup Calls:

Classified mission. Either nothing played or nothing released.

As with many of the early crew patches, this one was designed by the five crew members. The thirty-six stars symbolize the STS designation and were also part of a stylized American flag. The eagle symbolized our country's commitment to strength and vigilance. (NASA)

A combination of weather and the illness of John Creighton forced the postponement of this *Atlantis* mission from 22 February, to the 23rd, the 24th, and finally the 25th. NASA had long since stopped assigning backup crew members to all but the most important missions and this was the first time since Apollo 13 that a manned spaceflight had been affected by the illness of a crew member. The launch attempt on 25 February was scrubbed due to the failure of an Air Force range safety computer and NASA canceled an attempt the following day due to weather at the launch site. The classified four-hour window on 28 February opened at 05:00 UTC and *Atlantis* launched into the highest-inclination orbit of the flight campaign.

The launch trajectory exceeded the published maximum inclination of 57 degrees by a considerable margin. This required an inefficient dog-leg trajectory that saw *Atlantis* fly downrange toward 57 degrees and then turn toward 62 degrees once out over the Atlantic. The Air Force waived the normal flight rules that prohibited flying over land, with the trajectory taking the vehicle over or near Cape Hatteras, Cape Cod, and parts of Canada.

The primary payload remains officially classified, with NASA only acknowledging the deployment of a single satellite. At the time, *Aviation Week* described the payload as a large digital reconnaissance satellite. It appears that at least some parts of NASA endorse that view since the NASA National Space Science Data Center reported that a, "KH-11-10 was deployed from the orbiting STS-36 for the US Department of Defense. ... The satellite was reported to have malfunctioned after being placed in orbit." It is unclear if this is an official description or speculation on the part of whomever maintains the database. Most pundits believe the KH-11 Kennen resembled the Hubble Space Telescope in size and shape, mostly because both were built by Lockheed and transported in similar shipping containers. They also point to a NASA history of Hubble that, in discussing the reasons for switching from a 3-meter main Mirror to a smaller design, states, "In addition, changing to a 2.4-meter Mirror would lessen fabrication costs by using manufacturing technologies developed for military spy satellites." An unofficial history of the CIA mentions the primary Mirror on the initial KH-11s measured 2.34 meters, but that size reportedly increased in later versions.

Atlantis returned to Edwards on the first opportunity. Because of 15-knot head winds on the lakebed, and the lightweight of the orbiter, the 7,900-foot landing roll was shorter than expected. Despite the return from a high-inclination orbit, the TPS damage was "lower than average." This was the sixth "... successful ..." DoD flight.

Looking out of the internal airlock hatch into the middeck are, clockwise from the lower right, John Casper, Dave Hilmers, Pierre Thuot, John Creighton, and Mike Mullane. For the first time since Apollo 13, the illness of a crew member affected a launch. (NASA)

The crew, from bottom to top of stairs: John Creighton, John Casper, Pierre Thuot, Dave Hilmers, and Mike Mullane. At the right are Bill Lenoir (left), acting associate administrator for space flight, and Donald Puddy, director of flight crew operations at JSC. (NASA)

Personnel in Firing Room 1 in the KSC Launch Control Center catch a glimpse of Columbia during launch. Although all of the subsystem consoles faced the window, there was really very little view and most of the engineers watched launch on the small TVs in the consoles. (NASA)

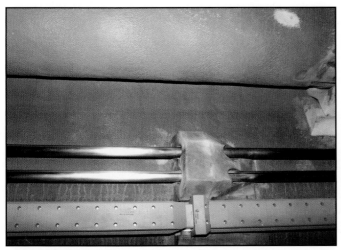

Details on ET-33. At the top is the 17-inch diameter LO_2 feedline. The two silver pipes are pressurization lines, while the item at the bottom is a cable tray that carried power and data for the sensors. (NASA)

A closeup of the starboard orbiter-ET attach point and LO_2 umbilical. The gray fitting is the attach point where a 2.5-inch-diameter bolt held the ET to the orbiter; the umbilical is to the left. (NASA)

John Creighton in his David Clark Company S1032 launch-entry suit on 25 February. The crew donned their partial-pressure suits in the O&C Building near the crew quarters. (NASA)

The crew exits the O&C Building for the second launch attempt. John Creighton (front left) and (in rear, from left), Dave Hilmers, Mike Mullane, and Pierre Thuot. (NASA)

Mike Mullane on the aft flight deck using a 70 mm Hasselblad camera. Each flight carried a variety of handheld still and motion picture (later, video) cameras. (NASA)

STS-31R (HST)

Mission:	35	NSSDC ID:		1990-037A
Vehicle:	OV-103 (10)	ET-34 (LWT)	SRB:	BI037 (LH)
				BI038 (RH)
Launch:	LC-39B	24 Apr 1990		12:33 UTC
Altitude:	333 nm	Inclination:		28.45 degrees
Landing:	EDW-22	29 Apr 1990		13:51 UTC
Landing Rev:	80	Mission Duration:		121 hrs 16 mins

Commander:	Loren J. Shriver (2)
Pilot:	Charles F. "Charlie" Bolden, Jr. (2)
MS1:	Bruce McCandless II (2)
MS2:	Steven A. "Steve" Hawley (3)
MS3:	Kathryn D. "Kathy" Sullivan (2)

Payloads:	Up: 28,643 lbs	Down: 0 lbs
	Hubble Space Telescope (23,905 lbs)	
	SRMS s/n 301	

Notes:
Advanced 18 Apr 1990 (to 12 Apr)
Advanced 12 Apr 1990 (to 10 Apr)
Scrubbed 10 Apr 1990 (APU oscillations)
Commander was to be John W. Young
Orbiter returned to KSC on 07 May 1990 (N905NA)

Wakeup Calls:

25 Apr	"Space is Our World"
26 Apr	"Shout" (*Animal House*)
27 Apr	"Kokomo"
28 Apr	"Cosmos"
29 Apr	"Rise and Shine"

NASA had scheduled *Discovery* for launch on 18 April 1990, a date that was moved forward to 12 April and finally 10 April. This was the first time NASA attempted to launch a space shuttle ahead of schedule. The 10 April attempt proceeded smoothly until problems with an auxiliary power unit caused engineers to scrub the launch at T-4 minutes. The mission management team rescheduled the launch for two weeks later to allow time to recharge the Hubble batteries. On 24 April, the MPS LO$_2$ outboard fill and drain valve did not close as expected and the launch team held the countdown while engineers shut the valve under manual control. The rest of the countdown proceeded smoothly.

The initial orbit after main engine cutoff was 330x48 nm and the OMS-2 burn raised this to 330x311 nm. A circularization burn (OMS-3) raised this to 333x332 nm, the highest mission flown to-date and the second highest of the flight campaign (STS-82, the second Hubble servicing mission, went to 335 nm).

This mission deployed the Hubble Space Telescope, the first of the Great Observatories. The crew used the monochrome camera at the end of the SRMS to examine Hubble for any damage that might have happened during launch, fortunately not finding any. During flight day 2, the crew prepared to support a contingency EVA to manually deploy the Hubble solar arrays if the automated sequence did not work as expected. After completing the in-suit pre-breathe period, Bruce McCandless and Kathy Sullivan entered the airlock, which was depressurized to 5.0 psia in preparation for the EVA. Following the unberthing of the telescope from the payload bay, one of the solar arrays failed to deploy on the first attempt but partially deployed on the second attempt. A third attempt resulted in the array opening fully. Once the solar arrays were successfully deployed, mission control canceled the contingency EVA, before the crew had exited the airlock. A similar exercise was conducted on flight day 4 in case the aperture door did not fully open.

The crew released the telescope on revolution 20 and ground controllers at the Space Telescope Operations Control Center at NASA Goddard completed activating the telescope. On flight day 4, Bruce McCandless and Kathy Sullivan again entered the airlock for a contingency EVA in case the aperture door did not fully open. Fortunately, it opened as expected and the EVA was again canceled. Of course, astronomers soon discovered a serious flaw in the main Mirror, largely negating the value of the telescope, at least initially.

Discovery landed on Runway 22 at Edwards, marking the first use of the new structural-carbon brakes. Post-landing inspections of the tires revealed only minor anomalies and minimal wear.

The crew patch featured Hubble against a background of the universe it would study. The cosmos included a stylistic depiction of galaxies in recognition of the contributions made by Sir Edwin Powell Hubble. Discovery trails a spectrum symbolic of red shift observations. (NASA)

The crew poses at NASA Dryden after landing. From the left are Steve Hawley, Charlie Bolden, Kathy Sullivan, Loren Shriver, and Bruce McCandless. All are wearing the S0132 launch-entry suits that became standard after the Challenger accident. (NASA)

Hubble was carefully transferred on 29 March from the surgically clean payload changeout room (PCR) at LC-39B into the payload bay of Discovery. The payload ground handling mechanism (PGHM, at left) extended the telescope into the payload bay (right), allowing technicians to close the active latches on the longeron bridges. (NASA)

The SRMS (at lower right) holds Hubble after lifting it from the payload bay. The solar arrays, rolled up on the arm extending from the middle of the telescope, have not yet been extended. Bruce McCandless and Kathy Sullivan began preparing for a contingency EVA in case the arrays could not be deployed. Fortunately, it was not necessary. (NASA)

Kathy Sullivan monitors student experiment 82-16, developed by Gregory S. Peterson at Utah State University, on the middeck. The intent was to observe the effects of microgravity on an electric arc. An Arriflex 16 mm camera photographed the experiment. (NASA)

The visual appearance of the STS-31R was typical of the period. Note the dark bare metal flipper doors on the upper wing surfaces of the orbiter; these would later become lightweight aluminum covered in white nomex felt (FRSI). (NASA)

STS-41

Mission:	36	NSSDC ID:	1990-090A
Vehicle:	OV-103 (11)	ET-39 (LWT)	SRB: BI040 (LH)
			BI037 (RH)
Launch:	LC-39B	06 Oct 1990	11:47 UTC
Altitude:	179 nm	Inclination:	28.45 degrees
Landing:	EDW-22	10 Oct 1990	13:58 UTC
Landing Rev:	66	Mission Duration:	98 hrs 10 mins

Commander:	Richard N. "Dick" Richards (2)
Pilot:	Robert D. "Bob" Cabana (1)
MS1:	Bruce E. Melnick (1)
MS2:	William M. "Shep" Shepherd (2)
MS3:	Thomas D. "Tom" Akers (1)

Payloads:	Up: 48,133 lbs	Down: 0 lbs
	Ulysses/IUS-PAM-S (44,024 lbs)	
	SSBUV (1,215 lbs)	
	SRMS s/n 301	

Notes: Partial SRB stack rollout 11 Jun 1990 to make
room in the VAB for STS-35 troubleshooting
Was to use Shuttle/Centaur upper stage
Orbiter returned to KSC on 16 Oct 1990 (N905NA)

Wakeup Calls:

07 Oct	"Rise and Shine, Discovery!"
08 Oct	The Coast Guard Hymn – "Semper Paratus"
09 Oct	"Fanfare for the Common Man"
10 Oct	"The Highwayman"

Ulysses, represented by the streaking silver teardrop passing over the Sun, would become the fastest man-made object, traveling more than 30 miles per second. The path around Jupiter was depicted by the bright red spiral originating from the payload bay. (NASA)

The launch team held this countdown several times for weather and technical issues, but *Discovery* still launched only 12 minutes into the 2.5-hour launch window.

The European Ulysses (formerly Solar-Polar) spacecraft was the primary payload. Originally, NASA had manifested Ulysses as STS-35/61F on the next flight of *Challenger* after STS-33/51L. The accident changed everything. A post-accident safety analysis determined the liquid-propellant Shuttle/Centaur upper stage was too dangerous to carry in the payload bay, so NASA switched Ulysses to a hastily assembled inertial upper stage (IUS) topped by a mission-specific payload assist module (PAM-S). This combination resulted in the heaviest space shuttle payload to-date.

Six hours after launch, on revolution 5, the crew deployed Ulysses. The spacecraft began its voyage to the Sun with a 16-month trip to Jupiter where gravitational energy was used to fling the spacecraft southward out of the orbital plane of the planets and on toward a solar south pole passage in 1994. The spacecraft crossed back over the orbital plane and made a pass through the solar north pole in 1995. By the time *Discovery* touched down at Edwards, Ulysses had already traveled more than 1 million miles on its five-year mission.

After deploying Ulysses, the crew began an ambitious schedule of science experiments. Understanding fire behavior in microgravity was part of the continuing research to improve space shuttle safety. Using a specially designed chamber, SSCE (solid surface combustion experiment) saw the crew burn a strip of paper to gain an understanding of the development of flame and its movement in the absence of convection currents. The SSBUV (shuttle solar backscatter ultraviolet) experiment carried an ozone detector identical to those on the NIMBUS-7 and TIROS weather satellites. By comparing measurements on *Discovery* with coordinated satellite observations, scientists could calibrate the satellite instruments. The goal of the physiological systems experiment (PSE) was to determine if drugs would be effective in reducing or eliminating some of the effects of osteoporosis. The crew also conducted the investigations into polymer membrane processing (IPMP) to determine the role convection currents played in the formation of membranes used for the purification of medicines, kidney dialysis, and water desalination.

Discovery landed on the concrete runway at Edwards and conducted a second test of the new structural-carbon brakes. When technicians opened the ET doors during the post-landing inspection, part of a 2.5-inch frangible nut dropped to the runway. This appeared to be an isolated anomaly and had no effect on the mission or subsequent turnaround processing.

An on-orbit portrait of, from the left, Tom Akers, Dick Richards, Bruce Melnick, Bob Cabana, and Bill Shepherd. Note the various flags and stickers from their alma maters on the original heavyweight middeck accommodations rack (MAR) in the background. (NASA)

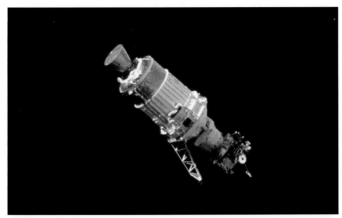

Ulysses, backdropped against the blackness of space, rapidly distances itself from Discovery and moves toward the beginning of its five-year mission to the Sun. The photo was taken from an aft flight deck window with a handheld Roliflex camera using 120 film. (NASA)

There were several ways for the crew to sleep in the orbiter. On long science (Spacelab) missions they used a sleepstation that looked like bunkbeds on the right side of the middeck. Shorter missions used sleeping bags like these hung on the wall. Or, and quite often, the crew just slept someplace quiet and out of the way. (NASA)

A view from the rotating service structure of Discovery at LC-39B with the crew cabin access arm at the left. Note the two red triangles on top of the cockpit warning about the explosive window that allowed the flight crew to egress a damaged orbiter on the ground. (NASA)

Ulysses with its cobbled-together IUS/PAM-S upper stage in the payload bay at LC-39B. The cancellation of Shuttle/Centaur had a dramatic effect on Ulysses, which now had to use a gravitational-assist route around Jupiter to accelerate to more than 100,000 mph. (NASA)

STS-38

Mission:	37	NSSDC ID:	1990-097A
Vehicle:	OV-104 (7)	ET-40 (LWT)	SRB: BI039 (LH)
			BI040 (RH)
Launch:	LC-39A	15 Nov 1990	23:48 UTC
Altitude:	142 nm	Inclination:	28.45 degrees
Landing:	KSC-33	20 Nov 1990	21:44 UTC
Landing Rev:	79	Mission Duration:	117 hrs 54 mins

Commander:	Richard O. "Dick" Covey (3)
Pilot:	Frank L. Culbertson, Jr. (1)
MS1:	Carl J. Meade (1)
MS2:	Robert C. "Bob" Springer (2)
MS3:	Charles D. "Sam" Gemar (1)

Payloads:	Up: 20,953 lbs	Down: 0 lbs
	USA-67 (AFP-658/SDS-B2) (20,953 lbs)	
	(Prowler)	
	No SRMS	

Notes: Originally stacked with ET-37 (later used by STS-37)
Hydrogen leaks during tanking
LH$_2$ tanking test 29 Jun 1990 (ET-37)
LH$_2$ tanking test 13 Jul 1990 (ET-37)
LH$_2$ tanking test 25 Jul 1990 (ET-37)
LH$_2$ tanking test 24 Oct 1990 (ET-40)
First flight with Air Force, Army, Marine Corps, and Navy crew members
Bad weather at Edwards for landing
First post-*Challenger* KSC landing

Wakeup Calls:

Classified mission so no details released but it is known that all four military hymns were played on one day.

The top orbiter, with the stylistic OMS burn, symbolized the dynamic nature of the Space Shuttle Program while the bottom orbiter, a monochrome Mirror image, acknowledged the thousands of individuals who worked behind the scenes. The other patch was a "secret." (NASA)

On 30 May 1990, NASA scrubbed the STS-35 launch because of an unusually high concentration of hydrogen around the orbiter-ET disconnect. A solution was elusive. To help understand the problem, NASA conducted a tanking test using the STS-38 stack on 29 June 1990. The test recorded high hydrogen concentrations and engineers conducted additional tanking tests on 13 July and 25 July to isolate the source of the leak. Afterward, NASA rolled the vehicle back to the VAB and demated the orbiter from the ET/SRB stack. During rollback, the vehicle remained parked outside the VAB as the STS-35 stack was rolled-back to the launch pad and *Atlantis* suffered minor hail damage during a thunderstorm.

Engineers replaced the original ET-37 with ET-40 and installed a new ET disconnect on *Atlantis*, leading to a successful (fourth) tanking test on 24 October. NASA scheduled the STS-38 launch for 9 November but issues with the DoD payload resulted in a further delay to 15 November. Liftoff occurred at the beginning of a four-hour classified launch window that began at 23:30 UTC.

Speculation is that the primary AFP-658 payload (USA-67) was an SDS-B communications relay satellite like those launched on STS-28R and, later, STS-53. The crew deployed the payload about seven hours after launch. Initially, *Aviation Week* reported the payload was a TRW Magnum/Orion electronic signals intelligence satellite similar to those launched by STS-20/51C and STS-33R. Given that published photos of the payload bay did not show the necessary aerospace support equipment for the inertial upper stage used on the other two Magnum launches, this seemed unlikely.

Other rumors include that *Atlantis* also deployed a stealth satellite called Prowler. In 1998, hobbyists discovered an unknown satellite, but they soon lost track of it. These observers speculate that Prowler was based on a modified Hughes HS-376 bus and was deployed 22 hours after the primary SDS-B2 payload. Again, published photos of the payload bay do not show the normal cradle and sunshields used by the Hughes satellites.

NASA canceled three landing opportunities on flight day 4 because of unacceptable weather at Edwards and extended the mission one day. Weather forecasts for Edwards the following day showed high winds on all runways, so the mission management team changed the primary landing site to KSC. NASA had been concerned about returning an orbiter to the Shuttle Landing Facility because the last mission to do so experienced tire failure because of the rough runway. However, the post-flight inspection showed no indication of excessive wear. This marked the seventh mission dedicated to the DoD, which "… successfully …" met all of its mission objectives.

From the left are Sam Gemar, Frank Culbertson, Bob Springer, Dick Covey, and Carl Meade. All of their launch-entry suits have a NASA "worm" logo and an American flag on the sleeves, and a traditional NASA "meatball," name tag, and crew patch on the front. (NASA)

Atlantis *hanging in the transfer aisle of the Vehicle Assembly Building. The orbiter has just come from the Orbiter Processing Facility and will soon be lifted into High Bay 1 to be mated with the waiting external tank and solid rocket boosters.* (NASA)

It's a long way up to clear the structural wall that separates the transfer aisle from the high bays. Here Atlantis *is entering High Bay 1 to be mated with the waiting stack. Note all of the platforms that allowed technicians to work on the vehicle while it is in the VAB.* (NASA)

The crew exits Atlantis *at the Shuttle Landing Facility. In this case they had changed out of their pressure suits for more comfortable clothes while they waited for the all-clear to egress the ship. Note the red carpet on the ground in front of the airstairs.* (NASA)

Columbia *(left) as STS-35 and* Atlantis *as STS-38 passing in the early morning of 9 August 1990 when engineers and technicians were trying to figure out why both vehicles were suffering hydrogen leaks. One issue was procedural, the other a contaminated seal.* (NASA)

STS-35

Mission:	38	NSSDC ID:	1990-106A
Vehicle:	OV-102 (10)	ET-35 (LWT)	SRB: BI038 (LH) BI039 (RH)
Launch:	LC-39B	02 Dec 1990	06:49 UTC
Altitude:	195 nm	Inclination:	28.45 degrees
Landing:	EDW-22	11 Dec 1990	05:55 UTC
Landing Rev:	144	Mission Duration:	215 hrs 05 mins

Commander:	Vance D. Brand (4)
Pilot:	Guy S. Gardner (2)
MS1:	Jeffrey A. "Jeff" Hoffman (2)
MS2:	John M. "Mike" Lounge (3)
MS3:	Robert A. Parker (2)
PS1:	Samuel T. "Sam" Durrance (1)
PS2:	Ronald A. "Ron" Parise (1)

Payloads:	Up: 29,720 lbs	Down: 0 lbs
	Astro-1 (27,760 lbs)	
	BBXRT (hitchhiker)	
	No SRMS	

Notes:
Scrubbed 30 May 1990 (hydrogen leak)
LH$_2$ tanking test 06 Jun 1990
Scrubbed 06 Sep 1990 (hydrogen leak)
Scrubbed 18 Sep 1990 (hydrogen leak)
Rolled-back to VAB 09 Oct 1990 (Hurricane Klaus)
LH$_2$ tanking test 30 Oct 1990
Backup payload specialists were
   Kenneth H. Nordsieck and John-David F. Bartoe
Orbiter returned to KSC on 20 Dec 1990 (N905NA)

Wakeup Calls:

Since crew worked two shifts around the clock,
mission control did not send any wakeup calls.

Designed by the crew, the patch symbolized Columbia flying above the atmosphere to better study the many celestial objects of the Universe, represented by the constellation Orion. The smaller patch was used by the Astro-1 team and showed the major instruments. (NASA)

Engineers detected an unusually high concentration of hydrogen around the orbiter-ET disconnect and in the orbiter aft compartment during propellant loading on 30 May 1990 and NASA grounded the space shuttle fleet while it investigated the issue. It took six months to work through the problems. Finally, on 2 December, *Columbia* was launched following short hold due to cloud cover at the launch site.

Prior to the *Challenger* accident, NASA had scheduled this mission for March 1986 as STS-34/61E commanded by Jon McBride. However, McBride retired from NASA in May 1989 and was replaced by Vance Brand. In addition, Guy Gardner and Mike Lounge replaced Dick Richards and Dave Leestma, respectively. The 59-year-old Brand was the oldest person to fly until Story Musgrave (61) on STS-80 in 1996 and John Glenn (77) on STS-95 in 1998.

The crew split into two teams to allow around-the-clock operations. The red team consisted of Guy Gardner, Robert Parker, and Ron Parise while the blue team included Jeff Hoffman, Sam Durrance, and Mike Lounge. Vance Brand was unassigned to either team and coordinated the activities.

The primary payload was Astro-1 and this was the first space shuttle mission controlled in part from the Spacelab Mission Operations Control Facility at MSFC. The Astro-1 observatory consisted of four telescopes to search the celestial sphere in the ultraviolet and X-ray wavelengths. These included the BBXRT (broadband X-ray telescope), HUT (Hopkins ultraviolet telescope), UIT (ultraviolet imaging telescope), and WUPPE (Wisconsin ultraviolet photo-polarimeter experiment).

The red team activated the payload and raised the telescopes about 11 hours after launch. Issues with the precision of the instrument pointing system (IPS) and the sequential overheating failures of both data display units used for controlling the telescopes forced ground teams at MSFC to aim the telescopes remotely with fine-tuning by the crew. However, BBXRT was directed from the outset by ground-based operators at NASA Goddard and was not affected. The payload specialists set up each instrument for the upcoming observation, identified the celestial target and provided the necessary pointing corrections for placing the object precisely in the field of view. They then started the observation sequences and monitored the data being recorded. Each observation took between ten minutes to a little over an hour.

Because weather forecasts indicated unacceptable conditions at Edwards, the mission management team decided to end the mission one day earlier than planned, albeit still at Edwards.

Vance Brand is at bottom center in this on-orbit portrait. Clockwise from Brand are Robert Parker, Ron Parise, Jeff Hoffman, Guy Gardner, Mike Lounge, and Sam Durrance, all sporting patriotic shirts. The crew used a self-timed 70 mm camera to expose the middeck image. (NASA)

On the middeck, Sam Durrance (left) and Jeff Hoffman teach "Space Classroom Assignment: The Stars" via television to classrooms on Earth. This gave students a lesson on the electromagnetic spectrum. Displayed behind them on the galley is a chart showing the range of light visible to each of the telescopes on Astro-1. (NASA)

KSC employees seldom missed an opportunity to watch a space shuttle, whether it be launch, landing, or a move. Here, the STS-35 stack is rolling out of the Vehicle Assembly Building heading toward LC-39B, a little more than four miles away. (NASA)

The various components of Astro-1 including the Hopkins ultraviolet telescope (HUT), ultraviolet imaging telescope (UIT), and the Wisconsin ultraviolet photo-polarimeter experiment (WUPPE) are visible on the Spacelab pallet long with the "igloo" control system housing. (NASA)

The orbiter transporter system (OTS) backs Columbia out of the OPF on her way to the VAB to be stacked for the second launch attempt after technicians replaced the 17-inch hydrogen disconnect with a new one borrowed from the not-yet-complete Endeavour. (NASA)

STS-37

Mission:	39		NSSDC ID:		1991-027A
Vehicle:	OV-104 (8)		ET-37 (LWT)		SRB: BI042
Launch:	LC-39B		05 Apr 1991		14:22 UTC
Altitude:	248 nm		Inclination:		28.45 degrees
Landing:	EDW-33		11 Apr 1991		13:56 UTC
Landing Rev:	93		Mission Duration:		143 hrs 32 mins

Commander:	Steven R. "Steve" Nagel (3)
Pilot:	Kenneth D. "Ken" Cameron (1)
MS1:	Linda M. Godwin (1)
MS2:	Jerry L. Ross (3)
MS3:	Jerome "Jay" Apt III (1)

Payloads:	Up: 40,561 lbs	Down: 0 lbs
	Gamma Ray Observatory (34,442 lbs)	
	CETA (803 lbs)	
	SRMS s/n 303	

Notes:	First EVA since STS-31/61B in 1985
	First flight of the upgraded AP-101S computer
	EVA1 07 Apr 1991 (Ross and Apt)
	EVA2 08 Apr 1991 (Ross and Apt)
	Orbiter returned to KSC on 18 Apr 1991 (N905NA)

Wakeup Calls:

06 Apr	Music by Marching Illini Band
07 Apr	The Marine Corps Hymn – "Halls of Montezuma"
08 Apr	"Hail Purdue"
09 Apr	"10,000 Men of Harvard Want Victory Today"
10 Apr	"La Bamba"
11 Apr	"Theme from Magnum PI"

The gamma connecting Atlantis and GRO symbolized the quest for gamma rays and the relationship between the manned and unmanned elements of the American space program. The two fields of three and seven stars refer to the STS designation. (NASA)

The launch team extended the T-9-minute hold on 5 April 1991 almost five minutes because of weather at KSC. Although both SRBs were retrieved after the launch, the impact with heavy seas damaged the left booster, which experienced more than 90-g during the impact. In particular, the forward segment case wall was dented and was subsequently scrapped. This was the first *Atlantis* flight using the AP-101S general-purpose computers and the structural-carbon brakes. At KSC, engineers installed a hydrogen dispersal system on the mobile launch platform in case the hydrogen leaks experienced by STS-35 and STS-38 reappeared. This system used gaseous nitrogen to dilute potential gaseous hydrogen leaks. However, no leaks were detected and the purge system was not used.

Atlantis carried the long-delayed Gamma Ray Observatory, the second of the Great Observatories. NASA expected GRO to spend two years searching for high-energy celestial gamma ray emissions that cannot penetrate the atmosphere. Its science instruments included BATSE (burst and transient source experiment), COMPTEL (imaging Compton telescope), EGRET (energetic gamma ray experiment telescope), and OSSE (oriented scintillation spectrometer experiment). GRO was the heaviest satellite to be deployed into low-earth orbit from space shuttle. It was also designed to be refueled on-orbit by space shuttle crews, although this never happened.

The crew used the SRMS to grapple GRO and raise it out of the payload bay, but the high-gain antenna failed to deploy despite shaking the satellite using the SRMS. Ultimately, Jerry Ross and Jay Apt freed the antenna during the first unscheduled EVA since April 1985. The antenna was released during the first 17 minutes of the EVA so the crew performed some of the tasks from a planned spacewalk the following day. During the second EVA, Apt reported the palm bar in his right glove punctured an area above the index finger, creating a minor leak. Even if it had come out of the hole, the leak rate would not have been great enough to endanger Apt.

About five months after it was deployed, GRO was renamed the Arthur Holly Compton Gamma Ray Observatory after the 1927 Nobel Laureate in Physics.

The mission management team waved-off the first landing opportunity at Edwards and rescheduled the landing for KSC the next day. That attempt was also waved off due to fog and the landing was shifted to Edwards once again. Due to incorrect winds aloft data, *Atlantis* landed 623 feet short of the intended touchdown spot on the lakebed runway, which is normally about 1,500 feet from the threshold, just in case things like this happened. The landing and rollout were otherwise normal.

The alternate crew portrait. From the left are Ken Cameron, Linda Godwin, Steve Nagel, Jay Apt, and Jerry Ross. The crew frequently posed for several portraits and one often featured them in uniform. There were also "gag" photos that were seldom released. (NASA)

Relaxing on the middeck. Ken Cameron is performing a "quick hands" feat with three tape cassettes, something a bit easier in microgravity. Jerry Ross later used the microgravity environment to have some fun with the bag of malted milk balls in his hands. (NASA)

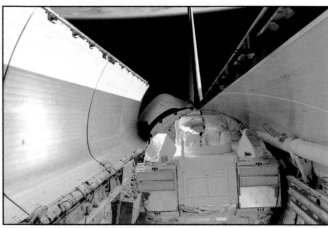

The crew opened the payload bay doors as soon as they were on-orbit, with the starboard door opening first. Note the SRMS arm on the right side of the photo with its elbow camera in its stowed position to clear the closed port-side door. (NASA)

Five David Clark S1032 launch-entry suits float in the middeck. Contrary to most reports, the suit was not a variant of the CSU-4/P partial-pressure suit, although the bladder system and other components were generally similar to various existing aircrew protective systems. (NASA)

Jay Apt works on the starboard side of the payload bay. Note the open hatch into the internal airlock and the two small windows on the aft flight deck. The pair of vertical brown tubes on the bulkhead were used on the ground to purge the payload bay. (NASA)

Atlantis being mated to the first Shuttle Carrier Aircraft (N905NA) in the Mate/Demate Device at NASA Dryden. Note the man-lift positioned so that workers can install the bolts that held the orbiter to the 747. Atlantis returned to KSC on 18 April 1991. (NASA)

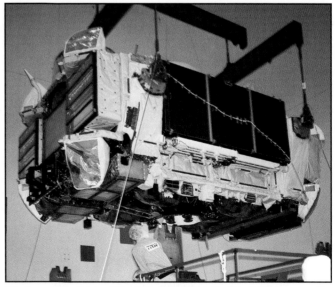

Workers at the Payload Hazardous Servicing Facility st KSC remove the Gamma Ray Observatory from its shipping container. The observatory was manufactured by TRW in Redondo Beach, California, and transported to KSC by truck. (NASA)

STS-39

Mission:	40	NSSDC ID:	1991-031A
Vehicle:	OV-103 (12)	ET-46 (LWT)	SRB: BI043
Launch:	LC-39A	28 Apr 1991	11:33 UTC
Altitude:	140 nm	Inclination:	57.00 degrees
Landing:	KSC-15	06 May 1991	18:57 UTC
Landing Rev:	134	Mission Duration:	199 hrs 22 mins

Commander:	Michael L. "Mike" Coats (3)
Pilot:	L. Blaine Hammond, Jr. (1)
MS1:	Gregory J. "Greg" Harbaugh (1)
MS2:	Donald R. "Don" McMonagle (1)
MS3:	Guion S. "Guy" Bluford, Jr. (3)
MS4:	Charles L. "Lacy" Veach (1)
MS5:	Richard J. "Rick" Hieb (1)

Payloads:	Up: 30,340 lbs	Down: 0 lbs
	AFP-675 (20,495 lbs)	
	SPAS-II (4,046 lbs)	
	STP-1 (hitchhiker)	
	SRMS s/n 301	

Notes:	First unclassified Department of Defense mission
	Delayed 09 Mar 1991 (ET umbilical doors)
	Rolled-back to the VAB 07 Mar 1991
	Scrubbed 23 Apr 1991 (SSME transducer)

Wakeup Calls:

Since crew worked two shifts around the clock, mission control did not send any wakeup calls.

The arrowhead shape of the crew patch represented a skyward aim to learn more about the atmosphere and space environment in support of the Department of Defense. The national symbol is represented by the star constellation Aguila (the eagle). (NASA)

Discovery was originally scheduled for launch on 9 March 1991, but during processing at LC-39A, engineers discovered cracks in all four hinges on the two ET umbilical doors. NASA opted to roll the vehicle back to the VAB on 7 March and then to the OPF on 15 March for repair. Hinges from Columbia were also found to have minor cracks, but were removed, reinforced, and installed on Discovery. Cracks were also found on Atlantis. The STS-39 stack was returned to LC-39A on 1 April for a scheduled launch on 23 April. The mission was again postponed when a main engine developed a problem while the ground team was loading the external tank. Mission management set a new launch date of 28 April.

This was the first unclassified DoD mission following seven classified flights. The payload bay contained AFP-675 consisting of five experiments on a Shuttle Pallet Satellite (SPAS-II). These included CIRRIS (cryogenic infrared radiance instrumentation for shuttle), Far UV (far ultraviolet), HUP (horizon ultraviolet program), QINMS (quadrupole ion neutral mass spectrometer), and URA (uniformly redundant array). The data recorders used by these experiments failed early in the flight. On flight day 4, the crew used the SRMS to deploy the SPAS pallet for about 39 hours. The following day, the crew again unberthed the SPAS, although this time it spent 23 hours attached the SRMS and not as a free-flyer. The crew was divided into two teams for around-the-clock operations. The red team included Blaine Hammond, Lacy Veach, and Rick Hieb while the blue team consisted of Greg Harbaugh, Don McMonagle, and Guy Bluford. Mike Coats kept his own hours, independent of the teams.

Secondary payloads included IBSS (infrared background signature survey) mounted on the SPAS pallet, STP-1 (space test program) mounted on a Hitchhiker-M carrier, CLOUDS-1A (cloud logic to optimize the use of defense systems), MPEC (multi-purpose experiment canister), and RME-III (radiation monitoring experiment). Only the contents of the MPEC were classified. According to a member of the crew, on flight day 8 Guy Bluford launched the classified payload by himself while, "the rest of us pretended not to notice."

After the crew closed the payload bay doors, mission control determined winds at Edwards would not subside and diverted the landing to the Shuttle landing Facility at KSC. After touchdown, Mike Coats and Blaine Hammond conducted a heavy braking test. The post-landing inspection of the tires revealed excessive wear on two tires, although the other two showed only the normal wear from a concrete runway landing.

In the front, from the left, are Don McMonagle, Mike Coats, Lacy Veach, and Greg Harbaugh; and in the back are Guy Bluford, Blaine Hammond, and Rick Hieb. Note that some of the shirts have their names embroidered. (NASA)

Taken from inside the crew cabin, this 35 mm photo shows the tops of canisters on the STP-1 payload on the cross-bay hitchhiker carrier (foreground) and the AFP-675 package that consisted of CIRRIS-1A (cryogenic infrared radiance instrumentation for shuttle), FAR-UV (far ultraviolet camera), HUP (horizon ultraviolet program), QINMS (quadruple ion neutral mass spectrometer), and URA (uniformly redundant array). Note the gold-covered cameras on the two corners of the aft bulkhead and on the elbow of the SRMS arm at right. (NASA)

The payload control station was on the port side of the aft cockpit and was reconfigured for every mission depending on the needs of the payload. By this time, laptop computers were also finding their way onto each mission to assist the crew. (NASA)

Discovery begins the long trek from the Vehicle Assembly Building to LC-39A. NASA had built the VAB to stack the Saturn V moon rocket and then reconfigured it to support space shuttle. Although a workable arrangement, it was far from ideal. (NASA)

STS-40

Mission:	41	NSSDC ID:	1991-040A
Vehicle:	OV-102 (11)	ET-41 (LWT)	SRB: BI044
Launch:	LC-39B	05 Jun 1991	13:25 UTC
Altitude:	161 nm	Inclination:	39.00 degrees
Landing:	EDW-22	14 Jun 1991	15:40 UTC
Landing Rev:	146	Mission Duration:	218 hrs 14 mins

Commander:	Bryan D. "OC" O'Connor (2)
Pilot:	Sidney M. "Sid" Gutierrez (1)
MS1:	James P. "Jim" Bagian (2)
MS2:	Tamara E. "Tammy" Jernigan (1)
MS3:	M. Rhea Seddon (2)
PS1:	F. Andrew "Drew" Gaffney (1)
PS2:	Millie E. Hughes-Fulford (1)

Payloads:	Up: 33,707 lbs	Down: 0 lbs
	Spacelab SLS-1 (28,114 lbs)	
	GAS (11)	
	No SRMS	

Notes:	Delayed 22 May 1991 (KSC workload)
	Delayed 27 May 1991 (MPS transducer)
	Scrubbed 01 Jun 1991 (IMU calibration error)
	Worked single-shift operations
	Backup payload specialist was
	Robert W. Phillips
	Orbiter returned to KSC on 21 Jun 1991 (N905NA)

Wakeup Calls:

06 Jun	"Great Balls of Fire"
07 Jun	Military medley
08 Jun	"Yakety Yak"
09 Jun	"Somewhere Out There" (*An American Tail*)
10 Jun	"Cow Patty"
11 Jun	"Shout", The Faber College Theme
	(*Animal House*)
12 Jun	"Twistin' the Night Away"
	(*Animal House*)
13 Jun	"Chain Gang"
14 Jun	"What a Wonderful World"

Against a background of the universe, seven silver stars, interspersed about the orbital path of Columbia, represented the seven crew members. The flight path formed a double-helix, designed to represent the DNA molecule common to all living creatures. (NASA)

This *Columbia* mission was postponed less than 48 hours before launch when engineers discovered a defective LH_2 transducer in the main propulsion system. Engineers feared if the transducer broke, pieces of it could cause a catastrophic failure of an SSME turbopump. The launch team scrubbed a launch attempt on 1 June prior to the T-20-minute hold after several attempts to calibrate an inertial measurement unit failed, so technicians replaced it. On 5 June, weather caused the launch team to hold at T-9 minutes, but weather conditions cleared and the launch proceeded smoothly. This was the first spaceflight that included three female crew members and the last flight of the original AP-101B general-purpose computers.

This was the fifth Spacelab mission and the first dedicated to life sciences. The mission featured the most detailed and interrelated physiological measurements in space since the three Skylab missions of 1973–74. A variety of experiments focused on the blood system; cardiovascular/cardiopulmonary system (heart, lungs and blood vessels); renal/endocrine system (kidneys and hormone-secreting organs and glands); immune system (white blood cells); musculoskeletal system (muscles and bones); and neurovestibular system (brains and nerves, eyes and inner ear). The experiments were conducted in the Spacelab long module and the orbiter middeck, focusing on the crew, 30 rodents, and several thousand tiny jellyfish. Of 18 experiments, ten involved humans, seven involved rodents, and one used jellyfish.

Shortly after the payload bay doors were opened, video of the aft bulkhead showed several thermal blankets were partially unfastened and a section of the aft bulkhead environmental seal was displaced on the left side of the bulkhead centerline. A team in Houston investigated the environmental seal anomaly and evaluated concerns for payload bay door closure, entry heating, and venting pressure. The results indicated a high level of confidence that normal door closure would yield a safe configuration for entry without requiring a contingency EVA. At the end of the mission, Bryan O'Conner and Sid Gutierrez oriented the orbiter nose-to-sun for 30 minutes to thermally condition the payload bay door seal prior to closing the left door. The left door closed and latched without incident, followed 30 minutes later by the right door. The deorbit burn, entry, and landing proceeded smoothly.

The post-flight runway inspection revealed thermal damage to the right ET umbilical door, with significant melting and erosion of the forward centerline latch fitting and adjacent tile. However, the internal bulb seal and thermal barrier were intact with no evidence of abnormal heating. Fortunately, there was no other damage.

In the front, from the left, are Drew Gaffney, Sid Gutierrez, Rhea Seddon, and Jim Bagian; in the back are Bryan O'Connor, Tammy Jernigan, and Millie Hughes-Fulford. Note the collection of military and college stickers and flags behind them on the middeck wall. (NASA)

The photographer disturbed Tammy Jernigan as she was eating on the middeck; the food tray for another crew member is attached to a locker behind her. Note the ever-present rolls of gray tape floating above her arm and below her food tray. (NASA)

The Spacelab pressurized module in the payload bay. Note that Columbia carried her American flag on the right side of the aft bulkhead, unlike the other orbiters that carried it on the left. (NASA)

A crawler-transporter backs away from MLP-3 after delivering the stack to LC-39B. The two crawlers were built by the Marion Power Shovel Company during the Apollo program and could carry loads weighing as much as 12.5 million pounds. (NASA)

Columbia returned to KSC on 21 June atop N905NA for down-mission processing, but was then ferried back across the country to Palmdale using N911NA on 10 August for major modifications, including the installation of the extended duration orbiter hardware. (NASA)

Columbia was easily identifiable by her black chines (upper forward part of the wings) and the shuttle infrared leeside temperature sensing (SILTS) experiment pod on top of her tail. The configuration of her thermal protection system tiles and blankets was also different. (NASA)

STS-43

Mission:	42		NSSDC ID:		1991-054A
Vehicle:	OV-104 (9)		ET-47 (LWT)		SRB: BI045
Launch:	LC-39A		02 Aug 1991		15:02 UTC
Altitude:	178 nm		Inclination:		28.46 degrees
Landing:	KSC-15		11 Aug 1991		12:24 UTC
Landing Rev:	142		Mission Duration:		213 hrs 21 mins

Commander: John E. Blaha (3)
Pilot: Michael A. "Mike" Baker (1)
MS1: Shannon W. Lucid (3)
MS2: G. David Low (2)
MS3: James C. "Jim" Adamson (2)

Payloads:	Up: 49,325 lbs	Down: 0 lbs
	TDRS-E/IUS (37,575 lbs)	
	SHARE-II (9,137 lbs)	
	No SRMS	

Notes: Delayed 23 Jul 1991 (master events controller)
Scrubbed 24 Jul 1991 (main engine controller)
Scrubbed 01 Aug 1991 (KSC weather)
First email from space (Lucid and Adamson)

Wakeup Calls:

03 Aug	"Back in the High Life"
04 Aug	"Dances With Wolves" (excerpt)
05 Aug	Medley from Rockwell-Downey employees
06 Aug	"Phantom of the Opera" (Clear Lake High School)
07 Aug	"What a Wonderful World"
08 Aug	"Cowboy in the Continental Suit"
09 Aug	"Washington and Lee" (fight song)
10 Aug	Sounds from Lucid's backyard (frogs, crickets, etc.)

The crew patch portrayed the evolution of the space program by highlighting thirty years of American manned space flight, from Mercury to space shuttle, commemorated by the emergence of Atlantis *from the outline of a Mercury capsule. (NASA)*

NASA originally scheduled this *Atlantis* launch for 23 July 1991, but moved it to the 24th to allow time to replace a faulty master events controller that commanded external tank separation. NASA scrubbed that attempt about five hours before liftoff because of a faulty main engine controller. Mission management rescheduled the launch for 1 August. Engineers extended the T-9-minute hold while technicians cycled the cabin pressure vent valve several times in an unsuccessful attempt to reset a bad position indication. The launch team concluded it was a faulty sensor and cleared the vehicle, but by then the weather was unacceptable. The count on 2 August proceeded smoothly.

The crew successfully deployed the fourth Tracking and Data Relay Satellite (TDRS-E) approximately six hours after launch and John Blaha and Mike Baker moved *Atlantis* safely away before the inertial upper stage ignited. The IUS first stage fired a little less than an hour after deployment and the second stage fired 11 hours later. The satellite separated from the IUS about 45 minutes after the last firing and drifted to its final position at 175 degrees west longitude. After on-orbit checkout, TDRS-5 became the fourth member of the constellation and was designated TDRS-West on 7 October 1991.

Just after the deployment of TDRS-E, the crew noted an object moving away from the aft end of the orbiter. The crews of STS-35 and STS-41 had noted similar objects. Engineers in Houston reviewed the imagery and concluded the object was solid oxygen that dislodged from one of the main engine nozzles. Data showed the main propulsion system contained 3,869 pounds of LO_2 at ET separation with 3,400 pounds of this vented through the engine nozzles. Some of the oxygen solidified on the nozzle.

The mission was notable for sending the first email from space. On 9 August, Shannon Lucid and Jim Adamson used AppleLink to send an email from a Macintosh Portable on the flight deck of *Atlantis* to CAPCOM Marsha Ivins at the Mission Control Center in Houston.

Landing at KSC occurred on the first opportunity. After landing, technicians found a piece of metal, which was the ET umbilical stud yoke, on the runway below the LO2 umbilical plate area. During the post-landing check, an audible leak was found in the main propulsion system at the liquid hydrogen 4-inch disconnect. A visual inspection of the disconnect showed a piece of the valve flapper seal had come loose and lodged in the flapper. In addition, the left inboard main landing-gear tire was worn through two cords on the inboard side of the tire and the wear was evenly spread around its circumference.

A composite photo showing the crew in front of what is likely a model of an orbiter superimposed over a striking sunrise. From the left are Shannon Lucid, Jim Adamson, John Blaha, David Low, and Mike Baker. All of the crew members are in their launch-entry suits. (NASA)

The crew deployed TDRS-E and its attached inertial upper stage about six hours after launch. After a few weeks of moving to geosynchronous orbit and on-orbit testing, the satellite became TDRS-5 and replaced TDRS-3 at 174 degrees west longitude. (NASA)

TDRS-E showing the business end of its solid-propellant inertial upper stage that would carry it to geosynchronous orbit. (NASA)

The black box is the space shuttle main engine controller that scrubbed the launch attempt on 24 July 1991. Large and bulky, the controller nonetheless proved generally reliable. (NASA)

In the white room at LC-39A, technicians make a final check of the launch-entry suit before Mike Baker climbs into Atlantis. Baker will don his gloves and helmet in the ship. (NASA)

TDRS-E as it was shipped to KSC. The satellite folded into a remarkably small package. The red panels with TRW written on them are protective covers for the solar arrays. (NASA)

STS-48

Mission:	43	NSSDC ID:	1991-063A
Vehicle:	OV-103 (13)	ET-42 (LWT)	SRB: BI046
Launch:	LC-39A	12 Sep 1991	23:11 UTC
Altitude:	308 nm	Inclination:	57.00 degrees
Landing:	EDW-22	18 Sep 1991	07:40 UTC
Landing Rev:	81	Mission Duration:	128 hrs 28 mins

Commander:	John O. Creighton (3)
Pilot:	Kenneth S. "Ken" Reightler, Jr. (1)
MS1:	Charles D. "Sam" Gemar (2)
MS2:	James F. "Jim" Buchli (4)
MS3:	Mark N. Brown (2)

Payloads:	Up: 21,569 lbs	Down: 0 lbs
	UARS (14,388 lbs)	
	SRMS s/n 301	

Notes: First N911A operational ferry flight
Orbiter returned to KSC on 26 Sep 1991 (N911NA)

Wakeup Calls:

13 Sep	"Hound Dog"
14 Sep	"Release Me" (for UARS)
15 Sep	"Bare Necessities" (*Jungle Book*)
16 Sep	"Are You Lonesome Tonight"
17 Sep	"Return to Sender"

The triangular shape of the crew patch represents the three atmospheric processes that determine upper atmospheric structure and behavior: chemistry, dynamics, and energy. The stars are those in the northern hemisphere as seen when UARS began its atmospheric study. (NASA)

During this *Discovery* countdown on 12 September 1991, air-to-ground radio noise between T-9 minutes and T-5 minutes led the launch director to hold the countdown at T-5 minutes to determine the cause. Engineers could not isolate the source of the noise, but since it appeared to be a ground problem, the launch team resumed the countdown. The noise disappeared just prior to launch and analysis after liftoff confirmed the problem originated in the ground equipment.

The crew successfully deployed the Upper Atmosphere Research Satellite (UARS) on flight day 3. During its planned 18-month mission the observatory would provide an increased understanding of the energy input into the upper atmosphere, global photochemistry of the upper atmosphere, dynamics of the upper atmosphere, the coupling among these processes, and the relationship between the upper and lower atmosphere. The UARS was the first major flight element of the NASA Mission to Planet Earth, a multi-year global research program that would use ground-based, airborne, and space-based instruments to study the Earth as a complete environmental system. Ultimately, the satellite greatly exceeded its planned mission duration and was finally retired after 14 years of service.

This was the first mission to test an electronic still (digital) camera in space. The modified Nikon F4 had a resolution of 1024x1024 pixels. The monochrome images contained 8 bits of digital information per pixel (256 levels of gray) and were stored on a removable hard disk. The crew could manipulate the images using a laptop computer before transmitting them to the ground via the orbiter downlink. Initially, the crew ran into a cable problem when they attempted to connect the camera to the laptop, but engineers on the ground talked the astronauts through the issue.

An on-orbit video from 15 September showed a flash of light and several objects appearing to fly in a controlled fashion; NASA explained the objects as small ice particles reacting to thruster firings.

The crew used a seven-second reaction control system firing to perform an evasive maneuver to avoid the spent upper stage from Cosmos 955. This was a Soviet ELINT satellite launched from the Plesetsk Cosmodrome on 20 September 1977 by a Vostok-2M (modified SS-6 Sapwood) launch vehicle. It ultimately reentered the atmosphere on 7 September 2000.

This was the second post-Challenger mission to have KSC as the planned end-of-mission landing site and the first to have a planned night landing at the Shuttle Landing Facility. An evaluation of weather conditions at KSC resulted in mission control delaying the landing one revolution, and ultimately in a change to Edwards.

The on-orbit crew portrait on the middeck of Discovery. From the left are Ken Reightler, Mark Brown, John Creighton, Sam Gemar, and Jim Buchli. Note the model of a space station truss, in the background, used in the MODE experiment. (NASA)

Jim Buchli poses with the structural test article (STA), a model of the space station truss structure, that was part of the middeck zero gravity dynamics experiment (MODE). This experiment studied the vibration characteristics of the jointed truss structure. The STA included four strain gages and eleven accelerometers. (NASA)

The SRMS has just released the Upper Atmosphere Research Satellite. The arm included an end effector that captured or released an object using a rotating ring to close three wire-snares around the payload-mounted grapple fixture. The grapple fixture included a 10-inch grapple pin and three alignment cams. (NASA)

The STS-48 stack making its way to LC-39A. The Upper Atmosphere Research Satellite was already in the payload changeout room at the pad, waiting to be installed into Discovery. The vertical installation at the launch pad was typical for deployable payloads. (NASA)

UARS on the SRMS just prior to deployment. Ultimately, the satellite greatly exceeded its planned eighteen-month mission duration and was finally retired after fourteen years of service. This was the first major flight element of the NASA Mission to Planet Earth. (NASA)

STS-44

Mission:	44	NSSDC ID:	1991-080A
Vehicle:	OV-104 (10)	ET-53 (LWT)	SRB: BI047
Launch:	LC-39A	24 Nov 1991	23:44 UTC
Altitude:	212 nm	Inclination:	28.45 degrees
Landing:	EDW-05R	01 Dec 1991	22:36 UTC
Landing Rev:	110	Mission Duration:	166 hrs 51 mins

Commander:	Frederick D. "Fred" Gregory (3)
Pilot:	Terence T. "Tom" Henricks (1)
MS1:	James S. "Jim" Voss (1)
MS2:	F. Story Musgrave (4)
MS3:	Mario "Trooper" Runco, Jr. (1)
PS1:	Thomas J. "Tom" Hennen (1)

Payloads:	Up: 47,235 lbs	Down: 0 lbs
	USA-75/IUS (DSP F16) (37,588 lbs)	
	No SRMS	

Notes: Unclassified Department of Defense mission
Delayed 19 Nov 1991 (IMU failure in IUS)
Backup payload specialist was
   Michael E. Belt
Orbiter returned to KSC on 08 Dec 1991 (N911NA)

Wakeup Calls:

25 Nov	Space, The Final Frontier (parody)
26 Nov	"Reveille" / "This is the Army, Mr. Jones"
27 Nov	"It's Time to Love (Put a Little Love in Your Heart)"
28 Nov	"Cheeseburger in Paradise"
29 Nov	"Twist and Shout" (*Ferris Bueller's Day Off*)
30 Nov	University of Alabama and Auburn fight songs
01 Dec	"In the Mood"

The black background of space, indicative of the mysteries of the universe, was illuminated by six large stars depicting the six crew members. The smaller stars represented all of the Americans who worked in support of the mission. (NASA)

N ASA had originally scheduled this *Atlantis* launch for 19 November 1991, but postponed it prior to tanking because of a faulty inertial measurement unit in the inertial upper stage. The launch was rescheduled for 24 November.

The primary payload on this unclassified DoD mission was the Defense Support Program F16 (Liberty) early warning satellite. The crew successfully deployed the spacecraft 7.5 hours after launch and then performed a 16.4-second two-engine OMS separation burn. The IUS then fired twice on its way to geosynchronous orbit. The DSP satellites operated by the Air Force Space Command could detect rocket launches and nuclear explosions using infrared emissions from these intense heat sources.

Secondary payloads included AMOS (Air Force Maui Optical Site calibration), CREAM (cosmic radiation effects and activation monitor), IOCM (interim operational contamination monitor), M88-1 (military man in space), RME-III (radiation monitoring experiment), SAMS (Space Acceleration Measurement System), UVPI (ultraviolet plume imager), VFT-1 (visual function tester), and Terra Scout. The objective of the Terra Scout experiment was to evaluate the ability of a specially trained crew member (Tom Hennen) to detect specific ground targets using a variety of visual aids, such as the spaceborne direct view optical system (SPADVOS). During the flight, Hennen attempted 29 observations and acquired 27 of them.

On flight day 4, Fred Gregory and Tom Henricks made a seven-second retrograde reaction control system maneuver to avoid space debris (Cosmos 851) that would pass near the orbiter during the crew sleep period. Cosmos 851 was a former Soviet electronics and signals intelligence satellite that had been launched on a Vostok-2M (modified SS-6 Sapwood) booster from the Plesetsk Cosmodrome on 27 August 1976. It ultimately reentered the atmosphere on 5 August 1989.

One of the three inertial measurement units on *Atlantis* failed on flight day 5. This failure invoked the flight rule requiring a minimum duration flight for loss of one IMU, so mission management shortened the ten-day mission by three days. The IMU remained powered up to allow the flight controllers to monitor it for the remainder of the mission, but was not used for onboard navigation.

The following day, the crew closed the payload bay doors and prepared for the deorbit burn. This mission was scheduled to land at KSC, but was diverted to Edwards as a safety precaution because of the IMU failure. Fred Gregory and Tom Henricks did not apply the brakes until the orbiter had slowed to 15 knots to accommodate a special lakebed runway bearing strength assessment, resulting in a long rollout of 11,191 feet.

In the front row, from the left, are Fred Gregory, Tom Hennen, and Jim Voss. Behind them are Mario Runco, Story Musgrave, and Tom Henricks. Hennen earned the nickname "Trash Man" during the flight after making a video about waste disposal. (NASA)

With the crew having a treadmill-like device onboard for exercise and biomedical testing, tennis shoes were in plentiful stock on the eight-day mission. Although not part of the original plan, the six crew members stored their shoes on the airlock hatch. (NASA)

Tom Henricks (foreground) "rows" on the hastily modified treadmill after the device experienced an anomaly that affected its ability to support the crew in "running" mode. Mario Runco is in the background awaiting his turn to exercise. (NASA)

Tom Henricks next to a water tank under the middeck floor after he installed a collection bag around the outlet of one of the humidity separators. This was a precautionary action after an anomaly within the environmental control and life support system. (NASA)

Atlantis in the Mate/Demate Device at NASA Dryden on 7 December 1991 being mated to N911NA for the return trip to Florida. The High Desert was a harsh environment but often presented some gorgeous sunsets (and sunrises). (NASA)

If only the DoD had been this cooperative on their other dedicated missions. Here is a gorgeous 70 mm photograph of the 33-foot-long, 14-foot-diameter Defense Support Program F16 Liberty spacecraft just prior to deployment. (NASA)

STS-42

Mission:	45	NSSDC ID:		1992-002A
Vehicle:	OV-103 (14)	ET-52 (LWT)		SRB: BI-048
Launch:	LC-39A	22 Jan 1992		14:53 UTC
Altitude:	162 nm	Inclination:		57.00 degrees
Landing:	EDW-22	30 Jan 1992		16:08 UTC
Landing Rev:	129	Mission Duration:		193 hrs 15 mins

Commander:	Ronald J. "Ron" Grabe (3)
Pilot:	Stephen S. "Oz" Oswald (1)
MS1:	Norman E. "Norm" Thagard (4)
MS2:	William F. "Reads" Readdy (1)
MS3:	David C. "Dave" Hilmers (4)
PS1:	Roberta L. Bondar (1)
PS2:	Ulf D. Merbold (2)

Payloads:	Up: 32,364 lbs	Down: 0 lbs
	Spacelab IML-1 (32,364 lbs)	
	GAS (10)	
	No SRMS	

Notes: Backup payload specialists were
                 Roger K. Crouch and Kenneth E. Money
             Orbiter returned to KSC on 16 Feb 1992 (N905NA)

Wakeup Calls:

    Since crew worked two shifts around the clock,
    mission control did not send any wakeup calls.

The four stars in the lower blue field and two stars in the upper blue field reflect the STS designation. The single gold star above the horizon on the right is in honor of Manley L. "Sonny" Carter, Jr. Note the Canadian and European Space Agency symbols in the border. (NASA)

N ASA delayed the 22 January 1992 launch of *Discovery* an hour due to a fuel cell anomaly and cloud cover at the launch site. Manley "Sonny" Carter was scheduled to fly as a mission specialist on this flight, but was killed in the crash of Atlantic Southeast Airlines Flight 2311 in Brunswick, Georgia, some seven months prior to launch. Dave Hilmers replaced him and the single gold star above the horizon on the right of the crew patch was in honor of Carter. The crew included the first Canadian female astronaut (Roberta Bondar) and the first West German astronaut (Ulf Merbold).

The hand-held ET separation imagery showed two possible divots in the intertank region that analysts estimated were about 14 inches in diameter, although the information was of only casual interest until after the *Columbia* accident 11 years in the future.

The primary payload was the International Microgravity Laboratory (IML-1), a pressurized Spacelab module, to explore the effects of weightlessness on living organisms and materials processing. The crew divided into two teams to allow around-the-clock monitoring of experiments. The red team consisted of Bill Readdy, Dave Hilmers, and Ulf Merbold, while the blue team included Ron Grabe, Steve Oswald, Norm Thagard, and Roberta Bondar. They began activating the Spacelab module about two and half hours after launch and entered the laboratory an hour later.

Secondary payloads included GOSAMR (gelation of sols applied microgravity research), IPMP (investigations into polymer membrane processing), RME-III (radiation monitoring experiment), student experiment 81-09 (convection in zero gravity), student experiment 83-02 (capillary rise of liquid through granular porous media), and ten GAS cans carried on a get-away special beam assembly.

An anomaly occurred in the waste collection system (WCS) when the crew was attempting to reconfigure the potty for commode use. The handle became disconnected from the control valve linkage and, as a result, they were unable to operate the commode. Engineers on the ground developed a repair procedure the crew, much to their relief, used to restore the potty to full operations.

With everything progressing smoothly and consumable usage being lower than planned, the mission management team extended the mission for one day and, as a result, the crew completed more than 100 percent of the planned science activities. The crew downlinked more than 100 hours of television and recorded 70 videotapes. The mission returned more than 100 crystals, billions of cells, and hundreds of plants. The entry and landing at Edwards on 30 January were uneventful.

Another photoshopped crew portrait (there is no location where this view is possible). From the left are Steve Oswald, Roberta Bondar, Norm Thagard, Ron Grabe, Dave Hilmers, Ulf Merbold, and Bill Readdy. All are in their launch-entry suits minus the gloves and helmets. (NASA)

Dave Hilmers, wearing a blue version of the launch-entry suit, floats in the water in the Weightless Environment Training Facility at JSC. His yellow and orange single-person life raft is behind him. This was part of the survival training in case of an emergency bailout over water. (NASA)

The Spacelab pressurized module in the payload bay of Discovery while still in the Orbiter Processing Facility. Note the manipulator positioning mechanism (MPM) pedestal on the longeron, despite the fact that the SRMS arm was not carried on this mission. (NASA)

A beautiful launch off LC-39A reflected in one of the tidal lagoons around the pad. Excepting the first two white tanks, the ETs were not painted. The various foams used to cover the tanks changed color, sometimes drastically, depending on age and how much sunlight that tank had been exposed to since it was manufactured. As sprayed at the Michoud Assembly Facility, the foam was a pale yellow; it turned increasing dark orange as time went on. So, essentially, there was no "correct" color for an external tank. (NASA)

STS-45

Mission:	46	NSSDC ID:		1992-015A
Vehicle:	OV-104 (11)	ET-44 (LWT)		SRB: BI049
Launch:	LC-39A	24 Mar 1992		13:14 UTC
Altitude:	160 nm	Inclination:		57.00 degrees
Landing:	KSC-33	02 Apr 1992		11:24 UTC
Landing Rev:	143	Mission Duration:		214 hrs 09 mins

Commander: Charles F. "Charlie" Bolden, Jr. (3)
Pilot: Brian Duffy (1)
MS1: Kathryn D. 'Kathy' Sullivan (3)
MS2: David C. "Dave" Leestma (3)
MS3: C. Michael "Mike" Foale (1)
PS1: Dirk D. Frimout (1)
PS2: Byron K. Lichtenberg (2)

Payloads: Up: 20,341 lbs   Down: 0 lbs
ATLAS-1 (20,341 lbs)
GAS (1)
No SRMS

Notes: Delayed 23 Mar 1992 (high H_2 concentrations)
Backup payload specialists were
    Michael L. Lampton and Charles R. Chappell

Wakeup Calls:

Since crew worked two shifts around the clock, mission control did not send any wakeup calls.

The colors of the setting Sun, measured by the ATLAS instruments, provide detailed information about ozone, carbon dioxide and other gases. The additional star in the ring is to recognize the alternate payload specialists and the entire ATLAS-1 team. (NASA)

This launch of *Atlantis* was originally scheduled for 23 March 1992, but was delayed one day because of high concentrations of gaseous oxygen and gaseous hydrogen in the orbiter aft compartment during tanking. Technicians could not recreate the leak during troubleshooting, leading engineers to theorize it was the result of the plumbing not being properly conditioned for the cryogenic propellants. No repairs were deemed necessary and NASA rescheduled the launch for the following day. Again, the oxygen concentration rose but quickly recovered to within acceptable limits. Engineers had anticipated this behavior and had already agreed to proceed. This attempt was briefly delayed by due to weather at the launch site.

Atlantis carried the Atmospheric Laboratory for Applications and Science (ATLAS-1) on two Spacelab pallets in the payload bay and the SSBUV (shuttle solar backscatter ultraviolet) experiment on the starboard payload bay wall. This was the first of ten ATLAS missions that NASA wanted to fly during the 11-year solar cycle where solar flares, sunspots, and magnetic activity vary from intense activity to relative calm. Regardless, there were only two additional ATLAS missions (STS-56 and STS-66) during the flight campaign.

The ATLAS payload, which was non-deployable, consisted of 12 instruments from America, Belgium, France, Germany, Japan, Switzerland, and The Netherlands. It supported 14 experiments in atmospheric chemistry, solar radiation, space plasma physics, and ultraviolet astronomy. The instruments included ACR (active cavity radiometer), AEPI (atmospheric emissions photometric imager), ALAE (atmospheric lyman-alpha emissions), ATMOS (atmospheric trace molecule spectroscopy), ENAP (energetic neutral atom precipitation), GRILLE (grille spectrometer), ISO (imaging spectrometric observatory), MAS (millimeter wave atmospheric sounder), SEPAC (space experiments with particle accelerators), SOLCON (measurement of solar constant), SOLSPEC (measurement of solar spectrum), SUSIM (solar ultraviolet spectral irradiance monitor).

The crew split into a red team consisting of Dave Leestma, Mike Foale, and Byron Lichtenberg and a blue team that included Charlie Bolden, Brian Duffy, Kathy Sullivan, and Dirk Frimout. The brief launch delay shifted the shadows approximately 1.5 degrees and the experiment times were adjusted accordingly. On the morning of flight day 6, the mission management team approved an additional day to collect more science data.

Atlantis returned to the Shuttle Landing Facility on the first landing opportunity.

On the flight deck of Atlantis. In the front row, from the left, are Kathy Sullivan and Charlie Bolden. Behind them are Dave Leestma, Brian Duffy, Byron Lichtenberg, Dirk Frimout, and Mike Foale. The "headpieces" worn by Sullivan and Bolden are actually shadows. (NASA)

Ever wonder what the little blue pieces of Velcro were meant to hold? The simple answer was everything. Here are the middeck lockers of Atlantis during the mission, with a collection of video cassettes, film canisters, and documentation affixed to the velcro. (NASA)

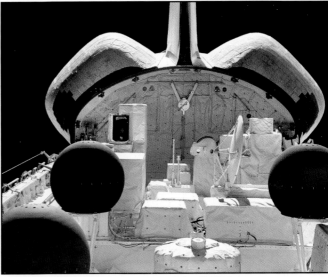

The three black spheres are part of the SEPAC (space experiments with particle accelerators) experiment. Note the EVA winch on the aft bulkhead to help a crew member close the payload bay doors during a contingency EVA if needed; thankfully, it never was. (NASA)

The ATLAS payload was mounted on two Spacelab pallets with an igloo support structure at the front (top in this photo). The forward bulkhead is at the top of the photo, with the airlock hatch cover clearly visible. Note the single GAS canister in front of the top right SEPAC sphere. (NASA)

Freedom Star tows the left solid rocket booster into Port Canaveral on her way to Hangar AF on Cape Canaveral Air Force Station. Freedom always retrieved the left booster while her sister ship Liberty Star retrieved the right. Freedom towed the booster (when "hipped") on the port side while Liberty hipped to starboard. (NASA)

A scenic view of the Atlantis stack on LC-39A. The 290-foot-high tower on the right held 300,000 gallons of water that was released just prior to main engine ignition to reduce the acoustic energy on the launch pad, preventing damage to the orbiter thermal protection system. (NASA)

STS-49

Mission:	47	NSSDC ID:	1992-026A
Vehicle:	OV-105 (1)	ET-43 (LWT)	SRB: BI050
Launch:	LC-39B	07 May 1992	23:40 UTC
Altitude:	198 nm	Inclination:	28.32 degrees
Landing:	EDW-22	16 May 1992	20:59 UTC
Landing Rev:	141	Mission Duration:	213 hrs 18 mins

Commander:	Daniel C. "Dan" Brandenstein (4)
Pilot:	Kevin P. "Chilli" Chilton (1)
MS1:	Richard J. "Rick" Hieb (2)
MS2:	Bruce E. Melnick (2)
MS3:	Pierre J. Thuot (2)
MS4:	Kathryn C. "Kathy" Thornton (2)
MS5:	Thomas D. "Tom" Akers (2)

Payloads:	Up: 37,444 lbs	Down: 0 lbs
	Orbus-21S Upper Stage (23,346 lbs)	
	ASEM (9,463 lbs)	
	SRMS s/n 303	

Notes:
Flight readiness firing on 06 Apr 1992
Delayed 04 May 1992 (photography requirements)
First flight of *Endeavour*
Retrieved and redeployed Intelsat VI F3
Only three-person EVA of the flight campaign
EVA1 10 May 1992 (Thuot and Hieb)
EVA2 11 May 1992 (Thuot and Hieb)
EVA3 13 May 1992 (Thuot, Hieb, and Akers)
EVA4 14 May 1992 (Thornton and Akers)
First use of drag chute during landing
Orbiter returned to KSC on 30 May 1992 (N911NA)

Wakeup Calls:

08 May	"God Bless the U.S.A."
09 May	"Rescue Me"
10 May	"Theme from Winnie the Pooh"
11 May	"Theme from Rocky"
12 May	"Kokomo"
13 May	(none)
14 May	"I Wake Up Every Morning With a Smile on My Face"
15 May	"Son of a Son of a Sailor"

The crew patch captures the spirit of exploration that has its origins in the early seagoing vessels that explored the uncharted reaches of the oceans. The ship depicted on the patch is HMS Endeavour, which James Cook commanded on his first Pacific expedition. (NASA)

In preparation for the maiden flight of *Endeavour*, NASA conducted a 22-second flight readiness firing (FRF) at LC-39B on 6 April 1992. Although the FRF successfully verified the overall operation of the orbiter and vehicle subsystems, engineers identified anomalies with all three SSMEs and replaced them on the pad. Initially, NASA scheduled the launch for 4 May, but moved it to 7 May to allow more daylight for documentary photography.

An ambitious plan called for this mission to rescue the Intelsat VI F3 satellite stranded in an unusable orbit since its launch by a Commercial Titan III on 14 March 1990. The attached Orbus-21S failed to separate from the satellite, preventing the upper stage from boosting it to geostationary orbit.

During the first EVA, Pierre Thuot and Rick Hieb were unable to attach a capture bar to the satellite from a position on the SRMS, resulting in the satellite being pushed away and wobbling. A second similar EVA the following day also failed. Finally, Thuot, Hieb, and Tom Akers made an unscheduled but ultimately successful hand capture as Dan Brandenstein maneuvered the orbiter to within a few feet of the 9,000-pound satellite. This was the first three-person EVA of the American space program and, somehow, three suited astronauts managed to fit in the internal airlock at the same time. The EVA team attached the capture bar and then began the previously planned satellite berthing activities. The astronauts attached a new Orbus-21S upper stage that later boosted Intelsat VI to its proper orbit. Because of the extra EVAs needed to capture Intelsat VI, the mission management team extended the mission two days.

A fourth EVA allowed Kathy Thornton and Tom Akers to practice assembly of station by EVA methods (ASEM) for the upcoming Space Station Freedom. Originally, this had been scheduled across two EVAs, but mission management shortened the activity due to the extra Intelsat VI efforts. The mission ultimately included 50 hours 54 minutes of EVA. The third spacewalk was the second longest of the flight campaign at 8 hours 29 minutes. The fourth EVA was the one-hundredth spacewalk in history.

Upon landing at Edwards, *Endeavour* tested the new drag chute, conservatively deploying it after nose-gear touchdown. Imagery showed the reefed parachute rode at a higher angle than expected, although it was positioned as expected after it disreefed.

In a harbinger of things to come, during the post-flight inspection, engineers noted a large debris strike on the right side of the vehicle immediately aft of the nose cap. The size and depth of this damage site was "indicative of an impact by a low-density material such as ET TPS foam."

The crew, still in their launch-entry suits, posing on the lakebed at Edwards AFB after the maiden flight of Endeavour. From the left are Rick Hieb, Kevin Chilton, Dan Brandenstein, Tom Akers, Pierre Thuot, Kathy Thornton, and Bruce Melnick. (NASA)

Endeavour sitting on LC-39B awaiting her first launch. As happened with each new orbiter, engineers conducted a flight readiness firing, this time on 6 April 1992, to verify all of the systems were working as expected. Note the NASA "worm" logo on the right wing. (NASA)

The crew was unable to capture the Intelsat VI F3 during the first two EVAs. Finally, Pierre Thuot, Rick Hieb, and Tom Akers managed to wrestle the satellite into the payload bay during the third EVA. This was the only three-person EVA from space shuttle. (NASA)

This landing marked the first use of the drag chute added in response to recommendations from the Rogers Commission investigation of the Challenger accident. The chute was a production feature on Endeavour and was retrofitted to the remaining orbiters. (NASA)

Kathy Thornton and Tom Akers are over the open storage unit that contained the ASEM (assembly of station by EVA methods) experiment. This was the 100th extravehicular activity (American and Russian) of the Space Age. Note Intelsat VI F3 in the background. (NASA)

STS-50

Mission:	48	NSSDC ID:	1992-034A
Vehicle:	OV-102 (12)	ET-50 (LWT)	SRB: BI051
Launch: Altitude:	LC-39A 163 nm	25 Jun 1992 Inclination:	16:12 UTC 28.46 degrees
Landing: Landing Rev:	KSC-33 221	09 Jul 1992 Mission Duration:	11:43 UTC 331 hrs 30 mins

Commander:	Richard N. "Dick" Richards (3)
Pilot:	Kenneth D. "Sox" Bowersox (1)
MS1:	Bonnie J. Dunbar (3)
MS2:	Ellen S. Baker (2)
MS3:	Carl J. Meade (2)
PS1:	Lawrence J. "Larry" DeLucas (1)
PS2:	Eugene H. "Gene" Trinh (1)

Payloads:	Up: 32,447 lbs	Down: 0 lbs

Spacelab USML-1 (24,305 lbs)
First flight of extended duration orbiter pallet
No SRMS

Notes: First flight of the EDO pallet
First KSC landing for *Columbia*
Backup payload specialists were
Joseph M. Prahl and Albert Sacco, Jr.

Wakeup Calls:

Since crew worked two shifts around the clock,
mission control did not send any wakeup calls.

The crew patch shows the space shuttle in the typical flying attitude for microgravity experiments. The spacelab module in the payload bay has the "µg" symbol for microgravity. The stars and stripes on the USML banner depict an all-American mission. (NASA)

The launch team delayed this *Columbia* countdown to wait for clouds around KSC to disperse. When this failed to happen, the launch director decided to proceed to the T-5-minute hold and wait for acceptable weather; however, the weather cleared at approximately T-7 minutes and the countdown continued.

It had been nine years since the first external tank bipod ramp debris incident on STS-7, but it happened again on STS-50 when the majority of the left ramp separated from ET-50, resulting in a piece of debris measuring 16x10 inches and weighing 0.98 pound. The only significant damage to *Columbia* was a 9x4-inch gash on one wing that was 0.5 inch deep; she would not be so lucky a decade later.

Experiments on the United States Microgravity Laboratory (USML-1) included CGF (crystal growth furnace), DPM (drop physics module), EDOMP (extended duration orbiter medical project), GBA (generic bioprocessing apparatus), GBX (glovebox facility), SAMS (space acceleration measurement system), SSCE (solid surface combustion experiment), and STDCE (surface tension driven convection experiment).

The science investigations fell into five basic areas of microgravity research: fluid dynamics (the study of how liquids and gases respond to the application or absence of differing forces), materials science (the study of materials solidification and crystal growth), combustion science (the study of the processes and phenomena of burning), biotechnology (the study of phenomena related to products derived from living organisms), and technology demonstrations that sought to prove experimental concepts for use in future space shuttle missions and on Space Station Freedom.

The crew split into red (Ken Bowersox, Larry DeLucas, Carl Meade, and Bonnie Dunbar) and blue (Dick Richards, Ellen Baker, and Gene Trinh) teams to operate around the clock. While most of the experiments were conducted in Spacelab, others operated on the middeck, including ASC-1 (astroculture), PCG (protein crystal growth), and ZCG (zeolite crystal growth). This mission marked the maiden flight of the extended duration orbiter pallet in the payload bay and regenerative carbon dioxide removal system (RCRS) in the crew module. The RCRS used regenerable solid amines to adsorb carbon dioxide and water from the crew module instead of the usual non-regenerable lithium hydroxide (LiOH) canisters.

Mission management added one day to allow the weather at Edwards to clear. When the weather did not cooperate, *Columbia* diverted to KSC and was the second orbiter to use a drag chute. Interestingly, this was the first time *Columbia* had landed at the Shuttle Landing Facility.

This on-orbit crew portrait was taken in the Spacelab pressurized module. Floating at the top is Ken Bowersox. In the blue shirts are, from the left, Carl Meade, Ellen Baker, and Gene Trinh. In the red shirts are Larry DeLucas, Dick Richards, and Bonnie Dunbar. (NASA)

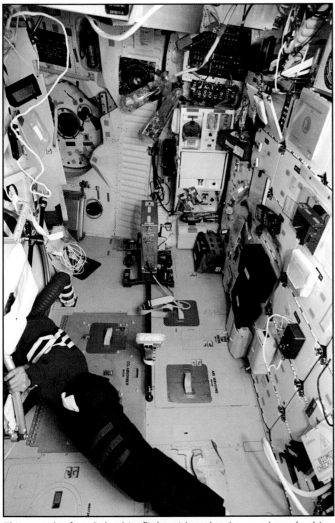

This was the first Columbia flight with a shuttle repackaged orbiter galley (SORG), shown at upper right. A SORG had previously flown on Endeavour for STS-49. Note the lack of a middeck accommodations rack (MAR) between the galley and hatch (at upper left). (NASA)

Artistic shot of Columbia going uphill on 25 June 1992. This was her 12th flight and the first launch after her extended duration orbiter modifications (called the J1 mod, although NASA did not technically consider it an orbiter maintenance and down period, OMDP). (NASA)

The extended duration orbiter (EDO) pallet being lowered into the payload bay. The 3,500-pound, 15-foot diameter wafer pallet carried four sets of cryogenic tanks with more propellant for the fuel cells, allowing the orbiter to fly longer missions. (NASA)

Columbia became the second orbiter to use a drag chute when she landed at the Shuttle Landing Facility at KSC. Interestingly, and a little surprisingly, this was the first time Columbia had landed in Florida during the eleven years she had been flying. (NASA)

STS-46

Mission:	49		NSSDC ID:		1992-049A
Vehicle:	OV-104 (12)		ET-48 (LWT)		SRB: BI052
Launch:	LC-39B		31 Jul 1992		13:57 UTC
Altitude:	231 nm		Inclination:		28.46 degrees
Landing:	KSC-33		08 Aug 1992		13:13 UTC
Landing Rev:	127		Mission Duration:		191 hrs 15 mins

Commander: Loren J. Shriver (3)
Pilot: Andrew M. "Andy" Allen (1)
MS1: Claude Nicollier (1)
MS2: Marsha S. Ivins (2)
MS3: Jeffrey A. "Jeff" Hoffman (3)
MS4: Franklin R. Chang-Diaz (3)
PS1: Franco E. Malerba (1)

Payloads: Up: 35,546 lbs   Down: 0 lbs
TSS-1 (18,594 lbs)
EURECA (9,901 lbs)
SRMS s/n 201

Notes: The 150th manned spaceflight to achieve orbit
Backup payload specialist was
   Umberto Guidoni

Wakeup Calls:

Since crew worked two shifts around the clock,
mission control did not send any wakeup calls.

The crew patch highlighted the two primary payloads: the Tethered Satellite System (TSS-1) and the European Retrievable Carrier (EURECA). The purple beam emanating from an electron generator in the payload bay spirals around Earth's magnetic field. (NASA)

Atlantis was launched 48 seconds late because the crew fell behind configuring switches on the flight deck, but this 31 July 1992 countdown was nevertheless regarded as the "cleanest" since the return-to-flight after the *Challenger* accident.

One of the primary objectives of the mission was to deploy the European Space Agency EURECA (European Retrievable Carrier) spacecraft. This was the largest satellite produced in Europe at the time and carried 15 major experiments, mostly in microgravity sciences. The crew unberthed the SRMS about six hours after launch and used it to support EURECA checkout and overnight park. A number of intermittent data problems occurred during the checkout and mission management waved-off the first opportunity to deploy the satellite. However, ground stations successfully communicated with EURECA while it was still attached to the SRMS, leading engineers to believe there was a compatibility problem with the orbiter systems. Therefore, the crew deployed EURECA on the first opportunity after the 24-hour delay. After deployment, EURECA used its thrusters to boost it into a 269-nm orbit. The satellite was ultimately recovered by STS-57 in June 1993.

Because of the 24-hour delay in the EURECA deployment, the mission management team extended the flight one day so all of the other mission objectives could be accomplished. Following the EURECA burn, Loren Shriver and Andy Allen moved to a 160-nm circular orbit in preparation the other primary objective, operating the Tethered Satellite System (TSS-1). This Italian satellite would be deployed on a 13-mile-long tether connected to the orbiter and researchers expected the effect of the tether passing through the far reaches of the atmosphere to generate 5,000 volts of electricity.

After some issues, the crew unreeled the satellite 587 feet before it stopped. Because of possible burned windings on the reel, the crew reeled the satellite in approximately 16 feet and then reeled out at a somewhat higher rate to 827 feet where the satellite again stopped. After conferring with engineers on the ground, the crew resumed deployment about 90 minutes later, but the satellite stalled immediately at 830 feet. The following day, the crew again reeled in the satellite to 733 feet where the tether became jammed and would not move in either direction. Engineers devised a workaround and the crew retrieved the satellite. Despite the problems, the satellite still managed to generate approximately 40 volts of power. NASA and the Italian Space Agency (ASI) would try again on STS-75.

After the TSS-1 satellite was safely stowed in the payload bay, Loren Shriver and Andy Allen lowered the orbit to conduct another experiment and then returned to the Shuttle Landing Facility at KSC.

Given the miles of tether for TSS-1, what better prop than silly string. This is one of several crew gag photos. From the left are Claude Nicollier, Franco Malerba, Masha Ivins, Jeff Hoffman, Franking Chang-Diaz, Andy Allen, and Loren Shriver. (Courtesy of Andy Allen)

The STS-46 payloads in the payload canister before being transfered into the payload bay. From the top are the Tethered Satellite System (TSS), European Retrievable Carrier (EURECA), and the EOIM-III (evaluation of oxygen interaction with materials) and TEMP2A-3 (thermal energy management) experiment packages. (NASA)

The Italian TSS-1 experiment did not go to plan. During deployment, the satellite reached a maximum distance of only 830 feet instead of the planned thirteen miles because of a jammed tether. Despite numerous attempts over several days to free the tether, the crew eventually gave up. NASA and the Italians would try again on STS-75. (NASA)

Atlantis heading from the Orbiter Processing Facility (OPF-1) to the Vehicle Assembly Building. Many KSC employees would often come out for roll-over, lining the short tow-way (which doubled as a normal road) or the nearby parking lots. (NASA)

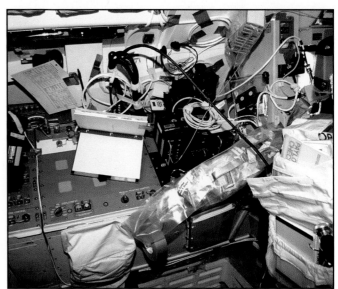

This area on the left side of the aft flight deck was referred to as "Marsha's work station" by the crew members, who usually added various descriptive modifiers such as "messy" or "cluttered." Note the ever-present roll of gray tape. (NASA)

STS-47

Mission:	50	NSSDC ID:	1992-061A
Vehicle:	OV-105 (2)	ET-45 (LWT)	SRB: BI-53
Launch:	LC-39B	12 Sep 1992	14:23 UTC
Altitude:	166 nm	Inclination:	57.00 degrees
Landing:	KSC-33	20 Sep 1992	12:54 UTC
Landing Rev:	126	Mission Duration:	190 hrs 30 mins

Commander:	Robert L. "Hoot" Gibson (4)
Pilot:	Curtis L. "Curt" Brown, Jr. (1)
MS1:	Mark C. Lee (2)
MS2:	Jerome "Jay" Apt III (2)
MS3:	N. Jan Davis (1)
MS4:	Mae C. Jemison (1)
PS1:	Mamoru "Mark" Mohri (1)

Payloads:	Up: 32,480 lbs	Down: 0 lbs
	Spacelab J (32,480 lbs)	
	GAS (9)	
	SRMS s/n 303	

Notes:	First female African-American astronaut (Jemison)
	First professional Japanese astronaut (Mohri)
	First married couple (Lee and Davis)
	First operational use of the drag chute
	Backup mission specialist was
	Stanley N. Koszelak
	Backup payload specialists were
	Chiaki Mukai and Takao Doi

Wakeup Calls:

Since crew worked two shifts around the clock, mission control did not send any wakeup calls.

The flags symbolize the side-by-side cooperation of America and Japan in the mission. The land masses of Japan and Alaska are represented on the crew patch, emphasizing the multi-national aspect of the flight as well as the high inclination 57-degree orbit. (NASA)

The second *Endeavour* mission celebrated the "golden flight" of the Space Shuttle Program on 7 May 1992 with the first on-time launch since STS-31/61B in 1985. The post-flight review of the orbiter umbilical well images revealed a large divot, approximately 16 inches in diameter with an indeterminable depth, on the ET intertank centered between the forward bipods.

During the 1980s, NASA instituted rules stipulating that husband/wife couples would not be launched together. But Mark Lee and Jan Davis had married in January 1991, well into the STS-47 training cycle; NASA allowed them to continue training and they became the first married couple in space. In addition, this mission included the first American mission specialist assigned a backup; Stanley Koszelak was the alternate for Mae Jemison.

Spacelab J included 24 materials science and 20 life sciences experiments, of which the National Space Development Agency of Japan (NASDA) sponsored 35, NASA 7, and collaborative efforts 2. Materials science investigations covered such fields as biotechnology, electronic materials, fluid dynamics and transport phenomena, glasses and ceramics, metals and alloys, and acceleration measurements. Life sciences included experiments on human health, cell separation and biology, developmental biology, animal and human physiology and behavior, space radiation, and biological rhythms. Test subjects included the crew, Japanese koi (carp), cultured animal and plant cells, chicken embryos, fruit flies, fungi and plant seeds, and frogs and frog eggs. To accommodate around-the-clock operations, the crew split into red (Curt Brown, Mark Lee, Mark Mohri) and blue (Jay Apt, Jan Davis, Mae Jemison) teams. Hoot Gibson could help either team as needed or tend to the orbiter.

Middeck experiments included AMOS (Air Force Maui Optical Site calibration), ISAIAH (Israeli Space Agency investigation about hornets), SAREX-II (shuttle amateur radio experiment), SSCE (solid surface combustion experiment), and UVPI (ultraviolet plume imager). There were also 12 GAS canisters (10 with nine experiments, 2 with ballast) on a GAS bridge in the payload bay. NASA planned STS-47 for seven days but mission management to extended the flight one day to allow additional science. Because of potentially unsatisfactory weather at KSC, mission controls waved off the first landing opportunity on flight day 8. Although the weather again threatened to divert the landing, it cleared enough to return to the Shuttle Landing Facility one revolution later.

Researchers considered the mission "an overwhelming success" since most of the experiment objectives were accomplished and the science return was greater than expected.

The on-orbit crew portrait, taken in the Spacehab module wearing very colorful attire. In the middle is Jay Apt. Clockwise from the center left are Jan Davis, Hoot Gibson, Curt Brown, Mae Jemison, Mamoru Mohri, and Mark Lee. Davis shows the effects of having long hair. (NASA)

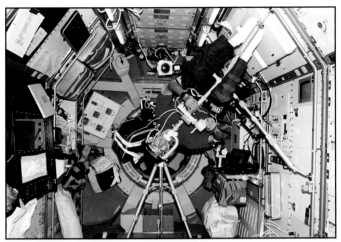

Mamoru Mohri participating in the comparative measurement of visual stability in Earth cosmic space experiment inside the pressurized Spacelab J module in the payload bay. The experiment was attempting to learn more about space adaptation syndrome. (NASA)

Hoot Gibson on the flight deck of Endeavour in his David Clark Company S1032 launch-entry suit. Note the parachute on his back and the life preserver under his arm and around his chest. (NASA)

An unusual angle on Endeavour as she heads uphill toward orbit. The solid rocket boosters provided about 85 percent of the initial thrust for the stack as it accelerated off the launch pad. (NASA)

Technicians install experiment racks into the pressured module. One of the supposed benefits of Spacelab was an ease of integrating different experiments; in practice it did not really work that well and Spacelab was eventually replaced by the less expensive Spacehab. (NASA)

The Spacelab pressurized module being lifted off its processing stand. The large hole in the forward bulkhead will be connected to a tunnel adapter that led to the internal airlock on Endeavour, providing a path for crew members to access the module on-orbit. (NASA)

STS-52

Mission:	51		NSSDC ID:		1992-070A
Vehicle:	OV-102 (13)		ET-55 (LWT)		SRB: BI054
Launch:	LC-39B		22 Oct 1992		17:10 UTC
Altitude:	169 nm		Inclination:		28.46 degrees
Landing:	KSC-33		01 Nov 1992		14:07 UTC
Landing Rev:	159		Mission Duration:		236 hrs 56 mins

Commander:	James D. "WxB" Wetherbee (2)
Pilot:	Michael A. "Mike" Baker (2)
MS1:	Charles L. "Lacy" Veach (2)
MS2:	William M. "Shep" Shepherd (3)
MS3:	Tamara E. "Tammy" Jernigan (2)
PS1:	Steven G. "Steve" MacLean (1)

Payloads:	Up: 26,862 lbs	Down: 0 lbs
	USMP-1 (14,555 lbs)	
	LAGEOS-II (6,209 lbs)	
	ASP (hitchhiker)	
	SRMS s/n 301	

Notes:	Delayed 15 Oct 1992 (SSME issues)
	Backup payload specialist was
	Bjarni V. Tryggvason
	Columbia carried ashes of Gene Roddenberry

Wakeup Calls:

23 Oct	"Wake Up, Columbia"
24 Oct	"Shake, Rattle, and Roll"
25 Oct	Unidentified
26 Oct	"The World is Waiting for the Sunrise"
27 Oct	"Birthday" (Baker's birthday)
28 Oct	Hawaiian music
29 Oct	"Mack the Knife"
30 Oct	"Bang the Drum"
31 Oct	"Monster Mash"
01 Nov	Notre Dame Victory March

The gold star symbolized the mission to explore the frontiers of space while the shape of the Greek letter lambda represents both the laser measurements to be taken from LAGEOS-II and the lambda point experiment as part of USMP-1. (NASA)

The target launch date in mid-October slipped when engineers decided to replace SSME-3 on *Columbia* because of possible cracks in the LH_2 coolant manifold on the nozzle. Changing the engine at the pad was less complex than continued X-ray analysis of the suspect area. Liftoff on 22 October 1992 was delayed due to crosswinds at the Shuttle Landing Facility and clouds at the Banjul TAL site but otherwise the countdown went smoothly. Imagery showed that an 8x4-inch piece of the left bipod ramp, weighing about 0.02-pound, separated from ET-55 during ascent. Since no probable cause could be reasonably established, no corrective action was taken.

On flight day 2, the crew deployed LAGEOS-II (laser geodynamic satellite). Built by the Italian Space Agency, Agenzia Spaziale Italiana (ASI), this was a passive laser ranging satellite that was 24 inches in diameter and had a dimpled appearance like a large golf ball because of 426 nearly equally spaced, cube-corner retroreflectors. The satellite obtained precise measurements of the crustal movements and gravitational field of Earth, as well as understanding the wobble in the planet's rotational axis.

The primary payload was the first United States Microgravity Payload (USMP-1) that consisted of three major experiments. The lambda-point experiment (LPE) studied fluid behavior in microgravity, the materials for the study of interesting phenomena of solidification on Earth and in orbit, MEPHISTO (materiel pour l'etude des phenomenes interessant la solidification sur terre et en orbite) studied metallurgical processes in microgravity, and SAMS (space acceleration measurement system) studied the microgravity environment onboard the orbiter. The MSFC Payload Operations Control Center monitored the USMP-1 experiments.

In addition, the crew positioned the SRMS in pre-designated configurations to expose witness plates attached to the arm to the atomic oxygen stream. The samples, installed along the length of the SRMS boom, were part of the MELEO (materials exposure to low-earth orbit) experiment that was investigating the suitability of several materials for space structures. The crew also used the SRMS to release the Canadian target assembly (CTA). Jim Wetherbee and Mike Baker performed two RCS separation maneuvers to provide some distance between the orbiter and satellite. The Canadian-designed SVS (space vision system) experiment used the cameras on the SRMS to monitor target dots of known size and spacing on the CTA. Using the video pixel count of the high-contrast dots, the SVS processor determined the distance and orientation to the target.

Columbia returned to the Shuttle Landing Facility at KSC on the first opportunity.

In the front, from the left, are Lacy Veach, Tammy Jernigan, and Bill Shepherd. In the back are Mike Baker, Jim Wetherbee, and Steve MacLean. Unusually, Veach and MacLean have leather name tags on their launch-entry suits rather than cloth versions. (NASA)

Columbia *returns to the Shuttle Landing Facility at KSC. Note the NASA UH-1 Huey hovering on the other side of the runway, undoubtedly filming the landing for the public affairs office. The SILTS pod on top of the tail was unique among the orbiters. (NASA)*

The crew captured this image of the onset of the summer monsoon over the Kalahari Desert, as illustrated by the thunderstorm towers poking up through the terminator. The USMP-1 payload is in front with the sunshield covering the LAGEOS-II satellite in the back. (NASA)

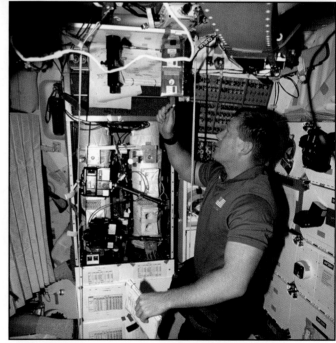

Bill Shepherd works with the crystals by vapor transport experiment (CVTE) on the middeck of Columbia. The experiment was in the location normally used by the middeck accommodations rack (MAR) next to the galley. Note the fire extinguisher at top left. (NASA)

STS-53

Mission:	52	NSSDC ID:	1992-086A
Vehicle:	OV-103 (15)	ET-49 (LWT)	SRB: BI055
Launch:	LC-39A	02 Dec 1992	13:24 UTC
Altitude:	204 nm	Inclination:	57.00 degrees
Landing:	EDW-22	09 Dec 1992	20:45 UTC
Landing Rev:	116	Mission Duration:	175 hrs 20 mins

Commander:	David M. "Dave" Walker (3)
Pilot:	Robert D. "Bob" Cabana (2)
MS1:	Guion S. "Guy" Bluford, Jr. (4)
MS2:	Michael R. "Rich" Clifford (1)
MS3:	James S. "Jim" Voss (2)

Payloads:	Up: 28,316 lbs	Down: 0 lbs
	USA-89 (SDS-B3) (20,953 lbs)	
	GCP (hitchhiker)	
	No SRMS	

Notes:	Last major payload for the Department of Defense
	First Dog Crew
	Orbiter returned to KSC on 18 Dec 1992 (N911NA)

Wakeup Calls:

03 Dec	"Jingle Bells"
04 Dec	"I Wanna be a Dog" (children's song)
05 Dec	Bagpipe medley
06 Dec	The Air Force Hymn – "Wild Blue Yonder"
07 Dec	The Navy Hymn – "Anchors Aweigh"
08 Dec	The Army Hymn – "Caissons Go Rolling Along"
09 Dec	The Marine Corps Hymn – "Halls of Montezuma"

The pentagonal shape of the crew patch represented the Department of Defense. The emblem shows Discovery rising to new achievements as it trails the astronaut symbol against a backdrop of the American flag. The five stars and three stripes symbolize the STS designation. (NASA)

The launch on 2 December 1992 was delayed to allow the Sun to melt ice that had formed on the ET due to overnight temperatures in the upper-40s. Initially, NASA planned this mission for six days, but subsequent changes led to a seven-day flight.

Redheaded Dave Walker had been known during his military service as "Red Dog" and those involved in STS-53 eventually became known as the "Dogs of War." So the first "Dog Crew" was born, although never acknowledged by NASA or the DoD. Walker was "Top Dog," Jim Voss was "Dog Face," and flight controller Wayne Hale became "Sled Dog." For their first wakeup call, mission control played a version of "Jingle Bells" performed by The Singing Dogs, a creation of Dr. Demento, a radio broadcaster specializing in strange and unusual recordings. CAPCOM Carl Meade said, "Crew dogs, wake up. We got work to do" to which Walker responded, "Good morning, Carl. Dogs of War are wide awake." The next morning the crew was greeted with the children's song "I Wanna be a Dog" by Nancy Cassidy. And so it continued.

Initially, the primary payload was simply known as DoD-1 and was the last major Department of Defense mission for space shuttle. The crew deployed the payload six hours after launch. Later identified cryptically as USA-89, most pundits believe this was the third Satellite Data System (SDS-B3) communications spacecraft to be deployed by space shuttle (the others were on STS-28R and STS-38). SDS satellites used a highly elliptical orbit, with a perigee of 162 nm and an apogee of 21,000 nm, to allow communications with polar areas unavailable to geosynchronous satellites. In a small switch from normal procedures, after the crew deployed classified payload, the remaining flight activities became unclassified. Nevertheless, as usual, DoD had little comment about the mission.

Middeck experiments included BLAST (battlefield laser acquisition sensor test), CLOUDS (cloud logic to optimize use of defense systems, CREAM (cosmic radiation effects and activation monitor), FARE (fluid acquisition and resupply experiment), HERCULES (hand-held, earth-oriented, real-time, cooperative, user-friendly, location-targeting and environmental system), MIS-1 (microcapsules in space), RME-III (radiation monitoring experiment), STL (space tissue loss), and VFT-2 (visual function tester).

Mission management diverted the landing to Edwards due to forecasted clouds at the Shuttle Landing Facility, but delayed the deorbit one revolution because of winds in the High Desert. Funny thing was, the clouds never arrived at KSC, but a cloud decided to sit over the arrival end of the runway at Edwards. It was a reality the weather office never attempted to explain.

In the front, from the left, are Guy Bluford and Jim Voss, while Dave Walker, Bob Cabana, and Rich Clifford are in the back. The photograph was taken at the new Space Center Houston (SCH) visitor complex adjacent to the Johnson Space Center. (NASA)

Discovery being lifted out of the transfer aisle of the Vehicle Assembly Building. The yellow device is the vertical sling that allowed the crane operator to translate the orbiter from horizontal to vertical. Note the red covers on the SSME and OMS engine nozzles. (NASA)

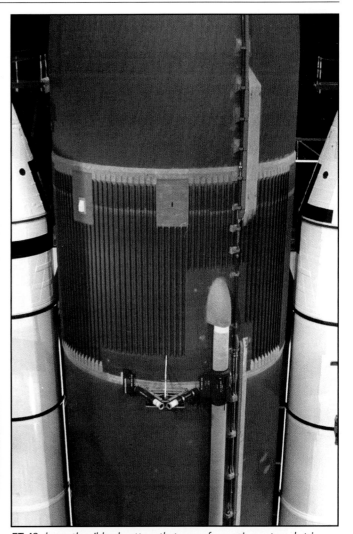

ET-49 shows the ribbed pattern that came from using external stringers to reinforce the unpressurized aluminum intertank. The 17-inch-diameter LO_2 feedline is at right. The smooth area at the top center of the intertank is a passive vent, while the range safety antenna is on the left. (NASA)

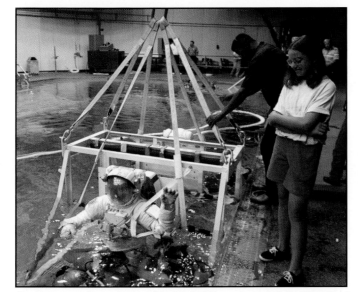

Jim Voss, wearing an extravehicular mobility unit spacesuit, is lowered into the Weightless Environment Training Facility (WET-F) pool at JSC while he waves to his daughter standing poolside. With no planned EVAs, Voss was practicing contingency procedures. (NASA)

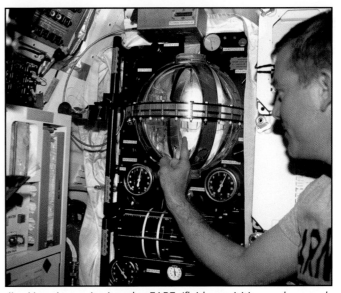

Jim Voss is monitoring the FARE (fluid acquisition and resupply equipment) experiment on the middeck. Voss points to one of two 12.5-inch spherical receiver tanks made of transparent acrylic. FARE investigated the dynamics of fluid transfer in microgravity. (NASA)

STS-54

Mission:	53	NSSDC ID:	1993-003A
Vehicle:	OV-105 (3)	ET-51 (LWT)	SRB: BI-056
Launch:	LC-39B	13 Jan 1993	13:59 UTC
Altitude:	173 nm	Inclination:	28.45 degrees
Landing:	KSC-33	19 Jan 1993	13:39 UTC
Landing Rev:	96	Mission Duration:	143 hrs 38 mins

Commander:	John H. Casper (2)
Pilot:	Donald R. "Don" McMonagle (2)
MS1:	Mario "Trooper" Runco, Jr. (2)
MS2:	Gregory J. "Greg" Harbaugh (2)
MS3:	Susan J. Helms (1)

Payloads:	Up: 49,039 lbs	Down: 0 lbs
	TDRS-F/IUS (37,397 lbs)	
	DXS (hitchhiker)	
	No SRMS	

Notes: EVA1 17 Jan 1993 (Harbaugh and Runco)
Shortest *Endeavour* mission

Wakeup Calls:

14 Jan	The Air Force Hymn – "Wild Blue Yonder"
15 Jan	"Hail Purdue"
16 Jan	(none)
17 Jan	"Centerfield"
18 Jan	"Stardust"
19 Jan	"Ain't Misbehavin'"

The crew patch depicted the American bald eagle placing a larger star among a constellation of four others, representing the placement of the fifth TDRS into orbit to join the four already in service. The blackness of space represents the Diffuse X-ray Spectrometer. (NASA)

The countdown proceeded smoothly, although unacceptable upper-level winds slightly delayed this *Endeavour* liftoff. Once on-orbit, the extended duration orbiter (EDO) waste collection system, a larger unit than the standard potty, failed during its first attempted use. After some troubleshooting, mission control advised the crew to cycle power to the potty before using it, and the unit operated satisfactorily for the remainder of the mission. The unit was making its first flight on *Endeavour* although it had previously flown on *Columbia*.

The crew successfully deployed the fifth Tracking and Data Relay Satellite (TDRS-F) and its inertial upper stage on flight day 1. The inertial upper stage fired one hour later to propel the satellite to an intermediate checkout orbit and then transferred it to geosynchronous orbit the following day. The satellite drifted at about 3 to 5 degrees per day to its checkout position at 150 degrees West longitude, where 5 to 6 weeks of detailed equipment calibration began. A secondary objective was operating the DXS (diffuse X-ray spectrometer) hitchhiker experiment that collected data from deep space to determine the wavelength and intensity of the strongest X-rays emitted by the hot stellar gases released by supernovas. Two DXS instruments were mounted on shuttle payload of opportunity carrier (SPOC) plates carried on get-away special adapter beams on each side of the payload bay. The DXS began scanning on revolution 7, but quickly experienced problems and automatically shut down. Researchers theorized these were caused by the high radiation regions of the South Atlantic Anomaly (SAA) and developed a bake-out procedure that restored most of the functionality for the remainder of the mission.

Middeck payloads included CGPA (commercial general bioprocessing apparatus), CHROMEX (chromosome and plant cell division in space experiment), PARE (physiological and anatomical rodent experiment), SAMS (space acceleration measurement system), and SSCE (solid surface combustion experiment).

On flight day 5, Greg Harbaugh and Mario Runco conducted the only EVA of the mission, performing a series of efforts to increase knowledge in preparation for assembling Space Station Freedom. They tested moving freely in the payload bay (although both were tethered to the payload bay sill safety line), climbing into foot restraints without using their hands, and simulated carrying large objects in the microgravity environment.

NASA rescheduled landing for one revolution earlier than planned due to an approaching weather front, but waved-off the attempt due to fog. This was the shortest mission for *Endeavour*.

Not the best copy, but of one of the more unusual crew gag photos. From the left are Don McMonagle, Mario Runco, John Casper, Susan Helms, and Greg Harbaugh. Although many crews posed for gag photos, the official NASA policy is not to release them. (NASA)

The ET was visible for quite a while after it was jettisoned, and early crews occasionally rolled the orbiter so that they could photograph the tank while it reentered the atmosphere. After the Columbia accident, the maneuver would become standard practice. (NASA)

Greg Harbaugh working in the back of the payload bay. The TDRS/IUS launch cradle is at the right. Note the yellow handholds just below the slidewire on the longeron sill, giving the EVA crew member something to grab onto while they moved around the payload bay. (NASA)

The mostly empty payload bay shows the aerospace support equipment used by the TDRS satellite and its inertial upper stage. Note the EVA slidewire running along each longeron sill, providing a location for crew members to tether themselves during spacewalks. (NASA)

All crews trained in the use of the escape baskets that allowed them to quickly egress the pad to an underground bunker some 1,200 feet away. There were seven baskets that could each hold three people, providing adequate capacity for the flight and ground crews. (NASA)

Another beautiful launch. The white smoke was actually steam from the 300,000 gallons of sound suppression water dumped on the pad shortly after main engine ignition to mitigate acoustic energy from the exhausts that could damage the thermal protection system. (NASA)

STS-56

Mission:	54	NSSDC ID:	1993-023A
Vehicle:	OV-103 (16)	ET-54 (LWT)	SRB: BI058
Launch:	LC-39B	08 Apr 1993	05:29 UTC
Altitude:	161 nm	Inclination:	57.00 degrees
Landing:	KSC-33	17 Apr 1993	11:38 UTC
Landing Rev:	148	Mission Duration:	222 hrs 08 mins

Commander:	Kenneth D. "Ken" Cameron (2)
Pilot:	Stephen S. "Oz" Oswald (2)
MS1:	C. Michael "Mike" Foale (2)
MS2:	Kenneth D. "Taco" Cockrell (1)
MS3:	Ellen Ochoa (1)

Payloads:	Up: 23,843 lbs	Down: 0 lbs
	ATLAS-2 (21,000 lbs)	
	SPARTAN-201-01 (2,840 lbs)	
	SRMS s/n 201	

Notes: Scrubbed 06 Apr 1993 (MPS LH$_2$ valve)

Wakeup Calls:

Since crew worked two shifts around the clock, mission control did not send any wakeup calls.

The payload bay contains ATLAS-2, SSBUV, and SPARTAN. ATLAS was part of the Mission to Planet Earth, so the planet features prominently. The atmosphere is depicted as a stylized visible spectrum and the sunrise is represented with an enlarged two-colored corona. (NASA)

The 6 April 1993 *Discovery* launch attempt proceeded smoothly until the T-9 minute hold when engineers had not completed discussions of a higher than normal temperature on one main engine. The launch team held the count for an hour while those discussions concluded with a decision to proceed with launch. However, the onboard general-purpose computers halted the countdown at T-11 seconds due to an anomaly in the main propulsion system and the launch director called for a 48-hour scrub turnaround. The final count on 8 April proceeded smoothly.

The primary payload was the second Atmospheric Laboratory for Applications and Science (ATLAS-2), designed to collect data on relationship between the energy output of the Sun and the middle atmosphere of Earth, and how these factors affect the ozone layer. The laboratory consisted of six atmospheric and solar instruments mounted on a Spacelab pallet and a seventh mounted in two GAS canisters. Atmospheric instruments included ATMOS (atmospheric trace molecule spectroscopy), MAS (millimeter wave atmospheric sounder), and SSBUV/A (shuttle solar backscatter ultraviolet). The solar science instruments were ACR (active cavity radiometer), SOLCON (measurement of solar constant), SOLSPEC (measurement of solar spectrum), and SUSIM (solar ultraviolet irradiance monitor). All seven ATLAS-2 instruments had flown on ATLAS-1 during STS-45 and flew a third time on STS-66.

On flight day 3, Ellen Ochoa used the SRMS to deploy SPARTAN-201 (Shuttle Pointed Autonomous Research Tool for Astronomy). This free-flying science platform studied the velocity and acceleration of solar wind and observed the corona using two telescopes, an ultraviolet coronal spectrometer (UVCS) and a white light coronagraph (YLC). The crew retrieved the free-flyer on flight day 5 and stowed the SRMS for the remainder of the mission.

The crew made radio contact with schools around world using the second shuttle amateur radio experiment (SAREX-II) and also talked to the Russian Mir space station—the first such contact between the two vehicles using amateur radio equipment. In addition, more than 30 investigations were flown as a part of the commercial materials dispersion apparatus (MDA) instrumentation technology associates (ITA) experiments.

The mission management team waved-off both landing opportunities on flight day 8 because of unacceptable weather at Shuttle Landing Facility. In the meantime, the crew reactivated five of the seven ATLAS-2 experiments to gather additional data. By flight day 9 the weather had cleared sufficiently and came back to KSC on the first landing opportunity.

The on-orbit portrait. In front are Ken Cameron and Mike Foale. In back, from the left, are Ellen Ochoa, Steve Oswald, and Ken Cockrell. The astronauts are posed against the ceiling of the aft flight deck, with the two small windows facing the payload bay to the right. (NASA)

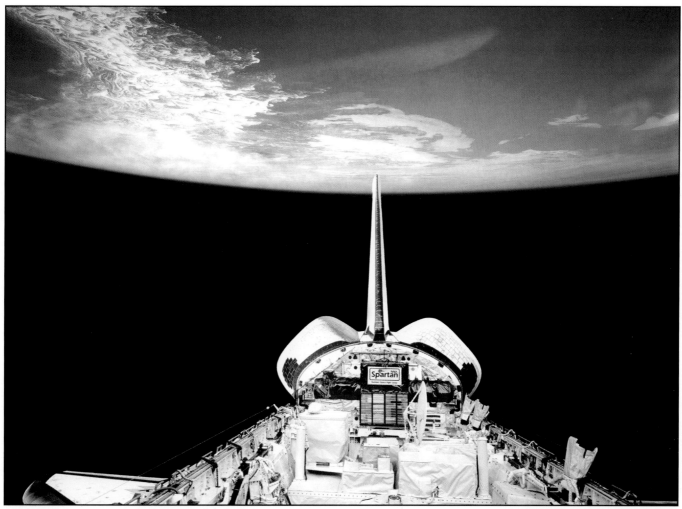

The ATLS-2 payloads are featured in this payload bay scene aboard Discovery, backdropped against an oblique view of the Kamchatka Peninsula. The orbiter was in the atmospheric monitoring attitude. The top of the igloo support container is partially visible in the foreground. Between the ATLAS pallet and the aft bulkhead of the payload bay is the SPARTAN-201 satellite, which was later released as a free-flyer and then reberthed for return to Earth. Also visible in frame are the MAS antenna, the active cavity radiometer irradiance monitor, and the SOLCON, SOLSPEC and SUSIM experiments. The empty MPMs at the right mean the SRMS arm was deployed and out of the frame. (NASA)

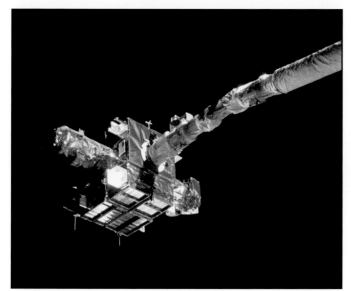

SPARTAN on the end of the SRMS while being deployed. After it was released, the platform operated independently, turning and pointing at the Sun, leaving the orbiter free for other activities. The free-flyer was also far enough away to avoid contamination by thruster firings. (NASA)

Astronauts Kevin Chilton (left) and Curtis Brown are seen at the spacecraft communicator (CAPCOM) console in the Flight Control Room (FCR-1) at the Mission Control Center. For the most part, only astronauts on the ground talked to the crews on-orbit. (NASA)

STS-55

Mission:	55	NSSDC ID:	1993-027A
Vehicle:	OV-102 (14)	ET-56 (LWT)	SRB: BI057
Launch:	LC-39A	26 Apr 1993	14:50 UTC
Altitude:	163 nm	Inclination:	28.45 degrees
Landing:	EDW-22	06 May 1993	14:31 UTC
Landing Rev:	160	Mission Duration:	239 hrs 40 mins

Commander:	Steven R. "Steve" Nagel (4)
Pilot:	Terence T. "Tom" Henricks (2)
MS1:	Jerry L. Ross (4)
MS2:	Charles J. "Charlie" Precourt (1)
MS3:	Bernard A. Harris, Jr. (1)
PS1:	Ulrich H. Walter (1)
PS2:	Hans W. Schlegel (1)

Payloads:	Up: 33,416 lbs	Down: 0 lbs
	Spacelab D2 (33,405 lbs)	
	No SRMS	

Notes:
Delayed 25 Feb 1993 (SSME issues)
Delayed 14 Mar 1993 (hydraulic leak)
Delayed 21 Mar 1993 (range conflict)
Aborted 22 Mar 1993 (SSME failure)
Delayed 24 Apr 1993 (IMU failure)
Backup payload specialists were
Gerhard P. J. Thiele and Renate L. Brümmer
Orbiter returned to KSC on 14 May 1993 (N905NA)

Wakeup Calls:

Since crew worked two shifts around the clock,
mission control did not send any wakeup calls.

The American and German flags on the crew patch represented the partnership of this mission. The two blue stars in the border represented the alternate payload specialists. The stars in the sky stand for each of the children of the primary crew members. (NASA)

This *Columbia* launch was first scheduled for 25 February 1993, but slipped to early March after questions arose about the turbopumps on all three main engines. A revised date of 14 March slipped again after a hydraulic flex hose burst in the aft compartment. Launch was set for 21 March but was pushed back 24 hours due to Eastern Range conflicts caused by a Delta II delay.

The launch attempt on 22 March was proceeding smoothly until it was aborted at T-3 seconds by the onboard general-purpose computers when a main engine failed to ignite completely. This was the first on-pad main engine abort since the return-to-flight and only the third of the flight campaign (STS-26/51F and STS-16/41D were the others). NASA decided to replace all three main engines as a precaution. The mission management team rescheduled the launch for 24 April, but scrubbed early when an inertial measurement unit failed its built-in test routine. The final countdown on 26 April proceeded smoothly. This was the last launch from LC-39A until February 1994 to allow for pad refurbishment and modification.

This was the second Spacelab mission sponsored by Germany (the first was STS-30/61A). The multi-discipline Spacelab D2 mission included experiments contributed by universities, research institutes, and industrial concerns in Germany and other countries. The crew split into a blue team consisting of Steve Nagel, Tom Henricks, Jerry Ross, and Ulrich Walter, and a red team that included Charlie Precourt, Bernard Harris, and Hans Schlegel.

They completed 88 experiments in astronomy, atmospheric physics, biological sciences, Earth observations, fluid physics, life sciences, and materials sciences. Many of the experiments advanced the research of the D1 mission by conducting similar tests, using upgraded hardware, or implementing methods that took advantage of the technical advancements since 1985. In addition, the D2 mission carried several new experiments. The mission also featured a unique support structure (USS) mounted in the payload bay near the Spacelab module that provided support for four additional experiment facilities that were connected to Spacelab for power and data, but ran independently. The experiments included AOET (atomic oxygen exposure tray), GAUSS (galactic ultrawide-angle Schmidt system camera), MAUS (material science autonomous payload), and MOMS (modular optoelectronic multispectral stereo scanner).

As a result of low clouds and humidity at the Shuttle Landing Facility, mission management delayed landing for one revolution and diverted *Columbia* to Edwards. This mission surpassed the 365th day in space for the space shuttle fleet and the 100th day in space for *Columbia*.

The seven primary and two alternate crew members pose in front of the Spacelab module. From the left (front) are Steve Nagel, Tom Henricks, Charlie Precourt, Bernard Harris, Ulrich Walter, Gerhard Thiele, and Hans Schlegel. In the back are Renate Brümmer and Jerry Ross. (NASA)

Columbia *in the Mate/Demate Device at NASA Dryden being mated to N905NA for the ferry flight back to KSC. (NASA)*

A space shuttle main engine ignites during the first launch attempt on 22 March 1993. The onboard general-purpose computers aborted the launch at T-3 seconds when SSME-3 did not ignite properly. This was the third on-pad main engine abort of the flight campaign. (NASA)

Hans Schlegel appears to be in a hurry to return to the crew quarters in the Operations and Checkout Building after the 22 March abort. Ulrich Walter, in the background, seems more casual about the situation. Note the German Space Agency (DLR) patches on their sleeves. (NASA)

STS-57

Mission:	56	NSSDC ID:	1993-037A
Vehicle:	OV-105 (4)	ET-58 (LWT)	SRB: BI059
Launch:	LC-39B	21 Jun 1993	13:07 UTC
Altitude:	258 nm	Inclination:	28.45 degrees
Landing:	KSC-33	01 Jul 1993	12:53 UTC
Landing Rev:	155	Mission Duration:	239 hrs 45 mins

Commander:	Ronald J. "Ron" Grabe (4)
Pilot:	Brian Duffy (2)
MS1:	G. David Low (3)
MS2:	Nancy J. "Smurf" Sherlock (1)
MS3:	Peter J. K. "Jeff" Wisoff (1)
MS4:	Janice E. Voss (1)

Payloads:	Up: 29,119 lbs	Down: 9,424 lbs
	Spacehab SM (19,119 lbs)	
	EURECA (9,424 lbs downmass)	
	SHOOT (hitchhiker)	
	GAS (10)	
	SRMS s/n 303	

Notes: Delayed 18 May 1993 (manifest)
Delayed 03 Jun 2993 (SSME issue)
Scrubbed 20 Jun 1993 (KSC and TAL weather)
Retrieved EURECA deployed by STS-46
EVA1 25 Jun 1993 (Low and Wisoff)

Wakeup Calls:

22 Jun	"Sitting on Top of the World"
23 Jun	"The Smurfs"
24 Jun	"Rendezvous"
25 Jun	"The Walk of Life"
26 Jun	"Holiday"
27 Jun	"I Got You (I Feel Good)"
28 Jun	"Catch a Falling Star"
29 Jun	(none)
30 Jun	(none)
01 Jul	"I'll Be Home For Christmas"

The characteristic flat-top shape of Spacehab is represented by the inner red border. The three gold plumes surrounded by the five stars trailing EURECA are suggestive of the astronaut symbol. The five stars and shape of the robotic arm symbolize the STS designation. (NASA)

NASA originally set this *Endeavour* launch for 18 May 1993, but rescheduled it to 3 June to allow liftoff and landing to occur during daylight. Mission management slipped the 3 June attempt to replace the high-pressure oxidizer turbopump on a main engine. The launch team scrubbed the 20 June attempt at T-5 minutes due to low clouds and rain at KSC and weather concerns at all three TAL sites. The following day, the Eastern Range called a short hold at T-5 minutes until an unidentified aircraft departed controlled airspace. The remainder of the countdown, the longest since the return-to-flight, proceeded smoothly.

This marked the first flight of the commercially developed Spacehab pressurized module. The crew studied body posture, the spacecraft environment, crystal growth, metal alloys, and the behavior of fluids in microgravity. The flight also included an evaluation of maintenance equipment that might be used on Space Station Freedom.

The other primary mission objective was retrieving the European Space Agency EURECA (European Retrievable Carrier) spacecraft deployed on STS-46 to study the long-term effects of microgravity. The crew used the SRMS to capture EURECA on 24 June, but ground controllers were unable to stow the spacecraft antennas so, on 25 June, David Low and Jeff Wisoff spent the beginning of a scheduled EVA manually folding them. The pair used the remainder of the spacewalk on planned tasks.

Secondary objectives included operating AMOS (Air Force Maui Optical Site calibration), FARE (fluid acquisition and resupply experiment), and SAREX-II (shuttle amateur radio experiment).

The payload bay also contained a GAS bridge assembly with ten GAS experiments and a single ballast can. These included the CONCAP-IV (consortium for materials development in space complex autonomous payload), CAN DO (G-324), and SHOOT (superfluid helium on-orbit transfer). To show how ambitious some of the GAS experiments were, consider CAN DO. The primary payload of CAN DO, known as GEOCAM, contained four Nikon 35 mm cameras equipped with 250-exposure film backs. The GEOCAM system closely matched the larger Skylab film format in both coverage and quality, allowing direct examination and comparison of the changes over the last 20 years. The 5-cubic-foot canister also contained 350 small, passive, student experiments.

The crew closed the payload bay doors before the landing attempt on 29 June was waved off; first for one revolution and then for one day due to bad weather at KSC, but unacceptable weather resulted in another 24-hour delay. This was the first flight with two waved-off landings since STS-32/61C in 1986.

From the left are Jeff Wisoff, Brian Duffy, Nancy Sherlock, Janice Voss, Ron Grabe, and David Low. Note the crew still has the NASA "worm" logo on their right sleeve, although the insignia was officially retired in mid-1992. The NASA "meatball" is on the chest of each suit. (NASA)

The first Spacehab single module to fly being prepared in the Operations and Checkout Building in the KSC industrial area on 4 February 1993. Spacehab was developed as a private venture and then leased to NASA to provide laboratory space and logistics volume. It was reportedly much less expensive to process than Spacelab. (NASA)

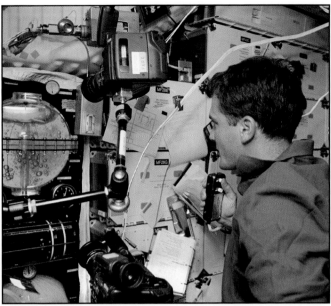

Jeff Wisoff monitors the fluid acquisition and resupply experiment (FARE-II), housed in four middeck lockers. The successor to the FARE-I effort flown on STS-53 in 1992, FARE-II was designed to demonstrate the effectiveness of a device to alleviate the problems associated with vapor-free liquid transfer in microgravity. (NASA)

Endeavour slows down on the Shuttle Landing Facility at KSC. The orbiter had a remarkably short nose landing-gear because engineers wanted to save weight and volume inside the fuselage. Although successful at achieving those goals, it also meant the orbiter "slapped" down hard as it rotated after main gear touchdown, causing a somewhat unpleasant jolt for the flight crew and placing unnecessary stress on the airframe. It was much the same phenomena experienced by the X-15 rocket plane, which had been designed by many of the same engineers. (NASA)

STS-51

Mission:	57	NSSDC ID:	1993-058A
Vehicle:	OV-103 (17)	ET-59 (LWT)	SRB: BI060
Launch:	LC-39B	12 Sep 1993	11:46 UTC
Altitude:	174 nm	Inclination:	28.45 degrees
Landing:	KSC-15	22 Sep 1993	07:57 UTC
Landing Rev:	157	Mission Duration:	236 hrs 11 mins

Commander:	Frank L. Culbertson, Jr. (2)
Pilot:	William F. "Reads" Readdy (2)
MS1:	James H. "Jim" Newman (1)
MS2:	Daniel W. "Dan" Bursch (1)
MS3:	Carl E. Walz (1)

Payloads:	Up: 54,006 lbs	Down: 0 lbs
	ACTS/TOS (24,030 lbs)	
	ORFEUS-SPAS-1 (7,321 lbs)	
	SRMS s/n 201	

Notes:	Scrubbed 17 Jul 1993 (SRB hold-down studs)
	Scrubbed 24 Jul 1993 (SRB HPU failure)
	Aborted 12 Aug 1993 (SSME-2 failure)
	First orbiter to fly a GPS receiver
	EVA1 16 Sep 1993 (Newman and Walz)

Wakeup Calls:

13 Sep	"Please Release Me" (ORFEUS-SPAS)
14 Sep	"Changes in Attitudes, Changes in Latitudes"
15 Sep	"Don't Let the Stars Get in Your Eyes"
16 Sep	"Walk, Don't Run"
17 Sep	(from crew) "Theme for the Common Man"
18 Sep	"Theme from Star Wars"
18 Sep	(from crew) "A Whole New World"
19 Sep	"Rendezvous"
20 Sep	"Heartbreak Hotel" (Max-Q)
21 Sep	"Surfin' Safari"
22 Sep	(none)

The gold star on the crew patch represents the ACTS and the stylized Shuttle Pallet Satellite represents the German-sponsored ASTRO-SPAS mission. The stars in Orion also commemorate the astronauts who have sacrificed their lives for the space program. (NASA)

The ground team scrubbed the first *Discovery* launch attempt during the T-20 minute hold on 17 July 1993 because of minor issues on the launch pad. The second attempt on 24 July was halted at T-19 seconds due to problems with one of two hydraulic power units on the right SRB. NASA rescheduled the launch for 4 August then changed it to 12 August to avoid the debris field of Comet Swift-Tuttle (Perseid meteor shower), which was expected to peak on 11 August. The third countdown was aborted at T-3 seconds due to a faulty sensor monitoring the fuel flow on a main engine. This was the fourth pad abort of the flight campaign and the second during 1993. All three main engines were changed at the pad.

NASA rescheduled launch for 10 September, then slipped it to 12 September to allow time to complete an ongoing review of Advanced Communications Technology Satellite (ACTS) following the loss of contact with the Mars Observer and issues with the NOAA-I satellite, both of which used a similar design. The 12 September launch proceeded smoothly with no unplanned holds.

This was the first mission to fly with a single GPS receiver but a fully operational triple-redundant three-string GPS would not fly for another 14 years, on STS-118.

The crew successfully deployed ACTS after a one-revolution delay because of S-band communications issues. The ACTS was a testbed for advanced communications satellite concepts. The transfer orbit stage (TOS) upper stage fired 45 minutes after deployment and boosted the satellite to geosynchronous orbit. However, during the deployment, two Super*Zip explosive cords in the aerospace support equipment cradle that released the spacecraft detonated incorrectly. The post-flight inspection found 36 tears in the payload bay thermal blankets, three gouges to metal cable trays, and one penetration through the aft bulkhead. All of the damage was easily repaired.

On flight day 2, the crew deployed the second primary payload, the Orbiting and Retrievable Far and Extreme Ultraviolet Spectrograph-Shuttle Pallet Satellite (ORFEUS-SPAS), the first in a series of ASTRO-SPAS astronomical missions. After six days as a free-flyer, the crew used the SRMS to retrieve ORFEUS-SPAS and return it to the payload bay. Jim Newman and Carl Walz conducted a successful EVA on flight day 3 to evaluate tools, tethers, and foot restrains for the upcoming Hubble Space Telescope servicing mission.

Mission control waved-off both landing opportunities on flight day 8 because of isolated rain showers in central Florida. The landing on the first opportunity the following day marked the first of 19 night landings at the Shuttle Landing Facility.

From the left are Frank Culbertson, Dan Bursch, Carl Walz, Bill Readdy, and Jim Newman. Surprisingly, NASA purchased only forty-nine launch-entry suits. Eleven of the suits used dark blue outer covers and were used for training, the remainder used orange covers like these. (NASA)

Carl Walz (red stripes) reaches for a power ratchet tool (PRT) while Jim Newman checks out mobility on the portable foot restraint (PFR). This was the third mission to include a preparatory EVA in response to the weaknesses in EVA training exposed during STS-49. (NASA)

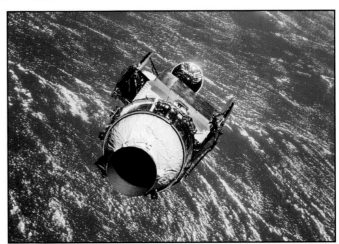

The Advanced Communications Technology Satellite (ACTS) with its unique transfer orbit stage (TOS) is backdropped over the blue ocean following its release from Discovery. (NASA)

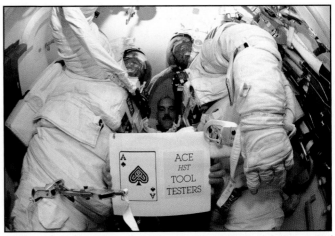

Bill Readdy holds a sign advertising the "Ace HST Tool Testers." Carl Walz (left) and Jim Newman, still in their extravehicular mobility unit suits, had just completed a lengthy EVA evaluating tools to be used on the first servicing mission to the Hubble Space Telescope. (NASA)

Discovery made the sixth night landing of the flight campaign. Note the drag chute silhouetted behind the orbiter and the small flame emanating from one of the auxiliary power unit exhaust beside the vertical stabilizer. (NASA)

Dan Bursch (left), Carl Walz, and Jim Newman watch as another crew member (out of frame) simulates a parachute jump into water during emergency bailout training at the Weightless Environment Training Facility (WET-F) at the Johnson Space Center. (NASA)

STS-58

Mission:	58	NSSDC ID:	1993-065A
Vehicle:	OV-102 (15)	ET-57 (LWT)	SRB: BI061
Launch:	LC-39B	18 Oct 1993	14:53 UTC
Altitude:	155 nm	Inclination:	39.00 degrees
Landing:	EDW-22	01 Nov 1993	15:07 UTC
Landing Rev:	225	Mission Duration:	336 hrs 13 mins

Commander:	John E. Blaha (4)
Pilot:	Richard A. "Rick" Searfoss (1)
MS1:	M. Rhea Seddon (3)
MS2:	William S. "Bill" McArthur, Jr. (1)
MS3:	David A. Wolf (1)
MS4:	Shannon W. Lucid (4)
PS1:	Martin J. Fettman (1)

Payloads:	Up: 32,011 lbs Down: 0 lbs
	Spacelab SLS-2 (32,011 lbs)
	Extended duration orbiter pallet
	No SRMS

Notes:	Scrubbed 14 Oct 1993 (range safety computer)
	Scrubbed 15 Oct 1993 (orbiter S-band failure)
	Backup payload specialists were
	Jay C. Buckey, Jr. and Laurence R. Young
	Last *Columbia* landing at Edwards
	Orbiter returned to KSC on 08 Nov 1993 (N911NA)

Wakeup Calls:

19 Oct	No music, just greeting
20 Oct	"Theme from 2001, a Space Odyssey"
21 Oct	(none)
22 Oct	"Jump In the Line" / "Doctor! Doctor!"
23 Oct	"I Know You're Out There, Somewhere"
24 Oct	"Back Home in Indiana"
25 Oct	"Shiny Happy People"
26 Oct	"Happy Trails"
27 Oct	"Look At Us Now"
28 Oct	(none)
29 Oct	"From A Distance"
30 Oct	"Theme from St. Elsewhere"
31 Oct	"Monster Mash"
01 Nov	(none)

The hexagonal shape of the patch depicts the carbon ring, a molecule common to all living organisms. Encircling the inner border of the crew patch is the double helix of DNA, representing the genetic basis of life. An EDO pallet is shown in the aft payload bay. (NASA)

Bad weather delayed the first launch attempt on 14 October 1993 for two hours during the T-9-minutes hold. When it cleared and the count resumed, a failure in the Air Force range safety system stopped the count at T-31-seconds. The Air Force was unable to correct the problem, forcing NASA to cancel this *Columbia* attempt. The launch team scrubbed the second attempt on 15 October when one of the two S-band transponders on the orbiter failed. In any case, it was unlikely the launch would have occurred because of unacceptable weather around the launch site. The 18 October countdown proceeded smoothly, delayed only ten seconds because of an unauthorized aircraft in the restricted area.

The primary goal of the second Spacelab Life Sciences (SLS-2) mission was to investigate the physiological response to microgravity and the subsequent re-adaptation to gravity. The payload consisted of 14 experiments, 8 using the crew as subjects and 6 using rats, focused on cardiovascular, regulatory, neurovestibular, and musculoskeletal systems. These experiments subjected the crew to rides on stationary bikes, rotating chairs, and rotating domes, and included taking a variety of blood, urine, and saliva samples. The crew used the astronaut science advisor (ASA), an early computer-based intelligent assistant (what is today called artificial intelligence), to help them work more efficiently and improve the quality of the science.

Secondary payloads included ITEPC (inter-mars tissue equivalent proportional counter), OARE (orbital acceleration research experiment), SAREX-II (space amateur radio experiment), and UMS (urine monitoring system) on the middeck. The orbiter was placed in an unusual attitude planned for the STS-74 United States Microgravity Laboratory (USML-2) mission for 15 hours to acquire OARE acceleration data that flight designers used to maximize the science return from the mission.

The crew also tested PILOT (portable inflight landing operations trainer), a laptop computer simulator that helped the commander and pilot maintain their proficiency for approach and landing during long duration flights. Similar simulators would be used through the end of the flight campaign in 2011.

Rhea Seddon sent down a special message to her husband, Hoot Gibson, when she surpassed his 632 hours 56 minutes in space (at the time). "He's still a really good guy, I still love him a lot, but I've got more hours in space than he does, so there!" she teased.

This mission had always been scheduled to land in California, and *Columbia* returned to Edwards on the first opportunity. As it turned out, this would be the last time *Columbia* would land in the High Desert.

The crew was inside the pressurized Spacelab module for their on-orbit portrait colorful shirts and the typical blue shorts. Clockwise from the bottom center are Rick Searfoss, John Blaha, Rhea Seddon, Shannon Lucid, Bill McArthur, Martin Fettman, and David Wolf. (NASA)

Bill McArthur eating on the middeck. Note that he is using a set of foot restraints to keep him in place and that he is "sitting" without the need for a chair. The seat behind him has its back folded down and a middeck locker tray strapped to it. (NASA)

ET-57 immediately after being jettisoned. There were cameras in the ET umbilical well on the orbiter that photographed the tank as it dropped away. The LO_2 umbilical is at lower left with the prominent 17-inch-diameter fill valve and a monoball electrical connector. (NASA)

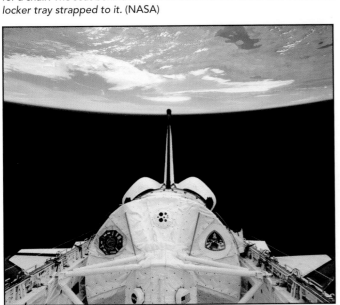

The Spacelab Life Sciences module in the payload bay. The truss structure in front of the pressurized module supported the long tunnel adapter that connected the orbiter crew compartment to the Spacelab module. Note the European Space Agency insignia on the left. (NASA)

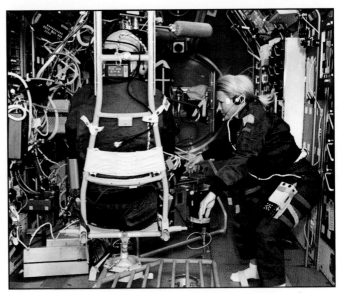

Rhea Seddon spins Martin Fettman in the rotating chair used to test ocular deviation and vestibular-ocular reflex (the changes in reflexive eye motions). One chair protocol required that the test subject be rotated about a vertical axis and stopped suddenly. (NASA)

STS-61 (HST-SM1)

Mission:	59	NSSDC ID:	1993-075A
Vehicle:	OV-105 (5)	ET-60 (LWT)	SRB: BI063
Launch: Altitude:	LC-39B 321 nm	02 Dec 1993 Inclination:	09:27 UTC 28.45 degrees
Landing: Landing Rev:	KSC-33 163	13 Dec 1993 Mission Duration:	05:26 UTC 259 hrs 59 mins

Commander:	Richard O. "Dick" Covey (4)
Pilot:	Kenneth D. "Sox" Bowersox (2)
MS1:	Kathryn C. "Kathy" Thornton (3)
MS2:	Claude Nicollier (2)
MS3:	Jeffrey A. "Jeff" Hoffman (4)
MS4:	F. Story Musgrave (5)
MS5:	Thomas D. "Tom" Akers (3)

Payloads:	Up: 24,363 lbs Down: 2,148 lbs Hubble Servicing Mission (30,961 lbs) SRMS s/n 303

Notes:	Scrubbed 01 Dec 1993 (KSC weather) EVA1 05 Dec 1993 (Hoffman and Musgrave) EVA2 06 Dec 1993 (Thornton and Akers) EVA3 07 Dec 1993 (Hoffman and Musgrave) EVA4 08 Dec 1993 (Thornton and Akers) EVA5 09 Dec 1993 (Hoffman and Musgrave) Backup mission specialist was Gregory J. Harbaugh

Wakeup Calls:

02 Dec	"Cosmos"
03 Dec	"Get Ready"
04 Dec	"Fanfare for the Common Man"
05 Dec	"With a Little Help From My Friends"
06 Dec	"Doctor My Eyes"
07 Dec	"I Can See Clearly Now"
08 Dec	Traditional Swiss Alpine song
09 Dec	"A Hard Day's Night"
10 Dec	"Mamas Don't Let Your Babies Grow Up to Be Cowboys"
11 Dec	"My Heroes Have Always Been Cowboys"
12 Dec	"I Can See For Miles"

The astronaut symbol is superimposed against the sky with the Earth underneath. The two circles represent the optical configuration of the Hubble Space Telescope where light is focused by reflections from a large primary Mirror and a smaller secondary Mirror. (NASA)

The Hubble Space Telescope, deployed from *Discovery* during STS-31R, had been designed from the beginning to be serviced by space shuttle while it was on-orbit. This capability became even more important after a significant flaw was discovered in the primary Mirror on the telescope that essentially eliminated any possibility of using the telescope as intended. Initially, NASA scheduled the launch of the first Hubble servicing mission (SM1) for 2 December 1993 from LC-39A. However, after rollout, engineers discovered contamination in the payload changeout room and decided to move *Endeavour* to LC-39B on 15 November. The contamination did not affect the payload package since it was well sealed.

The mission management team rescheduled the launch for 7 December, then moved it back to 2 December and then advanced it another day. The launch team scrubbed the attempt on 1 December due to weather at the launch site, although just before the scrub, the Eastern Range was also no-go due to a ship in the restricted area. The count proceeded smoothly the following day.

On 4 December, the crew used the SRMS to grapple Hubble and berth it in the flight service structure in the payload bay. During a then-record five EVAs, two teams of astronauts conducted the first servicing of Hubble. The crew completed most tasks sooner than expected and smoothly handled the few contingencies that did arise. In fact, the flight plan had initially included two additional EVAs, which were not needed. To complete the mission without too much fatigue, the five extravehicular activities were split between two pair of astronauts. The servicing tasks included replacing the high-speed photometer (HSP) with the corrective optics space telescope axial replacement (COSTAR) to correct the Mirror flaw discovered during on-orbit checkout, replacing the wide field/planetary camera (WF/PC) with the improved WF/PC-II, replacing the rate sensing units and electronic control units, installing new solar arrays, installing new magnetic sensing systems and fuse plugs, and repairing the Goddard high resolution spectrometer (GHRS). Astronomers generally praised the repairs as restoring the telescope to better than anticipated condition.

The crew released Hubble on flight day 9, about three hours later than planned to allow troubleshooting of an erratic subsystems monitor on the telescope; engineers had seen the anomaly before and it was not related to the servicing. Bill Clinton and Al Gore called to congratulate the crew and Flavio Cotti, the Swiss minister of internal affairs, called the following day to congratulate Claude Nicollier. Mission management cleared *Endeavour* to return to the Shuttle Landing Facility on the first opportunity.

The on-orbit portrait taken on the middeck of Endeavour with the crew in their embroidered shirts. In the front, from the left, are Claude Nicollier, Ken Bowersox, and Dick Covey. Behind them are Story Musgrave, Jeff Hoffman, Kathy Thornton, and Tom Akers. (NASA)

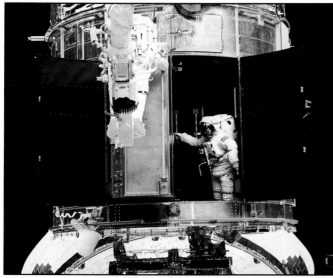

Tom Akers waits inside Hubble for Kathy Thornton, on the end of the SRMS with her back toward the camera, to hand him the COSTAR (corrective optics space telescope axial replacement) unit. (NASA)

Story Musgrave works near the end of the SRMS arm during the fifth EVA. The spacewalkers replaced the solar array drive electronics unit, installed a power supply redundancy kit, and manually deployed the primary drive mechanisms for both solar arrays. (NASA)

Jeff Hoffman standing on the SRMS foot restraint with the old wide field/planetary camera after installing the new instrument. (NASA)

Hubble after being released from Endeavour, about three hours later than panned due to some erratic behavior by one of the subsystems in the telescope. This was unrelated to the servicing mission and soon resolved by ground controllers. (NASA)

STS-60 (Near Mir)

Mission:	60	NSSDC ID:	1994-006A
Vehicle:	OV-103 (18)	ET-61 (LWT)	SRB: BI062
Launch: Altitude:	LC-39A 194 nm	03 Feb 1994 Inclination:	12:10 UTC 51.60 degrees
Landing: Landing Rev:	KSC-15 130	11 Feb 1994 Mission Duration:	19:20 UTC 199 hrs 09 mins

Commander: Charles F. "Charlie" Bolden, Jr. (4)
Pilot: Kenneth S. "Ken" Reightler, Jr. (2)
MS1: N. Jan Davis (2)
MS2: Ronald M. "Ron" Sega (1)
MS3: Franklin R. Chang-Diaz (4)
MS4: Sergei Konstantinovich Krikalev (3)

Payloads:	Up: 28,957 lbs	Down: 0 lbs
	Spacehab SM (9,452 lbs)	
	WSF-1 (3,785 lbs)	
	ODERACS-1 (4,406 lbs)	
	COB/GBA (hitchhiker)	
	GAS (3)	
	SRMS s/n 201	

Notes: First US/Russian Shuttle-Mir Program mission
Backup for Sergei Krikalev was
VladiMir Georgiyevich Titov

Wakeup Calls:

04 Feb	"Early Morning Riser"
05 Feb	"Rawhide" (by crew in simulator)
06 Feb	"The Bug"
07 Feb	"Let There Be Peace on Earth"
08 Feb	"Sweet Home Alabama"
09 Feb	Russian folk tunes
10 Feb	"I Get Around"
11 Feb	"Homeward Bound"

The countdown on 3 February 1994 had no unplanned holds and resulted in an on-time launch. This *Discovery* mission marked the first flight of a Russian cosmonaut (Sergei Krikalev) on space shuttle as one of the initial elements in implementing the Agreement on NASA/Russian Space Agency Cooperation in Human Space Flight, otherwise known as the Shuttle-Mir Program.

This mission marked the second flight of a Spacehab single module, which contained 13 major experiments. These included ASC-3 (astroculture), BPL (bioserve pilot laboratory, CGBA (commercial generic bioprocessing apparatus), CPCG (commercial protein crystal growth), ECLIPSE-Hab (equipment for controlled liquid phase sintering experiment-spacehab), IMMUNE-01 (immunology experiment), ORSEP (organic separation), PSB (Pennsylvania State biomodule), SAMS (space acceleration measurement system), SEF (space experiment facility), SOR/F (Stirling orbiter refrigerator/freezer), SRE (sample return experiment), and 3-DMA (three-dimensional microgravity accelerometer).

Also on board was the Wake Shield Facility (WSF-1), making the first in a planned series of flights. The experiment took advantage of the near vacuum of space to grow innovative thin film materials for use in electronics. The pre-deployment checkout of the WSF and the SRMS went well and, as a result, Jan Davis grappled and unberthed the WSF on flight day 3 but some issues caused it to be reberthed for the night. The major activity for flight day 4 was a second attempt to deploy the WSF. Davis again unberthed the payload, but problems with the WSF attitude control system caused mission control to wave-off the deployment. Instead, she positioned the SRMS wrist joint to point the WSF horizon sensor toward the Sun to warm it and left it in this configuration during the scheduled crew sleep period. The following day, mission management decided there was insufficient time remaining in the mission to deploy the WSF, so the payload remained attached to the end of the SRMS where it completed some of its planned science operations.

Good Morning America broadcast a live video hookup between *Discovery* and Mir on 8 February. The following day, the crew deployed six orbital debris radar calibration spheres (ODERACS) and all of the ground tracking stations successfully locked onto the spheres as predicted. They also deployed the Bremen Satellite (BREMSAT), although a spacecraft malfunction cause two experiments to lose data during the first 48 hours.

Mission control waved-off the first landing opportunity at KSC on flight day 9 because of bad weather, but the weather improved enough to allow landing on the second opportunity.

The American and Russian flags symbolize the partnership of the two countries and the open payload bay contains Spacehab and a get-away special bridge assembly. The Wake Shield Facility is shown on the SRMS prior to deployment. (NASA)

Five American astronauts and a Russian cosmonaut squeeze through the tunnel that connected Discovery and Spacehab. Charlie Bolden is at upper right and, clockwise from him are Ron Sega, Jan Davis, Franklin Chang-Diaz, Sergei Krikalev, and Ken Reightler. (NASA)

The Wake Shield Facility on the end of the SRMS. This was the first flight of the WSF, a 12-foot-diameter stainless steel disk that, in theory, generated an "ultra-vacuum" environment within which to grow thin semiconductor films for next-generation advanced electronics. (NASA)

Charlie Bolden and Ken Reightler on the flight deck of Discovery at KSC with checklists and procedures everywhere. The vehicle is not powered-up since the multifunction CRT display system displays are dark. There are four launch-entry suits plugged into the oxygen connections on the back of the center console. (NASA)

Leaving the Vehicle Assembly Building, Discovery heads for LC-39A on a crisp, clear winter day. The stack had been assembled in High Bay 3; other stacks were assembled in High Bay 1 (with the upper doors open). The Launch Control Center is in the background at left. (NASA)

Although it was decidedly dark outside, NASA did not consider this a night launch since it was not between fifteen minutes after nautical sunset and fifteen minutes before nautical sunrise. Nevertheless, it was quite a show for those in attendance. (NASA)

STS-62

Mission:	61		NSSDC ID:		1994-015A
Vehicle:	OV-102 (16)		ET-62 (LWT)		SRB: BI064
Launch:	LC-39B		04 Mar 1994		13:53 UTC
Altitude:	180 nm		Inclination:		39.00 degrees
Landing:	KSC-33		18 Mar 1994		13:11 UTC
Landing Rev:	224		Mission Duration:		335 hrs 17 mins

Commander: John H. Casper (3)
Pilot: Andrew M. "Andy" Allen (2)
MS1: Pierre J. Thuot (3)
MS2: Charles D. "Sam" Gemar (3)
MS3: Marsha S. Ivins (3)

Payloads:	Up: 30,016 lbs	Down: 0 lbs
	USMP-2 (9,606 lbs)	
	OAST-2 (5,789 lbs)	
	Extended duration orbiter pallet	
	SRMS s/n 301	

Notes: Delayed 03 Mar 1994 (KSC weather)
Landing was featured on Discovery Channel

Wakeup Calls:

05 Mar	"I Got You (I Feel Good)"
06 Mar	"Picky, Picky Head" (*Cool Runnings*)
07 Mar	Medley of Armed Forces Hymns
08 Mar	"Space Shuttle Boogie"
09 Mar	"Wake the World"
10 Mar	The Marine Corps Hymn – "Halls of Montezuma"
11 Mar	"Takin' Care of Business"
12 Mar	"Be Our Guest" (*Beauty and the Beast*)
13 Mar	Crew sent "Surfin' USA"
13 Mar	"I Get Around"
14 Mar	"Starship Trooper"
15 Mar	"View From Above"
16 Mar	"Travelin' Prayer"
17 Mar	"Living in Paradise"
18 Mar	"The Mermaid"

The varied hues of the rainbow on the horizon connote the varied, but complementary, nature of the payloads on this mission. The brilliant sunrise just beyond Columbia suggests the promise that research in space holds for the hopes and dreams of future generations. (NASA)

NASA originally scheduled this *Columbia* launch for 3 March 1994, but the launch director postponed it for 24 hours because of predicted unfavorable weather at KSC. The countdown on 4 March proceeded smoothly and the only deviation was a delay in deploying the SRB retrieval ships because of high seas. *Freedom Star* and *Liberty Star* left port on launch day and recovered the boosters and their parachutes on 6 March.

A divot was missing from the left bipod ramp on ET-62, but as with the previous incidents, engineers could not establish a conclusive cause, no damage was found on *Columbia*, and no corrective action was taken. It was all of only passing interest at the time.

The second United States Microgravity Payload (USMP-2) included five experiments to investigate materials processing and crystal growth in microgravity. The crew used the advanced automated directional solidification furnace (AADSF) to study the directional solidification of semiconductor materials. The major scientific goal of the material pour l'etude des phenomenes interessant la solidification sur terre et en orbite (MEPHISTO) experiment was to characterize the morphological transitions of faceted materials in the absence of gravity-induced thermosolutal convection. The isothermal dendritic growth experiment (IDGE) used succinonitrile (SCN) to study dendritic solidification of molten metals in microgravity. The critical fluid light scattering experiment-zeno (CFLSE-Zeno) studied the behavior of xenon at its critical point, which is where a fluid is simultaneously a gas and a liquid with the same density. The crew also conducted a number of biomedical activities to better understand the effects of prolonged space flight.

The NASA Office of Aeronautics and Space Technology OAST-2 payload consisted of six INSTEP (in-space technology program) experiments. Observations began with a three-minute release of nitrogen gas from a canister in the payload bay to study its effect on the glow of a plate constructed of materials that might be used on future satellites. Later, with the aft end of *Columbia* pointed toward Earth, John Casper and Andy Allen performed a 25-minute-long series of 360-degree spins to allow observations by the SKIT (spacecraft kinetic infrared test) instrument.

A stationary bike on the middeck had long been a staple of space shuttle flights to allow exercise to counter the effect of microgravity on the muscles. This mission marked the first flight of a new mounting system intended to keep vibrations from exercise from disturbing sensitive experiments, particularly on materials science missions.

Mission management cleared *Columbia* to return to the Shuttle Landing Facility on the first landing opportunity.

From the left are Sam Gemar, Andy Allen, Masha Ivins, John Casper, and Pierre Thuot. The crews frequently brought small items that were important to them or their family. In this case, Thuot brought a C. F. Martin backpacker guitar on the mission. (NASA)

The orbiter vibrated a lot during its entry from orbit, so photos tend to be fuzzy. This is Andy Allen in the pilot's seat. Note the monochrome green multifunction CRT display system. Eventually this would be replaced by the full-color multifunction electronic display system. (NASA)

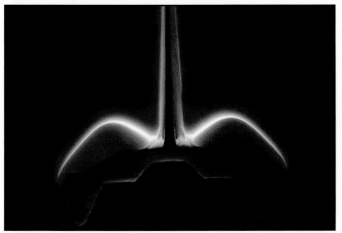

This was the "shuttle glow" phenomenon surrounding the vertical stabilizer and OMS pods. Scientists spent a considerable amount of time trying to understand this seemingly harmless phenomena. (NASA)

This view of a space shuttle launch was only reasonably seen from a boat, so decent photos of it are unusual since the waters were restricted to official use only. (NASA)

The limited duration space environment candidate materials exposure (LDCE) experiment exposed three identical sets of materials to the space environment using modified GAS canisters on the starboard side of the payload bay. (NASA)

John Casper crawls back in after practicing bailing out of the orbiter side hatch in the Spacecraft Mockup and Integration Laboratory at JSC. Barely visible in the hatchway of the crew compartment trainer (CCT) is the end of the crew escape pole. (NASA)

STS-59

Mission:	62		NSSDC ID:		1994-020A
Vehicle:	OV-105 (6)		ET-63 (LWT)		SRB: BI065
Launch:	LC-39A		09 Apr 1994		11:05 UTC
Altitude:	121 nm		Inclination:		57.00 degrees
Landing:	EDW-22		20 Apr 1994		16:55 UTC
Landing Rev:	183		Mission Duration:		269 hrs 49 mins

Commander:	Sidney M. "Sid" Gutierrez (2)
Pilot:	Kevin P. "Chilli" Chilton (2)
MS1:	Jerome "Jay" Apt III (3)
MS2:	Michael R. "Rich" Clifford (2)
MS3:	Linda M. Godwin (2)
MS4:	Thomas D. "Tom" Jones (1)

Payloads:	Up: 33,758 lbs	Down: 0 lbs
	SRL-1 (SIR-C) (33,758 lbs)	
	GAS (3)	
	SRMS s/n 303	

Notes:	Delayed 07 Apr 1994 (SSME inspections)
	Scrubbed 08 Apr 1994 (KSC weather)
	Launch was featured on Discovery Channel
	Orbiter returned to KSC on 02 May 1994 (N911NA)

Wakeup Calls:

Since crew worked two shifts around the clock, mission control only sent one wakeup call

18 Apr   "Freedom" (White Elementary School)

The gold astronaut symbol sweeps over Earth's surface from Endeavour, representing the operation of the spaceborne imaging radar (SIR-C) and MAPS sensors. The five stars on the left and nine stars on the right symbolize the STS designation. (NASA)

The launch team postponed the *Endeavour* launch attempt set for 7 April 1994 at T-27 hours to inspect vanes in the SSME high-pressure oxidizer preburner pumps after Rocketdyne found anomalies in similar units. The 8 April countdown proceeded smoothly up to the T-9 minute hold, which was extended because of clouds around the launch site. Late in the 2.5-hour launch window, the cloud conditions became acceptable; however, increased winds accompanied the clearing conditions. As the launch window closed, crosswinds at the Shuttle Landing Facility exceeded the 15-knot limit for an RTLS abort. As a result, NASA called a 24-hour scrub turnaround; the count proceeded smoothly the following day.

Once on-orbit, the crew split into two teams. The red team consisted of Sid Gutierrez, Kevin Chilton, and Linda Godwin while the blue team included Jay Apt, Rich Clifford, and Tom Jones. This allowed around-the-clock operation of the Space Radar Laboratory (SRL-1) sponsored by the German Space Agency (DARA) and Italian Space Agency (ASI). The payloads studied vegetation, hydrology, tectonics, topography, and global air pollution. The instruments included the third spaceborne imaging radar and the X-band synthetic aperture radar (SIR-C/X-SAR), as well as an atmospheric instrument called measurement of air pollution from satellites (MAPS). Thirteen countries contributed 49 principal investigators and more than 100 scientists to the SIR-C/X-SAR project. The instruments imaged more than 400 sites including swaths taken over 44 countries that covered more than 43.75 million square miles, the equivalent of 25 percent of the planet.

Secondary payloads included SAREX-II (shuttle amateur radio experiment), STL/NIH-C (space tissue loss/National Institute of Health-cells), and VFT-4 (visual function tester). New Mexico State University, Matra Marconi Space (France), and the Society of Japanese Aerospace Companies all sponsored GAS experiments. In addition, students at the University of Alabama-Huntsville developed CONCAP-IV (consortium for materials development in space complex autonomous payload) that flew in a GAS can.

Since there were sufficient consumables and the vehicle was in good shape, mission management extended the mission one day. The crew completed stowage activities in preparation for entry on the first landing opportunity, but mission control waved it off because of clouds around the Shuttle Landing Facility. The second opportunity was also waved-off because of unfavorable weather in the landing area as well as potential crosswind violations. On 20 April, mission control waved-off the first KSC landing opportunity due to weather concerns and diverted *Endeavour* to Edwards.

The crew on a camping trip in the Brazos Bend State Park, Texas, a few months after the mission. From the left, Sid Gutierrez and Tom Jones are crouching while Kevin Chilton, Jay Apt, Linda Godwin, and Rich Clifford are standing. (Courtesy of Tom Jones)

Longer missions frequently carrier a sleepstation along the starboard wall of the middeck. Each bunk had its own reading light and air vent, and a curtain across the opening allowed a modicum of privacy. Here are Sid Gutierrez, Linda Godwin, and Kevin Chilton. (NASA)

Unlike the first launch attempt on 8 April that was frustrated by clouds and high winds at KSC, the weather on 9 April was acceptable and resulted in some beautiful launch photography. The Sun was beginning to peek above the eastern horizon. (NASA)

Sid Gutierrez (left) and Kevin Chilton (right) with Rich Clifford between them on the flight deck of Endeavour. Note parts of the orange launch-entry suit paraphernalia are still on the seats and the early laptop computer on the glareshield above the instrument panel. (NASA)

The Space Radar Laboratory included the third spaceborne imaging radar, the X-band synthetic aperture radar, and the measurement of air pollution from satellite experiment. An area of the Pacific Ocean northeast of Hawaii forms the backdrop for the image. (NASA)

STS-65

Mission:	63	NSSDC ID:	1994-039A
Vehicle:	OV-102 (17)	ET-64 (LWT)	SRB: BI066
Launch:	LC-39A	08 Jul 1994	16:43 UTC
Altitude:	163 nm	Inclination:	28.45 degrees
Landing:	KSC-33	23 Jul 1994	10:39 UTC
Landing Rev:	235	Mission Duration:	353 hrs 55 mins

Commander:	Robert D. "Bob" Cabana (3)
Pilot:	James D. "Jim" Halsell, Jr. (1)
MS1:	Richard J. "Rick" Hieb (3)
MS2:	Carl E. Walz (2)
MS3:	Leroy Chiao (1)
MS4:	Donald A. "Don" Thomas (1)
PS1:	Chiaki Mukai (1)

Payloads:	Up: 32,880 lbs	Down: 0 lbs
	Spacelab IML-2 (25,043 lbs)	
	Extended duration orbiter pallet	
	No SRMS	

Notes:	First space shuttle test of the second TDRS ground station at White Sands, New Mexico
	Backup payload specialist was Jean-Jacques Favier

Wakeup Calls:

Since crew worked two shifts around the clock, mission control did not send any wakeup calls.

The second IML mission is reflected in the crew patch by two gold stars shooting toward the heavens behind the lettering. Columbia is reaching into space on an international quest for a better understanding of the effects of space flight on materials processing and life sciences. (NASA)

The countdown on 8 July 1994 proceeded smoothly with no unplanned holds. This *Columbia* mission marked the second flight of the International Microgravity Laboratory (IML-2), carrying more than twice the number of experiments as the first mission on STS-42. The IML concept applied results from one mission to the next to broaden investigations between missions.

The crew split into two teams to allow around-the-clock operations. The red team consisted of Bob Cabana, Jim Halsell, Rick Hieb, and Chiaki Mukai while the blue team included Leroy Chiao, Don Thomas, and Carl Walz. Mukai became the first Japanese woman to fly in space and set a record for the longest flight to-date by a female astronaut. As had become normal for Spacelab missions, the crew augmented the regenerative carbon dioxide removal system (RCRS) with lithium hydroxide (LiOH) canisters.

More than 80 primary experiments, representing 200 principal investigators from 13 countries and 6 space agencies, were located in the pressurized Spacelab module. The agencies included the European Space Agency (ESA), French Space Agency (CNES), German Space Agency (DARA), Canadian Space Agency (CSA), National Space Development Agency of Japan (NASDA), and NASA. The experiments covered material science, fluid science, microgravity environment and countermeasure, bioprocessing, space biology, human physiology, and radiation biology.

Secondary payloads included CPCG (commercial protein crystal growth), OARE (orbital acceleration research experiment), and SAREX-II (shuttle amateur radio experiment). In addition, the flight supported the AMOS (Air Force Maui Optical Site calibration) and MAST (military application of ship tracks) experiments that were manifested as payloads of opportunity.

The OARE made highly accurate measurements, on the order of one-billionth of the acceleration of gravity (10 nano-g), of very low frequency (steady state to 1 Hz) orbiter accelerations. The experiment acquired near steady-state microgravity acceleration data in support of IML-2 operations from ten minutes after launch until approximately ten minutes after entry interface.

Mission control waved-off the first landing attempt on 22 July because of low clouds, winds, and heavy offshore rains around KSC. They also waved-off the second opportunity and extended the mission one day because they wanted to land at the Shuttle Landing Facility. After the second wave-off, the crew performed two OMS burns to provide the ability to remain on-orbit an extra day if needed. In the end, it didn't matter since the weather at KSC cleared sufficiently to land on the first opportunity the following day.

The crew poses in the Spacelab pressurized module used for the second International Microgravity Laboratory. In the front, from the left, are Rick Hieb, Chiaki Mukai, Bob Cabana, and Jim Halsell. Behind them are Don Thomas, Carl Walz, and Leroy Chiao. (NASA)

Carl Walz floats through the Spacelab module. Most of the experiment facilities were mounted in 12 racks along the sides of the laboratory. In addition, one experiment facility and other support equipment were in the middeck, which was connected to Spacelab by a tunnel. (NASA)

Columbia flies over Lake Nyasa, located between Malawi, Mozambique, and Tanzania. The Spacelab tunnel included an awkward S-shaped "joggle" section that translated the low-mounted height of the orbiter airlock to the center-mounted height of the Spacelab hatch. (NASA)

Leroy Chiao (top) places a sample in the biorack incubator as Don Thomas handles another sample inside the biorack glovebox. The glovebox was used to prepare samples for the biorack and slow rotating centrifuge microscope (NIZEMI) experiments. (NASA)

The normal potty could only accommodate a crew of seven for 14 days, so NASA built this extended duration orbiter waste management system that had additional capacity and more redundancy. Most importantly, the crews preferred its 8-inch opening to the normal 4-inch opening. (NASA)

STS-64

Mission:	64	NSSDC ID:	1994-059A
Vehicle:	OV-103 (19)	ET-66 (LWT)	SRB: BI068
Launch:	LC-39B	09 Sep 1994	22:23 UTC
Altitude:	141 nm	Inclination:	57.00 degrees
Landing:	EDW-04	20 Sep 1994	21:14 UTC
Landing Rev:	176	Mission Duration:	262 hrs 50 mins

Commander:	Richard N. "Dick" Richards (4)
Pilot:	L. Blaine Hammond, Jr. (2)
MS1:	Jerry M. Linenger (1)
MS2:	Susan J. Helms (2)
MS3:	Carl J. Meade (3)
MS4:	Mark C. Lee (3)

Payloads:	Up: 28,463 lbs	Down: 0 lbs
	LITE (22,679 lbs)	
	SPARTAN-201-02 (2,840 lbs)	
	SPIFEX	
	ROMPS (hitchhiker)	
	GAS (12)	
	SRMS s/n 201	

Notes:	First ACES suits (Linenger and Meade)
	EVA1 16 Sep 1994 (Meade and Lee)
	First test of SAFER backpack
	First untethered EVA since STS-19/51A
	Last untethered EVA of the flight campaign
	Orbiter returned to KSC on 27 Sep 1994 (N905NA)

Wakeup Calls:

10 Sep	Parody of "Fun, Fun, Fun" (Max-Q)
11 Sep	Parody of "My Girl" (Max-Q)
12 Sep	"Ace in the Hole"
13 Sep	Parody of "I Get Around" (Max-Q)
14 Sep	Parody of "Theme from Green Acres" (Max-Q)
15 Sep	"Hound Dog"
16 Sep	"EVA Surfing" (Max-Q)
17 Sep	"Another Saturday Night" (Max-Q)
18 Sep	"This Is the Time"
19 Sep	"Yakety Yak"
20 Sep	Sounds of chirping birds and a crowing rooster

LITE used a three-wavelength laser symbolized by the three gold rays emanating from the star in the payload bay that formed part of the astronaut symbol. SPARTAN was shown on the SRMS arm and the two untethered EVA crew members represented the SAFER tests. (NASA)

NASA scheduled the 2.5-hour late afternoon launch window on 9 September 1994 to allow night operation of the LITE laser early in the mission. The launch team extended the T-9-minute hold due to heavy cloud cover around KSC and at T-5 minutes because of potential rain over the Shuttle Landing Facility.

This *Discovery* mission marked the first flight of LITE (LIDAR in-space technology experiment) that used a laser optical radar to study the atmosphere as part of the Mission to Planet Earth. LITE operated for 53 hours, yielding more than 43 hours of high-rate data, providing unprecedented views of cloud structures, storm systems, dust clouds, pollutants, forest burning, and surface reflectance. The instrument studied the atmosphere above Africa, northern Europe, Indonesia, Russia, and the south Pacific. Data images clearly indicated high cloud cover and dust storms over West Africa and dramatically outlined the structure of the Super Typhoon Melissa, including details of the eye of the storm. The mission management team extended the mission by one day to allow additional LITE operations.

The crew used the SRMS to deploy SPARTAN-201 (Shuttle Pointed Autonomous Research Tool for Astronomy) that collected data about the acceleration and velocity of the solar wind and measured the corona of the Sun. The crew retrieved the free-flyer two days later. This mission also included the shuttle plume impingement flight experiment (SPIFEX), a 33-foot-long extension for the SRMS arm that was carried on the starboard longeron sill. This was one of the few times the orbiter carried anything on that sill until the advent of the orbiter boom sensor system (OBSS) after the *Columbia* accident. SPIFEX collected data on RCS plumes to better understand the potential effects of the plumes on large space structures. ROMPS (robot operated processing system) was the first American robotics system operated in space, split between two GAS cans mounted on the payload bay wall. In addition, there was a GAS bridge with 12 cans.

The crew completed all preparations for deorbit and landing on flight day 9, but the weather was unacceptable for both landing opportunities. While the crew was reconfiguring the orbiter for an extra day on-orbit, the backup flight system (BFS) exhibited an issue that had recently been seen in the simulator in Houston. Fortunately, all the crew needed to do was reboot the general-purpose computer and the problem cleared itself.

On flight day 10, the crew again completed all preparations for deorbit and closed the payload bay doors. Mission control waved-off both KSC landing opportunities because of unacceptable weather and directed the crew to land at Edwards.

Clockwise from the upper right are Dick Richards, Carl Meade, Susan Helms, Blaine Hammond, Mark Lee, and Jerry Linenger. During launch and entry, Linenger and Meade wore the advanced crew escape suits while the rest of the crew wore the older launch-entry suits. (NASA)

Soon after landing, technicians connected a purge unit to the right-hand T-0 umbilical and a cooling unit to the left-hand umbilical. These supported the orbiter until it was completely powered down and safed. Here is a convoy moving Discovery from the concrete runway at Edwards AFB to the Mate/Demate Device at the NASA Dryden Flight Research Center. (NASA)

Presaging the much later OBSS, Discovery carried a 33-foot-long extension for the SRMS arm, called SPIFEX, on the starboard payload bay longeron sill. (NASA)

Mark Lee (red stripes) tests the SAFER (simplified aid for EVA rescue) backpack, a device designed for use in the event a crew member became untethered while conducting an EVA. These marked the first untethered American EVAs in ten years, since the original manned maneuvering units were retired prior to the Challenger accident. (NASA)

Discovery lifts-off from LC-39B. At this point the pad still had the original hammerhead crane on top of the fixed service structure as well as the large lightning mast that protected the vehicle from the frequent Florida storms. The fixed service structure had been created, largely, by using parts of the original Apollo-era launch umbilical towers that serviced the Saturn V moon rockets. Space shuttle had recycled as much of the ground infrastructure as possible to keep initial costs down, often to the detriment of future operational costs. (NASA)

STS-68

Mission:	65	NSSDC ID:	1994-062A
Vehicle:	OV-105 (7)	ET-65 (LWT)	SRB: BI067
Launch:	LC-39A	30 Sep 1994	11:16 UTC
Altitude:	120 nm	Inclination:	57.00 degrees
Landing:	EDW-22	11 Oct 1994	17:03 UTC
Landing Rev:	182	Mission Duration:	269 hrs 46 mins

Commander:	Michael A. "Mike" Baker (3)
Pilot:	Terrence W. "Terry" Wilcutt (1)
MS1:	Steven L. "Steve" Smith (1)
MS2:	Daniel W. "Dan" Bursch (2)
MS3:	Peter J. K. "Jeff" Wisoff (2)
MS4:	Thomas D. "Tom" Jones (2)

Payloads:	Up: 34,525 lbs	Down: 0 lbs
	SRL-2 (SIR-C) (33,758 lbs)	
	GAS (3)	
	SRMS s/n 303	

Notes:	Aborted on 18 Aug 1994 (SSME-3 turbopump)
	Orbiter returned to KSC on 20 Oct 1994 (N911NA)

Wakeup Calls:

Since crew worked two shifts around the clock, mission control only sent one wakeup call.

11 Oct   "Tiny Bubbles"

The world's land masses and oceans dominate the center field of the crew patch, with Endeavour circling the globe. The SRL-2 letters span the width and breadth of Earth, symbolizing worldwide coverage of the SIR-C/X-SAR radars and MAPS instruments. (NASA)

The *Endeavour* launch attempt on 18 August 1994 was aborted at T-1.9 seconds when the onboard general-purpose computers shut down all three main engines after detecting a high discharge temperature in a main engine high-pressure oxidizer turbopump. This was the fifth on-pad abort of the flight campaign. *Endeavour* returned to the VAB and technicians replaced all three engines with the ones that had been installed on *Atlantis* for STS-66. The countdown for the second launch attempt on 30 September proceeded smoothly to an on-time liftoff.

This marked the second flight in 1994 of the Space Radar Laboratory (SRL-2) as part of the Mission to Planet Earth. Flying SRL during different seasons allowed researchers to compare changes between this flight and the first SRL mission aboard STS-59 in April. The overall goal of the mission was to better understand the changes caused by natural processes and compare them to changes brought about by human activity. The crew split into two teams to support around-the-clock observations. The red team included Mike Baker, Terry Wilcutt, and Jeff Wisoff while the blue team consisted of Dan Bursch, Tom Jones, and Steve Smith

Besides observing the same locations imaged during the first flight, SRL-2 took advantage of several unusual events, including an erupting volcano on the Kamchatka Peninsula in Russia. The volcano had begun erupting a couple of weeks earlier, but the latest burst from Kliuchevskoi occurred about eight hours after the STS-68 launch. Researchers also evaluated the ability of the SIR-C/X-SAR imaging radars to discern difference between such human-induced phenomena as an oil spill in the ocean and naturally occurring film. Mike Baker and Terry Wilcutt demonstrated their skills, and the maneuverability of the orbiter, by piloting *Endeavour* to within 30 feet of where it had flown during SRL-1. The mission management team approved a one-day extension to gather additional science data.

The mission took advantage of an opportunity to study fires set in British *Columbia* for forest management, with the MAPS (measurement of air pollution from satellites) instrument providing a better understanding of carbon monoxide emissions from burning forests. *Endeavour* also carried five GAS cans, two sponsored by university student groups, one by Swedish Space Corporation, and two by the U.S. Postal Service that carried 500,000 stamps commemorating the 25th anniversary of Apollo 11.

Mission control waved-off the first landing opportunity because of weather at KSC that was trending toward unacceptable conditions. Ultimately, mission management team decided to land at Edwards, which proceeded without incident.

Clockwise from bottom right are Tom Jones, Mike Baker, Dan Bursch, Terry Wilcutt, Steve Smith, and Jeff Wisoff. In this case, the crew wore shirts representing which science team they were on. Jones had flown only six months earlier on STS-59 that carried the SRL-1 payload. (NASA)

Tom Jones floating on the aft flight deck with a video camera and a Linhof large format still camera (left). Note the two overhead windows that were frequently used when photographing Earth. (NASA)

Terry Wilcutt takes advantage of the microgravity environment to juggle five cameras on the aft flight deck. A large format Linhof camera is at left, joining a battery of handheld Hasselblads. (NASA)

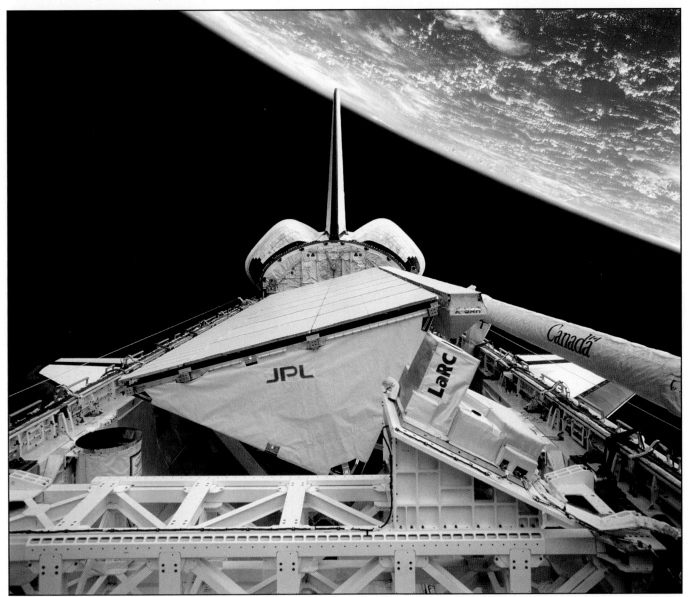

The large angled antenna is part of the spaceborne imaging radar (SIR-C) and X-band synthetic aperture radar (X-SAR). The multipurpose experiment support structure (MPESS) carrying the measurement of air pollution from satellites (MAPS) is at bottom frame. One of the three get-away special canisters is at the left. This was G-316, sponsored by North Carolina A&T State University in Greensboro, that studied the effects of microgravity on the survival, mating, and development of the milkweed bug. The SRMS arm is at right. (NASA)

STS-66

Mission:	66	NSSDC ID:	1994-073A
Vehicle:	OV-104 (13)	ET-67 (LWT)	SRB: BI069
Launch:	LC-39B	03 Nov 1994	17:00 UTC
Altitude:	165 nm	Inclination:	57.00 degrees
Landing:	EDW-22	14 Nov 1994	15:35 UTC
Landing Rev:	174	Mission Duration:	262 hrs 34 mins

Commander: Donald R. "Don" McMonagle (3)
Pilot: Curtis L. "Curt" Brown, Jr. (2)
MS1: Ellen Ochoa (2)
MS2: Joseph R. "Joe" Tanner (1)
MS3: Jean-François A. Clervoy (1)
MS4: Scott E. Parazynski (1)

Payloads: Up: 30,714 lbs   Down: 0 lbs
ATLAS-3 (18,001 lbs)
CRISTA-SPAS-01 (7,194 lbs)
SRMS s/n 202

Notes: Delayed 27 Oct 1994 (manifest)
Orbiter returned to KSC on 22 Nov 1994 (N911NA)

Wakeup Calls:

Since crew worked two shifts around the clock,
mission control did not send any wakeup calls.

NASA had scheduled the launch for 27 October 1994, but used the SSMEs from *Atlantis* to replace the ones in *Endeavour* following the 18 August abort of STS-68, causing a one-week delay. The launch director held the count on 3 November at T-5 minutes to evaluate the winds at the TAL sites, which finally settled enough for launch.

The primary payloads were the third Atmospheric Laboratory for Applications and Science (ATLAS-3) and the Cryogenic Infra-red Spectrometers and Telescopes for the Atmosphere-Shuttle Pallet Satellite (CRISTA-SPAS). The ATLAS-3 instruments were mounted on a Spacelab pallet and included ACRIM (active cavity radiometer irradiance monitor), ATMOS (atmospheric trace molecule spectroscopy), ESCAPE-II (experiment of the sun complementing ATLAS payload and education), SOLCON (measurement of solar constant), SOLSPEC (measurement of solar spectrum), SSBUV/A (shuttle solar backscatter ultraviolet), and SUSIM (solar ultraviolet spectral irradiance monitor). ATLAS-3 made the first detailed middle atmosphere measurements of the northern hemisphere during late fall.

Ellen Ochoa released CRISTA-SPAS from the SRMS on flight day 2. The satellite provided data that complemented those obtained by the Upper Atmosphere Research Satellite (UARS) during STS-48. CRISTA was the first instrument to provide detailed information on the conditions in the upper atmosphere including the dynamics of winds, temperature changes, and atmospheric movements that distribute the gases that influence ozone chemistry.

For the retrieval of CRISTA-SPAS, Don McMonagle and Curt Brown used an R-bar approach (radius vector; i.e., from above or below) to evaluate the technique that would be used for the upcoming Mir docking missions. This technique saved propellant and reduced the risk of orbiter RCS firings contaminating the target. The rendezvous required one OMS and four RCS maneuvers. During the rendezvous with CRISTA-SPAS, McMonagle and Brown performed a "MAHRSI football" maneuver that allowed the middle atmosphere high-resolution spectrograph investigation (MAHRSI) instrument to make observations of the orbiter and the area immediately around it. The CRISTA instrument collected 180 hours of data and MAHRSI acquired about 200 hours of data.

Mission control diverted the landing to Edwards because of high winds, rain, and clouds at the Shuttle Landing Facility caused by Tropical Storm Gordon. This was the fourth diverted landing of 1994.

This was the last solo flight for *Atlantis* for more than 14 years since all future missions, except one, were to Mir or the ISS. The exception was the STS-125 Hubble servicing mission (SM4) in 2009.

Atlantis is trailed by gold plumes representing the astronaut symbol and is superimposed over Earth, much of which is visible from the high inclination orbit. The gaze of instruments on ATLAS and CHRISTA-SPAS is illustrated by the stylized sunrise and visible spectrum. (NASA)

In their official portrait, from the left, are Jean-Francois Clervoy, Scott Parazynski, Curt Brown, Joe Tanner, Don McMonagle, and Ellen Ochoa. Three of them, Clervoy, Parazynski, and Tanner, were members of the 1992 astronaut class and were making their initial flights in space. (NASA)

Atlantis *in the Mate/Demate Device at NASA Dryden. The black tiles turned gray as the result of repeated exposures to the entry environment, so the darker tiles were newer ones: The tiles around the main landing-gear doors were frequently replaced since opening and closing to doors tended to break the adjacent tiles. (NASA)*

The Atmospheric Laboratory for Applications and Science (ATLAS-3) payload in the payload bay. The cylindrical object in front is the pressurized Spacelab igloo that provided power, data handling, and communications equipment for the instruments mounted on the Spacelab pallets behind it. SSBUV was in the GAS cans at left. (NASA)

Most damage to the orbiter thermal protection system generally looked something like the white areas shown here; gouges in the tile caused by debris hitting the vehicle. A small amount of this sort of damage occurred on every flight and was easily repaired after landing. (NASA)

Don McMonagle (left) and Curt Brown in the crew compartment trainer (CCT) at the JSC Shuttle Mockup and Integration Laboratory. The CCT was a fixed-base trainer (i.e., it did not replica flight movement) but included an exact duplicate of the crew module for training. (NASA)

STS-63 (Near Mir)

Mission:	67		NSSDC ID:		1995-004A
Vehicle:	OV-103 (20)		ET-68 (LWT)		SRB: BI070
Launch:	LC-39B		03 Feb 1995		05:22 UTC
Altitude:	214 nm		Inclination:		51.60 degrees
Landing:	KSC-15		11 Feb 1995		11:52 UTC
Landing Rev:	129		Mission Duration:		198 hrs 28 mins

Commander:	James D. "WxB" Wetherbee (3)
Pilot:	Eileen M. Collins (1)
MS1:	Bernard A. Harris, Jr. (2)
MS2:	C. Michael "Mike" Foale (3)
MS3:	Janice E. Voss (2)
MS4:	VladiMir Georgiyevich Titov (3)

Payloads:	Up: 27,554 lbs Down: 0 lbs
	Spacehab SM (9,427 lbs)
	SPARTAN-204 (2,651 lbs)
	ODERACS-2 (4,406 lbs)
	CGP/O2 (hitchhiker)
	SRMS s/n 201

Notes:	Scrubbed 02 Feb 1995 (IMU-2 failure)
	First rendezvous (not docking) with Mir
	First female space shuttle pilot (Collins)
	Carried Coca-Cola fountain dispenser (FGBA)
	EVA1 09 Feb 1995 (Foale and Harris)
	Backup for VladiMir Titov was
	Sergei Konstantinovich Krikalev

Wakeup Calls:

04 Feb	"On Orbit is the Place to Be" (Max-Q)
05 Feb	"Another Saturday Night" (Max-Q)
06 Feb	"Make New Friends"
07 Feb	"Blue Danube Waltz" from 2001, A Space Odyssey
08 Feb	"Opening of Time" / "Dark Side of the Moon"
09 Feb	"Surfing EVA" (Max-Q)
10 Feb	"Theme from Monty Python's Flying Circus"
11 Feb	"The End"

Spacehab and SPARTAN are in the payload bay and Mir is printed in Cyrillic on the side of the station. The American and Russian flags at the bottom represent the cooperative nature of the mission. The six rays of the Sun and the three stars symbolize the STS designation. (NASA)

The launch team scrubbed the *Discovery* attempt on 2 February 1995 after an inertial measurement unit failed. The countdown the following day proceeded smoothly with the launch team making minor adjustments to accommodate the short five-minute launch window required for the first rendezvous with Mir.

Beginning on flight day 1, a series of RCS burns brought *Discovery* in line with Mir. The original plan called for the orbiter to approach no closer than 32.8 feet (10 meters) and then complete a fly-around of the Russian space station. However, shortly after MECO, a reaction control system jet began leaking and the crew could not correct the problem. This led to Russian concerns about a lack of RCS redundancy during rendezvous. After negotiations and technical exchanges between American and Russian officials, the Russians agreed a close approach could still be safely accomplished and mission control directed the crew to proceed with the rendezvous.

Jim Wetherbee manually flew *Discovery* within 39 feet of Mir, saying "As we are bringing our spaceships closer together, we are bringing our nations closer together," after the orbiter was at the point of closest approach. He continued, "The next time we approach, we will shake your hand and together we will lead our world into the next millennium." After station keeping for about 15 minutes, Wetherbee backed-off 400 feet and performed a fly-around of the station.

The crew also worked with payloads aboard the Spacehab single module that carried 11 biotechnology experiments, 3 advanced materials development experiments, 4 technology demonstrations, and 2 pieces of supporting hardware measuring on-orbit accelerations. The experiments included the fluids generic bioprocessing apparatus (FGBA) to determine if carbonated beverages could be produced from separately stored carbon dioxide, water, and flavored syrups. The unit held 1.65 liters of Coca-Cola and Diet Coke. The company had flown a previous experiment on STS-26/51F and would fly a further improved dispenser on STS-77.

On flight day 2, the crew deployed ODERACS-II (orbital debris radar calibration system) and used the SRMS to unberth SPARTAN-204 (Shuttle Pointed Autonomous Research Tool for Astronomy). The payload remained suspended on the arm to observe the shuttle glow phenomenon. They later released SPARTAN-204 for 48 hours as a free-flyer so its far-ultraviolet imaging spectrograph could study targets in the interstellar medium.

Preparations for landing began on flight day 8 and *Discovery* returned to the Shuttle Landing Facility on the first opportunity.

On the aft flight deck, from the left, are Janice Voss, Bernard Harris, VladiMir Titov, Jim Wetherbee, Mike Foale, and Eileen Collins. The crew members are wearing shirts that match the colors in both the American and Russian flags. (NASA)

The fluids generic bioprocessing apparatus (FGBA-2), otherwise known as the Coke dispenser, in the Spacehab module. There were separate dispensers for Coke and Diet Coke. Despite the hype at the time, it was a serious experiment. (NASA)

Valeri VladiMirovich Polyakov looks out a window on Mir while Discovery was maneuvering nearby. Polyakov is the holder of the record for the longest single stay in space, living aboard Mir for more than 14 months during a single mission and more than 22 months in total. (NASA)

Bernard Harris records data from experiments located in middeck lockers. The commercial protein crystal growth (CPCG) experiment was sponsored by the University of Alabama at Birmingham. (NASA)

Mir was the first continuously inhabited station on-orbit and long held the record for the longest continuous human presence in space at 3,644 days, until the ISS surpassed it on 23 October 2010. (NASA)

Mike Foale (red stripes) grabs SPARTAN-204 as Bernard Harris looks on. Before the Mir docking, the crew held SPARTAN with the SRMS to observe the "shuttle glow" phenomena. After undocking, SPARTAN became a free-flyer observing various celestial targets. (NASA)

A photo snapped by an EVA crew member from the end of the SRMS looking back into the payload bay. The Spacehab single module is at the right wearing both American and Russian flags. The SRMS end effector blocks the view of the external airlock. (NASA)

STS-67

Mission:	68	NSSDC ID:	1995-007A
Vehicle:	OV-105 (8)	ET-69 (LWT)	SRB: BI071
Launch:	LC-39A	02 Mar 1995	06:38 UTC
Altitude:	193 nm	Inclination:	28.45 degrees
Landing:	EDW-22	18 Mar 1995	21:48 UTC
Landing Rev:	262	Mission Duration:	399 hrs 09 mins

Commander:	Stephen S. "Oz" Oswald (3)
Pilot:	William G. "Borneo"Gregory (1)
MS1:	John M. Grunsfeld (1)
MS2:	Wendy B. Lawrence (1)
MS3:	Tamara E. "Tammy" Jernigan (3)
PS1:	Samuel T. "Sam" Durrance (2)
PS2:	Ronald A. "Ron" Parise (2)

Payloads:	Up: 28,528 lbs	Down: 0 lbs
	Astro-2 (19,450 lbs)	
	Extended duration orbiter pallet	
	GAS (2)	
	SRMS s/n 303	

Notes:	First space shuttle flight connected to the Internet
	Only Endeavour flight using EDO pallet
	Backup payload specialist was
	Scott D. Vangen
	Longest Endeavour mission
	Orbiter returned to KSC on 27 Mar 1995 (N905NA)

Wakeup Calls:

    Since crew worked two shifts around the clock,
    mission control did not send any wakeup calls.

The three sets of rays, diverging from the payload bay represent the three Astro-2 telescopes. The spiral galaxy, Jupiter, and the four moons (for a total of six space objects) as well as the seven stars of the crew patch symbolize the STS designation. (NASA)

This *Endeavour* liftoff was briefly delayed due to concerns about a feedline heater on the flash evaporator system but launch proceeded smoothly. This was the first mission NASA announced was connected to internet. Users of more than 200,000 computers in 59 countries visited the Astro-2 home page at MSFC and crew members answered some of the questions while they were on-orbit. The crew split into two teams to support around-the-clock operations. The red team consisted of Steve Oswald, Bill Gregory, John Grunsfeld, and Ron Parise while the blue team included Wendy Lawrence, Tammy Jernigan, and Sam Durrance.

Astro-2 marked the second flight of the same three ultraviolet telescopes flown by Astro-1 aboard STS-35 in December 1990. The HUT (Hopkins ultraviolet telescope), developed at the Johns Hopkins University, performed spectroscopy in the far ultraviolet spectrum to identify physical processes and chemical composition in objects. Researchers had made improvements to HUT after Astro-1 that increased its sensitivity three-fold. The UIT (ultraviolet imaging telescope), sponsored by NASA Goddard, took wide-field ultraviolet photographs of objects. These experiments selected targets from a list of more than 600 objects ranging from some inside the solar system to individual stars, nebulae, supernova remnants, galaxies, and active extragalactic objects. The WUPPE (Wisconsin ultraviolet photo-polarimeter experiment), built at the University of Wisconsin, measured the photometry and polarization of ultraviolet radiation from objects.

The HUT completed more than 200 observations of more than 100 celestial objects. Investigators believed the telescope collected enough data to meet its primary mission objective: detecting the presence of intergalactic helium, a telltale remnant of theoretical big bang explosion that began the universe. The UIT imaged about two-dozen large spiral galaxies and took the first ultraviolet images of the entire moon. The instrument also studied rare stars that were 100 times as hot as Sun, elliptical galaxies, and some of faintest galaxies in universe. WUPPE yielded a "treasure chest of data," according to its principal investigator, greatly expanding the database on ultraviolet spectra-polarimetry. Targets included dust clouds in Milky Way and the nearby Large Magellanic Cloud.

On 17 March, the mission management team extended the flight by one day due to poor weather at both KSC and Edwards. The planned return to the Shuttle Landing Facility the following day was waved off due to winds, clouds, and potential rain. Ultimately, *Endeavour* diverted to Edwards. This was the longest mission for *Endeavour*.

In front, from the left, are Steve Oswald, Tammy Jernigan, and Bill Gregory. In the back are Ron Parise, Wendy Lawrence, John Grunsfeld, and Sam Durrance. Both payload specialists flew on the Astro-1 mission aboard STS-35 in December 1990. (NASA)

The Astro-2 payload consisted of three telescopes mounted atop a sophisticated instrument pointing system (IPS): the HUT (Hopkins ultraviolet telescope), UIT (ultraviolet imaging telescope), and WUPPE (Wisconsin ultraviolet photo-polarimeter experiment). (NASA)

Endeavour sitting in the Mate/Demate Device at NASA Dryden. The NASA facility was located in a corner of Edwards AFB and did not have its own runways, using the natural lakebeds or the Air Force concrete runways as needed to support its operations. (NASA)

Speedbrake open wide and drag chute deployed, Endeavour slows down on Runway 22 at Edwards AFB. The vehicle rolled for 9,962 feet from main gear touchdown to wheels stop, about average for an orbiter weighing 217,450 pounds at touchdown. (NASA)

N905NA, the original Shuttle Carrier Aircraft, takes off from Edwards heading toward the Kennedy Space Center. The ferry flight made stops at Dyess AFB, Texas, and Columbus AFB, Mississippi, on its way across the country, arriving at KSC on 27 March 1995. (NASA)

STS-71 (S/MM-01)

Mission:	69	NSSDC ID:	1995-030A
Vehicle:	OV-104 (14)	ET-70 (LWT)	SRB: BI072
Launch:	LC-39A	27 Jun 1995	19:32 UTC
Altitude:	215 nm	Inclination:	51.60 degrees
Landing:	KSC-15	07 Jul 1995	14:55 UTC
Landing Rev:	154	Mission Duration:	235 hrs 22 mins

Commander:	Robert L. "Hoot" Gibson (5)
Pilot:	Charles J. "Charlie" Precourt (2)
MS1:	Ellen S. Baker (3)
MS2:	Gregory J. "Greg" Harbaugh (3)
MS3:	Bonnie J. Dunbar (4)
Up:	Anatoly Yakovlevich Solovyev (4)
Up:	Nikolai Mikhailovich Budarin (1)
Down:	Gennadi Mikhailovich Strekalov (5)
Down:	VladiMir Nikolayevich Dezhurov (1)
Down:	Norman E. "Norm" Thagard (5)

Payloads:	Up: 26,577 lbs	Down: 476 lbs
	Spacelab-Mir (24,557 lbs)	
	No SRMS	

Notes:	Delayed late-May 1995 (Russian work)
	Scrubbed 23 Jun 1995 (KSC weather)
	Scrubbed 24 Jun 1995 (KSC weather)
	First flight of external airlock (Mir configuration)
	First rendezvous with Mir
	First space shuttle crew rotation flight
	Atlantis carried eight crew members down

Wakeup Calls:

28 Jun	"I Got You Babe"
29 Jun	"From a Distance"
30 Jun	"Your Wildest Dreams"
01 Jul	"Kuca, Kuca, Kuca" (Russian pop song)
02 Jul	"Changes in Latitudes, Changes in Attitudes"
03 Jul	Florida State Seminole Fight Song
04 Jul	"America the Beautiful"
05 Jul	"I Love My Moon (children's song)
06 Jul	Parody of "Hello, Goodbye"
07 Jul	"Take the Long Way Home"

The rising Sun symbolizes the dawn of a new era of cooperation between America and Russia. Atlantis and Mir are shown in separate circles converging at the center of the crew patch, symbolizing the union of the two separate space programs. (NASA)

NASA planned this launch of *Atlantis* for late May 1995, but slipped it into June to accommodate Russian activities necessary for the first Mir docking, including the launch of the Spektr module and a series of spacewalks to reconfigure the space station. The launch team scrubbed the 23 June attempt when rain and lightning at the launch site prevented loading the ET. They scrubbed the second attempt on 24 June at T-9 minutes, again due to stormy weather in central Florida. Mission management called a three-day stand-down since the forecast was unfavorable and the launch team needed a break. The 27 June countdown proceeded smoothly.

On flight day 2, the crew powered-up the androgynous peripheral docking system (APDS) and extended the guide ring to the ready-for-docking position. Rendezvous occurred at 13:00 UTC on 29 June using the R-bar (radius vector; i.e., from above or below) approach practiced on STS-66 with *Atlantis* closing in on Mir from directly below. R-bar approaches allowed natural forces to slow the orbiter more than would occur with a V-bar (velocity vector; i.e., from in front or behind) approach, reducing the need for RCS firings. The final docking began with *Atlantis* about 2,500 feet below Mir and Hoot Gibson at the controls on the aft flight deck. The orbiter stayed 250 feet away from Mir until American and Russian flight controllers approved the docking. Gibson then maneuvered the orbiter to a point about 30 feet from Mir before beginning his final approach. Docking occurred 218 nm above the Lake Baykal region of the Russian Federation. The orbiter docking system androgynous port served as the actual connection to a similar interface on the docking port on the Kristall module. When linked, *Atlantis* and Mir formed largest spacecraft on-orbit to-date, with a total mass of nearly 500,000 pounds.

After the crews opened the hatches, the STS-71 crew passed into Mir for a welcoming ceremony. For the next 100 hours or so, the crews transferred equipment to and from Mir. They also conducted 15 separate biomedical and scientific experiments in the pressurized Spacelab module in the payload bay using the EO-18 crew as test subjects. Just prior to the undocking on 4 July, the EO-19 crew flew their Soyuz TM-21 spacecraft to a position where they could photograph *Atlantis* and Mir separating, providing the first fairly long-distance on-orbit photography of a space shuttle orbiter.

The returning crew of eight equaled the largest crew of the flight campaign (STS-30/61A). To ease their entry into gravity environment after more than 100 days in space, EO-18 crew members Norm Thagard, VladiMir Dezhurov, and Gennadi Strekalov lay supine in custom-made Russian seats on the middeck.

The crews in the Spacehab module. In the top row, from the left, are Charlie Precourt, Hoot Gibson, Greg Harbaugh, Ellen Baker, Norman Thagard, and Gennadi Strekalov. The bottom row includes Nikolai Budarin, Bonnie Dunbar, Anatoly Solovyev, and VladiMir Dezhurov. (NASA)

The orbiter docking system on Atlantis ready to capture Mir. The ODS had been developed by the Russians but was used extensively by space shuttle. Interestingly, this was the first electronically recorded color still image to be downlinked during the American space program. (NASA)

Charlie Precourt floats from Atlantis into the Mir Kristall module. Judging by Mir and the later ISS, space stations cannot be the neat, tidy platforms depicted in most science fiction movies since equipment and supplies are stored wherever there is available room. (NASA)

Atlantis commander Hoot Gibson (foreground) offers a wide smile and a handshake to Mir 18 commander VladiMir Dezhurov. This was the first time since the Apollo-Soyuz Test Project mission in 1975 that Americans and Russians had met each other in space. (NASA)

An unusual view of SRB separation from the Sony DSC-V1 Cyber-shot digital camera in the left orbiter umbilical. Normally this camera was not turned on until tank separation, but on STS-71 it captured four frames during SRB sep. It later captured its normal ET sep images. (NASA)

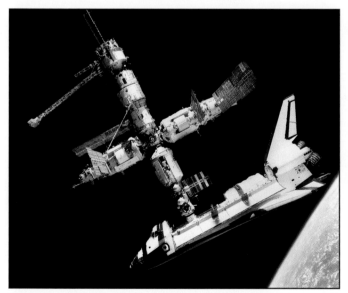

The newly arrived EO-19 crew undocked their Soyuz TM-21 temporarily to photograph the departure of the STS-71 and EO-18 crews aboard Atlantis. Note how far aft the external airlock and orbiter docking system are in the payload bay. (NASA)

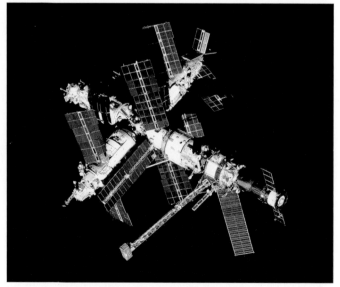

Mir as photographed from Atlantis on 29 June 1995. The Soviet Union started launching parts of Mir on 20 February 1986 using a Proton K booster. The last major piece, the Docking Module, would be delivered to the now-Russian station by Atlantis during STS-74. (NASA)

STS-70

Mission:	70	NSSDC ID:	1995-035A
Vehicle:	OV-103 (21)	ET-71 (LWT)	SRB: BI073
Launch:	LC-39B	13 Jul 1995	13:42 UTC
Altitude:	166 nm	Inclination:	28.45 degrees
Landing:	KSC-33	22 Jul 1995	12:03 UTC
Landing Rev:	143	Mission Duration:	214 hrs 20 mins

Commander:	Terence T. "Tom" Henricks (3)
Pilot:	Kevin R. Kregel (1)
MS1:	Donald A. "Don" Thomas (2)
MS2:	Nancy J. S. Currie (2)
MS3:	Mary Ellen Weber (1)

Payloads:	Up: 46,799 lbs	Down: 0 lbs
	TDRS-G/IUS (37,575 lbs)	
	No SRMS	

Notes:
Delayed 08 Jun 1995 (woodpeckers)
Rolled back to the VAB for repairs
Seventh and last space shuttle TDRS deployment
Last large satellite deployed by space shuttle
First use of the new Flight Control Room
First flight of a Block I space shuttle main engine

Wakeup Calls:

14 Jul	"Theme from Woody Woodpecker"
15 Jul	"Beautiful Ohio"
16 Jul	"God Bless the USA"
	(Ferguson Elementary School)
17 Jul	"Cleveland Indian Talkin' Tribe" (fight song)
18 Jul	"Beer Barrel Polka"
19 Jul	(unknown)
20 Jul	"Eyes of Texas"

The TDRS constellation was depicted by three gold stars representing the triad of spacecraft transmitting data to Earth. The stylized red, white, and blue ribbon represented the American goal of linking space exploration to the advancement of all mankind. (NASA)

This launch was first targeted for 22 June 1995; however, due to various Russian delays affecting the Spektr module and STS-71, NASA opted to swap the STS-70 and STS-71 launch dates and accelerated processing to ready *Discovery* and her payloads for a launch no earlier than 8 June, with *Atlantis* to follow as STS-71 later in the month. This schedule was abandoned following the Memorial Day weekend when Northern Flicker Woodpeckers poked 181 holes in the spray-on foam insulation that covered the external tank. The holes ranged from large excavations about 4 inches in diameter to single pecks and claw marks. Attempts to repair the damage at the pad were unsuccessful and the stack returned to the VAB on 8 June. After technicians repaired the insulation, the stack returned to the pad on 15 June. The countdown on 13 July proceeded smoothly. Don Thomas and Mary Ellen Weber deployed the seventh Tracking Data and Relay Satellite (TDRS-G) about six hours after launch and the attached inertial upper stage placed the satellite into geosynchronous orbit at 179.88 degree. This became the sixth TDRS placed in operational use since TDRS-B was lost with *Challenger*. It was the last large communications satellite deployed by space shuttle.

Secondary payloads included BDS (bioreactor demonstration system), BRIC (biological research in canisters), CPCG (commercial protein crystal growth), HERCULES-B (hand-held, earth-oriented, real-time, cooperative, user-friendly, location-targeting and environmental system), MAST (military applications of ship tracks), MIS-B (microencapsulation in space), PARE/NIH-R (physiological and anatomical rodent experiment/national institutes of health-rodents), RME-III (radiation monitoring experiment), STL/NIH-C (space tissue loss/national institutes of health-cells), SAREX-II (shuttle amateur radio experiment), VFT-4 (visual function tester), and WINDEX (windows experiment). HERCULES-B was the third generation of a Department of Defense experiment using a space-based geolocating system that tagged every frame of video with latitude and longitude with an accuracy of 3 miles.

The crew completed all entry stowage and deorbit preparations in preparation for entry on the nominal end-of-mission landing day. However, mission management waved-off both KSC opportunities on flight day 7 because of ground fog. The crew reopened the payload bay doors to provide cooling for the orbiter. The first opportunity the following day as also waved off due to weather. The weather finally cooperated on the second opportunity and *Discovery* returned to the Shuttle Landing Facility. Despite some minor issues, engineers considered STS-70 the "most trouble-free" mission to-date.

From the left are Kevin Kregel, Nancy Currie, Tom Henricks, Mary Ellen Weber, and Don Thomas. Note the outer covers of their launch-entry suits have a round velcro area where the crew patch normally affixes, despite the STS-70 patch being a vertical rectangle. (NASA)

This mission marked the first use of new sleep restraints, demonstrated here by Kevin Kregel on the middeck. After the flight, the crew commented the new restraints were more comfortable than the original ones and they became standard for future missions. (NASA)

Discovery is lowered into place on ET-71 in VAB High Bay 3. The red pieces over the various engine nozzles and around the ET attach points are protective covers used during ground processing and were removed after the elements were mated. (NASA)

A wide angle view from the back shows activity in the new White Flight Control Room at the Mission Control Center, which was dedicated during the STS-70 mission. Discovery was just passing over Florida at the time this photo was taken (note mercator map and TV scene on screens). It cost NASA about $50 million to develop the new MCC that replaced the original mainframe based, NASA-unique design with an industry-standard workstation-based, local area network system. The room is actually much smaller than it appears in photographs. (NASA)

STS-69

Mission:	71	NSSDC ID:	1995-048A
Vehicle:	OV-105 (9)	ET-72 (LWT)	SRB: BI074
Launch:	LC-39A	07 Sep 1995	15:09 UTC
Altitude:	201 nm	Inclination:	28.45 degrees
Landing:	KSC-33	18 Sep 1995	11:39 UTC
Landing Rev:	170	Mission Duration:	250 hrs 29 mins

Commander:	David M. "Dave" Walker (4)
Pilot:	Kenneth D. "Taco" Cockrell (2)
MS1:	James S. "Jim" Voss (3)
MS2:	James H. "Jim" Newman (2)
MS3:	Michael L. "Mike" Gernhardt (1)

Payloads:	Up: 38,855 lbs	Down: 0 lbs
	SPARTAN-201-03 (2,842 lbs)	
	WSF-2 (4,358 lbs)	
	IEH-1 (hitchhiker)	
	CAPL-2 (hitchhiker)	
	GAS (3)	
	SRMS s/n 303	

Notes:	Delayed 03 Aug 1995 (Hurricane Erin)
	Delayed 08 Aug 1995 (SRB concerns)
	Scrubbed 31 Aug 1995 (fuel cell)
	Second Dog Crew
	First time two spacecraft had been deployed and retrieved on the same mission
	EVA1 16 Sep 1995 (Voss and Gernhardt)

Wakeup Calls:

08 Sep	"Hound Dog"
09 Sep	"Theme from Scooby Doo"
10 Sep	"Bingo"
11 Sep	"Theme from Rin Tin Tin"
12 Sep	"A Hard Day's Night"
13 Sep	"Theme from Patton"
14 Sep	"Theme from Underdog"
15 Sep	"He's a Tramp" (*Lady and the Tramp*)
16 Sep	"Walk Like a Man"
17 Sep	"Theme from Peanuts (Snoopy's Theme)"

The original 3 August 1995 launch date was delayed by several events. On 1 August, NASA rolled *Endeavour* back to the VAB because Hurricane Erin was threatening central Florida. Subsequently, the stack returned to LC-39A on 8 August, but engineers became concerned over o-ring erosion on the solid rocket boosters found after STS-70 and STS-71 and delayed the flight until the issue could be further evaluated. After engineers decided the erosion was not a concern, NASA scheduled the launch for 31 August, but scrubbed it prior to tanking when a fuel cell failed during start-up. A smooth countdown preceded the liftoff on 7 September.

This was the second "Dog Crew," following STS-53, on which both Dave Walker and Jim Voss flew. Each STS-69 crew member adopted a nickname: Walker was "Red Dog," Ken Cockrell was "Cujo," Jim Voss was "Dogface," Jim Newman was "Pluto," and Mike Gernhardt was "Underdog." The crew posed with dog bowls at their preflight breakfast and the wakeup calls were decidedly dog related. Comically, somebody had smuggled dried dog food into the pantry on *Endeavour*, to which Walker said, "We haven't been up here long enough yet to be this hungry. But if we were able to cajole our management to let us stay up long enough, we would eat this."

The SPARTAN-201-03 (Shuttle Pointed Autonomous Research Tool for Astronomy) free-flyer studied the outer atmosphere of the Sun and its transition into solar wind that constantly flows past Earth. The timing of the SPARTAN flight coincided with passage of Ulysses over the north polar region of the Sun, expanding the range of data collected about the interaction between the Sun and its outflowing wind of charged particles. The crew released SPARTAN on flight day 2, but during rendezvous on flight day 4, the attitude of the free-flyer was not as expected since the spacecraft had gone into safe mode. The crew performed a fly-around and then successfully grappled SPARTAN with the SRMS and brought it back into the payload bay.

The other primary payload was the Wake Shield Facility (WSF-2). This was a 12-foot-diameter stainless steel disk designed to generate an "ultra-vacuum" environment to grow thin films for advanced electronics. The experiment had first flown, mostly unsuccessfully, aboard STS-60. This time the crew used the SRMS to deploy the WSF on flight day 5 and the free-flyer used its onboard cold gas propulsion to remain at a station-keeping distance of 23 to 35 miles. The free-flyer was retrieved on flight day 8. Despite several problems during the flight, it had completed four successful thin film growth runs.

The crew performed the deorbit burn for the first opportunity at the Shuttle Landing Facility.

The Wake Shield Facility (WSF) was represented by the astronaut symbol against a flat disk. The two stylized space shuttles highlight the ascent and entry phases of the mission and the two spiral plumes symbolize the deployment and retrieval of two spacecraft. (NASA)

In the front, from the left, are Ken Cockrell and Dave Walker while in the back are Jim Voss, Mike Gernhardt, and Jim Newman. Note the open airlock hatch at the right and the variety of photos taped to the middeck wall behind the crew members. (NASA)

Workers at Hangar AE on Cape Canaveral AFS lower the disk-shaped Wake Shield Facility onto its payload bay carrier. The carrier had a containment vessel that protected the spacecraft from possible contamination prior to deployment. (NASA)

The payload canister at the Vertical Processing Facility at KSC. At the top is the Wake Shield Facility, followed by the international extreme ultraviolet hitchhiker (IEH-1) and the capillary pumped loop (CAPL-2) hitchhiker. SPARTAN is out of frame above the WSF. (NASA)

Jim Voss during the only EVA of the mission. He was standing on a mobile foot restraint attached to the SRMS arm and was tethered to the SRMS as well. The EVA evaluated tools and enhancements to the space suits for the future space station. (NASA)

The Wake Shield Facility (WSF-2) just prior to being released by the SRMS. The three WSF flights aboard space shuttle proved the vacuum wake concept by growing gallium arsenide (GaAs) and aluminum gallium arsenide (AlGaAs) semiconductor thin films in space. (NASA)

The SRMS (at extreme right) moving into position to grapple SPARTAN-201-03. The spacecraft studied the outer atmosphere of the Sun and its transition into the solar wind that flows past the Earth. This was the third flight of SPARTAN-201 on space shuttle. (NASA)

STS-73

Mission:	72	NSSDC ID:		1995-056A	
Vehicle:	OV-102 (18)	ET-73 (LWT)		SRB: BI075	
Launch:	LC-39B	20 Oct 1995		13:53 UTC	
Altitude:	151 nm	Inclination:		39.00 degrees	
Landing:	KSC-33	05 Nov 1995		11:46 UTC	
Landing Rev:	255	Mission Duration:		381 hrs 52 mins	

Commander: Kenneth D. "Sox" Bowersox (3)
Pilot: Kent V. "Rommel" Rominger (1)
MS1: Catherine G. "Cady" Coleman (1)
MS2: Michael E. "LA" López-Alegría (1)
MS3: Kathryn C. "Kathy" Thornton (4)
PS1: Fred W. Leslie (1)
PS2: Albert "Al" Sacco, Jr. (1)

Payloads: Up: 33,705 lbs   Down: 0 lbs
USML-2 (27,398 lbs)
Extended duration orbiter pallet
No SRMS

Notes: Delayed 25 Sep 1995 (KSC work load)
Scrubbed 28 Sep 1995 (SSME-1 leak)
Delayed 05 Oct 1995 (Hurricane Opal)
Delayed 06 Oct 1995 (hydraulic system issue)
Scrubbed 07 Oct 1995 (SSMEC failure)
Delayed 14 Oct 1995 (SSME inspections)
Scrubbed 15 Oct 1995 (KSC weather)
First flight of a Block IA space shuttle main engine
Backup payload specialists were
   R. Glynn Holt and David H. Matthiesen

Wakeup Calls:

Since crew worked two shifts around the clock,
mission control did not send any wakeup calls.

In the foreground are the five "natural element" polyhedrons that were investigated by Plato and, later, Euclid representing fire, earth, air, and water. The shape of the emblem is a dodecahedron, a fifth element the Pythagoreans thought represented the cosmos. (NASA)

The launch of *Columbia* was scheduled for 25 September 1995 but was delayed due to work constraints at KSC. A launch attempt on 28 September was scrubbed prior to crew ingress when the main fuel valve on a main engine began leaking. NASA rescheduled the launch for 5 October, but weather from Hurricane Opal resulted in a one-day delay. The launch team scrubbed a 6 October attempt prior to ET loading when engineers determined hydraulic fluid had been inadvertently drained during the main engine replacement. The count on 7 October was halted by a failure in a space shuttle main engine controller. NASA rescheduled the launch for 14 October, but this slipped to allow time to inspect the main engine oxidizer ducts for cracks. An attempt on 15 October was first delayed and then scrubbed due to weather. The next attempt was tentatively scheduled for 19 October based on the launch of an Atlas at Cape Canaveral, but that slipped a day when the Atlas did not launch on time. This mission tied STS-32/61C for the most scrubs.

The crew split into two teams to allow around-the-clock operations. The red team consisted of Ken Bowersox, Kent Rominger, Kathy Thornton, and Al Sacco while the blue team included Cady Coleman, Michael López-Alegría, and Fred Leslie.

The second United States Microgravity Laboratory (USML-2) consisted of 14 experiments, 7 investigations, 1 demonstration, and 1 evaluation. Some of these were suggested by the results from USML-1 aboard STS-50 in 1992. For instance, that mission had provided new insights into theoretical models of fluid physics, the role of gravity in combustion and flame spreading, and how gravity affects the formation of semiconductor crystals. USML-2 built on that foundation and gathered data in five major scientific areas, including biotechnology, combustion science, commercial space processing technologies, fluid physics, and materials science. This was the longest Spacelab mission dedicated to microgravity research.

As had become customary for laboratory missions, science activities were directed from the Spacelab Mission Operations Control facility at MSFC. In addition, science teams at several NASA centers and universities monitored and supported the experiments.

The secondary payload was the orbital acceleration research experiment (OARE). This experiment, which flew on several missions, measured microgravity levels caused by atmospheric drag, changes in orbiter velocity due to solar and atmospheric effects, and vibrations of onboard equipment. It also provided a baseline environment definition for other experiments.

The weather at KSC was good and *Columbia* returned to the Shuttle Landing Facility on the first opportunity.

Michael López-Alegría has his arms folded at front center. Clockwise from him are Kathy Thornton, Cady Coleman, Al Sacco, Kent Rominger, Fred Leslie, and Ken Bowersox. The crew members are in shirts representing which of the science teams they were assigned to. (NASA)

Ken Bowersox retrieves a crow bar from the inflight maintenance (IFM) tool set onboard the Spacelab module. Every flight carried a variety of tools to fix whatever might go wrong during the mission; the orbiter was well outside the normal AAA service area. (NASA)

Kent Rominger is obviously confused since he is sitting in the commander's seat in the crew compartment trainer at JSC. Rommel was the pilot for this mission and should have been in the right-hand seat. The CCT was essentially identical to a flight vehicle. (NASA)

Fred Leslie working with the crystal growth furnace (CGF) in the pressurized Spacelab module. (NASA)

Al Sacco is assisted by two scuba divers as he hangs by his parachute harness during emergency egress training at JSC. (NASA)

Al Sacco rappels using a Sky-Genie during emergency egress training at JSC. Note the yellow markings saying "cut here"—telling emergency ground crews they could safely enter the vehicle at that location since there were no electrical cables or hydraulic lines in that area. (NASA)

Crew members in their launch-entry suits practice evacuating the orbiter using the crew compartment trainer at JSC. The inflatable slide was essentially the same type used by commercial airliners and allowed the crew to abandon the orbiter after an emergency landing. (NASA)

STS-74 (S/MM-02)

Mission:	73	NSSDC ID:	1995-061A
Vehicle:	OV-104 (15)	ET-74 (LWT)	SRB: BI076
Launch:	LC-39A	12 Nov 1995	12:31 UTC
Altitude:	185 nm	Inclination:	51.60 degrees
Landing:	KSC-33	20 Nov 1995	17:02 UTC
Landing Rev:	128	Mission Duration:	196 hrs 31 mins

Commander:	Kenneth D. "Ken" Cameron (3)
Pilot:	James D. "Jim" Halsell, Jr. (2)
MS1:	Chris A. Hadfield (1)
MS2:	Jerry L. Ross (5)
MS3:	William S. "Bill" McArthur, Jr. (2)

Payloads:	Up: 23,687 lbs	Down: 690 lbs
	Russian Docking Module (9,066 lbs)	
	GPP (hitchhiker)	
	SRMS s/n 301	

Notes:	Scrubbed 11 Nov 1995 (TAL weather)

Wakeup Calls:

13 Nov	"Dance of the Flowers"
14 Nov	"Yeager's Triumph" (*The Right Stuff*)
15 Nov	"Somewhere Over the Rainbow" (*Wizard of Oz*)
16 Nov	"Blue Danube"
17 Nov	"Northwest Passage"
18 Nov	"Do Wah Diddy Diddy"
19 Nov	"Theme from The Dream is Alive"

NASA scrubbed the first launch attempt on 11 November at T-5 minutes because of poor weather at the Zaragoza and Moron TAL sites. The count the following day proceeded smoothly with *Atlantis* lifting-off at the beginning of a ten-minute window with no unplanned holds. This was the fourth of nine Mir missions and the second docking. This marked the first time astronauts from America, Canada, ESA, and Russia were in space on the same complex at the same time.

This mission delivered the Russian Docking Module and two solar arrays to Mir. The module, built by RCS-Energia, simplified future space shuttle dockings to the space station by providing additional clearance between the orbiter and Mir. The Mir crew installed the solar arrays on the Docking Module after the orbiter departed. Secondary payloads included GLO-4 (spacecraft glow experiment), ICBC (IMAX cargo bay camera), PASDE (photogrammetric appendage structural dynamics experiment), and SAREX-II (shuttle amateur radio experiment).

On flight day 1, the crew depressurized the crew module to 10.2 psia to prepare for a contingency EVA in the event the Docking Module could not be mated to the androgynous peripheral docking system (APDS). Ultimately, the EVA was not needed and the pressure was restored to normal. On flight day 3, Chris Hadfield used the SRMS to grapple the Docking Module and move it within 5 inches of the APDS capture ring in preparation for the thrusting sequence designed to force capture. Ken Cameron and Jim Halsell fired six downward RCS jets and succeeded to capture the ring. Once locked together, the APDS docking ring retracted and a series of hooks and latches engaged to complete docking.

The crew pressurized the Docking Module using gases from *Atlantis* and then opened the hatch to the Docking Module and inspected it. Ken Cameron and Jim Halsell then maneuvered *Atlantis* toward Mir using three OMS firings and docked to the Kristall module using the top androgynous unit on the Docking Module. The success of this operation provided confidence for the STS-88/ISS-2A assembly flight that would use the same technique to install Node 1 (Unity) to the Russian Functional Cargo Block (FGB, Zarya). *Atlantis* spent three days docked with Mir.

On 18 November, *Atlantis* undocked from the bottom androgynous unit, leaving the Docking Module attached to Kristall where it provided clearance between the Mir solar arrays and the orbiters during future dockings. The crew completed the deorbit preparations for entry and landed at the Shuttle Landing Facility on the first opportunity.

The central focus of the crew patch was the Russian-built Docking Module. The rainbow across the horizon represented the atmosphere, the thin membrane protecting all nations, while the three flags across the bottom show those nations participating in the mission. (NASA)

The crew took their on-orbit portrait in the Docking Module, which would greatly simplify future Mir dockings and also taught procedures for the eventual ISS. Clockwise from the lower left are Jim Halsell, Chris Hadfield, Jerry Ross, Ken Cameron, and Bill McArthur. (NASA)

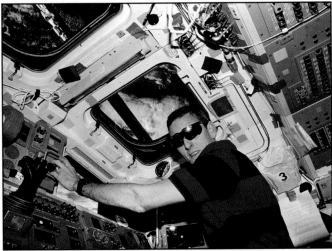

Bill McArthur at the SRMS controls on the aft flight deck. The two overhead windows and two windows into the payload bay, along with cameras on the arm itself, allowed the operator to see pretty much everything the SRMS could reasonable reach. (NASA)

Chris Hadfield makes his way among supplies and logistics hardware aboard Mir. Even more so than the future ISS, Mir had very limited storage for materiel. Note the latches and double pressure seals (gaskets) along the outer rim of the hatch frame. (NASA)

The STS-74 and EO-20 crews pose for a group portrait in the Mir Base Block. The EO-20 crew included, from the left, Sergei Avdeyev, Thomas Reiter, and Yuri Gidzenko. The STS-74 crew is in the two-tone shirts while the Russians wore their normal orbital uniforms. (NASA)

Chris Hadfield used the SRMS to place the Docking Module on top of the orbiter docking system, which itself was on top of the external airlock. After the Docking Module was securely attached to the ODS, Atlantis maneuvered to mate it to the Kristall module on Mir. (NASA)

Mir as seen from Atlantis as she headed home. The new Docking Module, with its distinctive orange color, is at the bottom of the photo. The Russians called the module the Stykovochnyy Otsek (SO) and assigned it a designation of GRAU index 316GK. (NASA)

STS-72

Mission:	74	NSSDC ID:		1996-001A	
Vehicle:	OV-105 (10)	ET-75 (LWT)		SRB: BI077	
Launch:	LC-39B	11 Jan 1996		09:41 UTC	
Altitude:	257 nm	Inclination:		28.45 degrees	
Landing:	KSC-15	20 Jan 1996		07:43 UTC	
Landing Rev:	142	Mission Duration:		214 hrs 01 mins	

Commander: Brian Duffy (3)
Pilot: Brent W. Jett, Jr. (1)
MS1: Leroy Chiao (2)
MS2: Winston E. Scott (1)
MS3: Koichi Wakata (1)
MS4: Daniel T. "Dan" Barry (1)

Payloads: Up: 23,661 lbs   Down: 7,670 lbs
OAST-Flyer (2,579 lbs)
Space Flyer Unit (7,670 lbs downmass)
SLA-01 (hitchhiker)
(GAS (3)
SRMS s/n 303

Notes: SFU launched by an H-II on 18 March 1995
EVA1 14 Jan 1995 (Chiao and Barry)
EVA2 16 Jan 1995 (Chiao and Scott)
Crew and families featured in PBS "Astronauts"

Wakeup Calls:

12 Jan	"Theme from Star Wars"
13 Jan	"Sea in Springtime" (traditional Japanese song)
14 Jan	Theme from the original *Godzilla*
15 Jan	"Theme from Star Trek Next Generation"
16 Jan	"Smallest Astronaut"
17 Jan	"Heigh-Ho, Heigh-Ho, It's Off To Work We Go"
18 Jan	"All I Wanna Do"
19 Jan	"Darth Vader's Them" (*Star Wars*)
20 Jan	"Theme from The Dream is Alive"

The inner gold border represents the distinct octagonal shape of the Space Flyer Unit and the OAST-Flyer is shown just after release from the SRMS. The stars represent the hometowns of the crew in America and Japan. Notably, the EVA astronaut is wearing a crew patch. (NASA)

The launch window on 11 January 1996 was 49 minutes long, based on the position of the Space Flyer Unit (SFU) that Japan had paid NASA $50 million to recover. The National Space Development Agency of Japan (NASDA) had launched the SFU from the Tanegashima Space Center on 18 March 1995 aboard an H-II booster. The spacecraft spent ten months conducting research in astronomy, biology, engineering, and materials science.

Endeavour rendezvoused with the SFU on flight day 3 and Japanese ground controllers made several attempts to verify the solar panels were completely retracted; all failed and the panels were jettisoned as a precaution. NASA had incorporated this contingency into preflight training, but the procedure delayed capture by 93 minutes. Koichi Wakata then grappled the SFU with the SRMS and berthed it in the payload bay.

The other primary objective was the operation of the NASA Office of Aeronautics and Space Technology OAST-Flyer. This was the seventh in a series of missions using the reusable SPARTAN (Shuttle Pointed Autonomous Research Tool for Astronomy) spacecraft. Experiments aboard the spacecraft included REFLEX (return flux experiment) to test accuracy of computer models predicting spacecraft exposure to contamination, GADACS (global positioning system attitude determination and control experiment) to demonstrate GPS technology in space, SELODE (solar exposure to laser ordnance device) to test laser ordnance devices, SPRE (spartan packet radio experiment), and W3EAX (Amateur Radio Association at the University of Maryland amateur radio communications experiment). Koichi Wakata again operated the SRMS to deploy the OAST-Flyer and retrieve it after 46 hours of free flight. Oddly, as *Endeavour* maneuvered to rendezvous with OAST-Flyer, the crew noted it was 63 miles from the predicted retrieval position.

On flight day 5, Leroy Chiao and Dan Barry evaluated a new portable work platform and a rigid umbilical that might be used to route various fluid and electrical lines on the space station. Chiao and Winston Scott continued the evaluation during a second EVA on flight day 7. The crew members were favorably impressed with the new EMU lights and the body restraint tether, but did not use a pair of new actively heated gloves. During the second EVA, both crew members evaluated new electronic cuff checklists to replace the traditional flip cards used to provide task directions. The electronic cuffs did not prove particularly successful since they were difficult to read in direct sunlight and used a comparatively small font.

The crew closed the payload bay doors and performed the deorbit burn for the first KSC landing opportunity.

In front, from the left, are Dan Barry, Brian Duffy, and Leroy Chiao. In the rear are Koichi Wakata, Brett Jett, and Winston Scott. Flying in the bright sunlight, the crew had put shades in the aft flight deck windows to prevent heavy shadowing. (NASA)

This was the eighth night landing of the flight campaign, using the same definition as for launch. Despite the appearance, the main engines were not operating and the bright spots are simply a reflection off the injector plate that normally supplied propellants into the engines. (NASA)

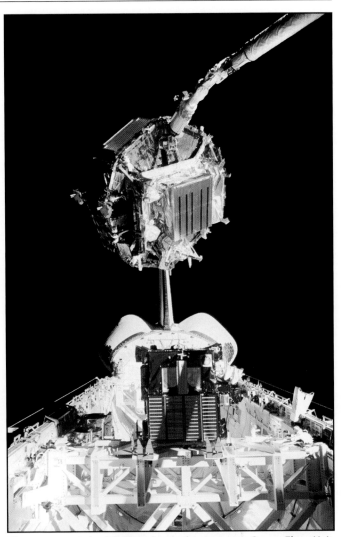

The crew used the SRMS to berth the Japanese Space Flyer Unit. The SRMS could only capture objects equipped with the appropriate grapple fixture, making it impossible to retrieve uncooperative or unknown targets (such as Russian spy satellites). (NASA)

This was the twelfth night launch of the flight campaign and the third for Endeavour. NASA defined a night launch as one between 15 minutes after nautical sunset and 15 minutes before nautical sunrise. Ultimately there would be 34 night launches, mostly to support Mir and ISS. (NASA)

The rotating service structure is rolled back at LC-39B to reveal the full Endeavour stack. The "Go Endeavour" banner was provided by the Space Flight Awareness program that attempted to instill pride and dedication in the workforce. (NASA)

STS-75

Mission:	75	NSSDC ID:	1996-012A
Vehicle:	OV-102 (19)	ET-76 (LWT)	SRB: BI078
Launch: Altitude:	LC-39B 173 nm	22 Feb 1996 Inclination:	20:18 UTC 28.46 degrees
Landing: Landing Rev:	KSC-33 252	09 Mar 1996 Mission Duration:	13:59 UTC 377 hrs 40 mins

Commander:	Andrew M. "Andy" Allen (3)
Pilot:	Scott J. "Doc" Horowitz (1)
MS1:	Jeffrey A. "Jeff" Hoffman (5)
MS2:	Maurizio Cheli (1)
MS3:	Claude Nicollier (3)
MS4:	Franklin R. Chang-Diaz (5)
PS1:	Umberto Guidoni (1)

Payloads:	Up: 32,006 lbs	Down: 0 lbs
	USMP-3 (30,515 lbs)	
	TSS-1R (1,396 lbs)	
	Extended duration orbiter pallet	
	No SRMS	

Notes:	Fictional subject of an internet hoax on microgravity sex experiments

Wakeup Calls:

Since crew worked two shifts around the clock, mission control did not send any wakeup calls.

The Tethered Satellite is passing through Earth's magnetic field and the tether is crossing Earth's terminator signifying the dawn of a new era for space tether applications and in mankind's knowledge of the ionosphere, material science, and thermodynamics. (NASA)

The countdown for this *Columbia* mission on 22 February 1996 resulted in an on-time launch. However, approximately six seconds after liftoff, Andy Allen and Doc Horowitz reported the left SSME chamber pressure tape meter on the flight deck was reading incorrectly, indicating approximately 40 percent thrust instead of 104 percent thrust. The crew saw the meter tracked the other chamber pressure meters throughout ascent, but with a 60 percent bias. Telemetry showed the engine was operating normally.

The primary payloads were the third United States Microgravity Payload (USMP-3) and the reflight of the Tethered Satellite System (TSS-1R). This was the second attempt to conduct the TSS experiment, the first ending poorly aboard STS-46 in 1992. The crew began operating TSS-1R on flight day 4. All went to plan until, after deploying 64,627 of an expected 68,640 feet of cable, the 0.1-inch-diameter tether broke, resulting in the loss of the satellite. Before the tether broke, instruments recorded 3,500 volts and 500 milliamps, largely validating the experiment. Although the objectives were not fully met, all science instruments were operating and gathering data during the five-hour period prior to the tether break.

The USMP-3 consisted of four major payloads mounted on two mission peculiar experiment support structures (MPESS) in the payload bay. These included AADSF (advanced automated directional solidification furnace), IDGE (isothermal dendritic growth experiment), MEPHISTO (material pour l'etude des phenomenes interessant la solidification sur terre et en orbite), and ZENO (critical fluid light scattering experiment). There were also three smaller experiments on the middeck: CPCG-IV (commercial protein crystal growth), OARE (orbital acceleration research experiment), and SAMS (space acceleration measurement system).

UFO enthusiasts, and various conspiracy theorists, widely circulated excerpts of a video shot during STS-75 claiming that visual anomalies in the footage represent an unexplained paranormal phenomenon. The crew identified the UFOs as small particles of debris filmed out of focus and journalist James Oberg wrote an analysis of the footage that debunked most of the claims, to no avail. In addition, "STS-75" was described in the fictional NASA document 12-571-3570 that became available several years before this mission was launched. The document purports to report on experiments to determine effective sexual positions in microgravity. It was pure fiction and bore no resemblance to anything that happened on this mission (or any other).

Mission control waved-off the first KSC landing opportunity due to cloud cover, but the second opportunity was successful.

On the left, front to back, are Franklin Chang-Diaz, Maurizio Cheli, and Scott Horowitz. On the right are Jeff Hoffman, Umberto Guidoni, Andy Allen, and Claude Nicollier. The crew is posed in front of the crew hatch while the vehicle was at LC-39B. (NASA)

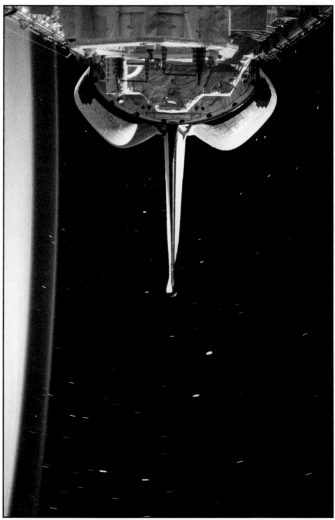

Columbia's vertical stabilizer appears to point to the four stars of the Southern Cross. The empty cradle for the Tethered Satellite System can be seen closest to the camera, with USMP-3 and the extended duration orbiter pallet behind it. (NASA)

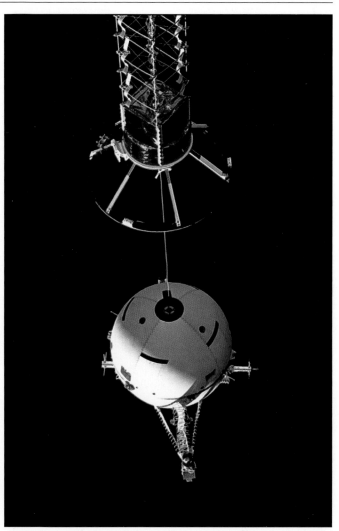

The Tethered Satellite System and part of its supportive boom device prior to deployment operations. After deploying 64,627 of an expected 68,640 feet of cable, the 0.1-inch-diameter tether broke, resulting in the loss of the Italian satellite. (NASA)

Jeff Hoffman (left) and Franklin Chang-Diaz hold up a sign to celebrate the fact that each had surpassed the 1,000-hour mark in space during the mission. Both of the astronauts had flown five times. They were posed on the forward flight deck of Columbia. (NASA)

The Tethered Satellite System showing the round satellite and its cradle. The tether reel, holding some 69,000 feet (13 miles) of cable, is below the cradle. The entire assembly would be mounted on a Spacelab pallet and placed in the payload bay. (NASA)

STS-76 (S/MM-03)

Mission:	76	NSSDC ID:	1996-018A
Vehicle:	OV-104 (16)	ET-77 (LWT)	SRB: BI079
Launch:	LC-39B	22 Mar 1996	08:13 UTC
Altitude:	216 nm	Inclination:	51.60 degrees
Landing:	EDW-22	31 Mar 1996	13:30 UTC
Landing Rev:	145	Mission Duration:	221 hrs 16 mins

Commander:	Kevin P. "Chilli" Chilton (3)
Pilot:	Richard A. "Rick" Searfoss (2)
MS1:	Ronald M. "Ron" Sega (2)
MS2:	Michael R. "Rich" Clifford (3)
MS3:	Linda M. Godwin (3)
Up:	Shannon W. Lucid (5)

Payloads:	Up: 24,605 lbs	Down: 736 lbs
	Spacehab SM (10,387 lbs)	
	GAS (1)	
	No SRMS	

Notes:
Delayed 21 Mar 1996 (KSC weather)
Last flight of a Phase II space shuttle main engine
First operational use of SAFER devices
EVA1 26 Mar 1996 (Godwin and Clifford)
Backup Mir crew member was
  John E. Blaha
Orbiter returned to KSC on 12 Apr 1996 (N905NA)

Wakeup Calls:

23 Mar	"Anywhere is ..."
24 Mar	"When the Roll is Called Up Yonder"
25 Mar	"Awake, the Harp"
26 Mar	"Another Saturday Night" (Max-Q)
27 Mar	"Free Flying"
28 Mar	"Jumpin' at the Woodside"
29 Mar	"Stars and Stripes Forever"

The "Spirit of 76," an era of new beginnings, is represented by the orbiter rising through the circle of 13 stars in the Betsy Ross flag. The three gold trails and the ring of stars in union form the astronaut symbol. The patch was partly designed by Brandon Clifford, age 12. (NASA)

The launch director called a 24-hour scrub prior to loading the ET on 21 March 1996 because of high winds and rough seas around the launch site. The count on 22 March resulted in an on-time *Atlantis* launch. Post-launch inspections of MLP-3 revealed a 63-foot long crack on one of the steel plates running from the north end of the left flame hole to the north side of the MLP surface. Cracks were sometimes found and easily repaired on the MLPs following launch, but this one was particularly large.

Atlantis carried a Spacehab single module with a large quantity of equipment slated for transfer to Mir, as well as the European Space Agency Biorack experiment. The Biorack shared a double rack in Spacehab with the LSLE (life sciences laboratory equipment) freezer and contained 11 experiments from America, France, Germany, Switzerland, and The Netherlands. These studied the effect of microgravity and cosmic radiation on plants, tissues, cells, bacteria, and insects and the effects of microgravity on bone loss.

The crew rendezvoused with Mir on flight day 3, following the same R-bar approach used during STS-74. During five days of docked operations, the crew transferred 1,506 pounds of water and 4,787 pounds of materiel to Mir. They also transferred the MBGX (Mir glovebox) to replenish a glovebox already on the station, and the QUELD (Queen's University experiment in liquid diffusion and LPS (high-temperature liquid phase sintering) experiments.

On flight day 6, Linda Godwin and Rich Clifford conducted the first American EVA around two mated spacecraft. The major activity was attaching four MEEP (Mir environmental effects payload) experiments to the Docking Module. These experiments were designed to characterize the environment around Mir over an 18-month period. Godwin and Clifford wore the SAFER (simplified aid for EVA rescue) propulsive devices first flight tested during STS-64, marking their first operational use.

On 28 March, the mission management team shortened the flight by one day due to weather concerns at the Shuttle Landing Facility on 31 March and 1 April. But the weather deteriorated sooner than expected and mission management waved-off the two opportunities on 30 March. They waved-off the first KSC opportunity on 31 March because of thunderstorm activity and *Atlantis* landed at Edwards.

Unusually, there was a problem on the ferry flight to KSC. *Atlantis* departed Edwards atop of N905NA on 6 April, but a fire warning indicator for the right inboard engine on the 747 convinced Gordon Fullerton and Thomas McMurtry to shut down the engine and return to Edwards. The engine was replaced and the SCA carrying *Atlantis* arrived at KSC on 12 April.

In the front, from the left, are Ron Sega, Kevin Chilton, and Rick Searfoss. In the back are Rich Clifford, Shannon Lucid, and Linda Godwin. During the mission, Clifford and Godwin performed the first extravehicular activity during Shuttle-Mir docked operations. (NASA)

This view, taken from an overhead window in Atlantis, shows most of the elements of Mir. The Base Block is in the near foreground and a small piece of the Docking Module delivered by STS-74 is at top. Note Soyuz TM-23 docked at bottom left. (NASA)

Taken by the EO-21 crew aboard Mir on 23 March, this photo shows the payload bay configuration. Atlantis initially carried her external airlock further aft than would become standard for future missions to the ISS. Note the long tunnel connecting the Spacehab module. (NASA)

Rich Clifford preparing to move a gyrodyne from the Spacehab module aboard Atlantis to the Mir space station. The gyrodyne was later installed in the Mir Base Block. As their name implied, the spinning gyrodynes provided attitude control for Mir. (NASA)

Powerful Xenon lights attempted to illuminate the landing area as Atlantis approached the concrete runway at Edwards AFB. Each Xenon light emitted 1,000 million candlepower. Note the Gulfstream II Shuttle Training Aircraft barely visible at the top of the frame. (NASA)

Kevin Chilton (left) and Rick Searfoss practicing in Atlantis on 6 March 1996 at the Kennedy Space Center. Both are wearing the new advanced crew escape suits, although the rest of the crew would wear the older launch-entry suits during the mission. (NASA)

STS-77

Mission:	77	NSSDC ID:	1996-032A
Vehicle:	OV-105 (11)	ET-78 (LWT)	SRB: BI080
Launch: Altitude:	LC-39B 154 nm	19 May 1996 Inclination:	10:30 UTC 39.03 degrees
Landing: Landing Rev:	KSC-33 161	29 May 1996 Mission Duration:	11:10 UTC 240 hrs 39 mins

Commander:	John H. Casper (4)
Pilot:	Curtis L. "Curt" Brown, Jr. (3)
MS1:	Andrew S. "Andy" Thomas (1)
MS2:	Daniel W. "Dan" Bursch (3)
MS3:	Mario "Trooper" Runco, Jr. (3)
MS4:	J. Marc Garneau (2)

Payloads:	Up: 37,042 lbs	Down: 0 lbs
	Spacehab SM (8,948 lbs)	
	SPARTAN-207 (1,878 lbs)	
	PAMS/STU (115 lbs)	
	TEAMS (hitchhiker)	
	GAS (11)	
	SRMS s/n 301	

Notes:	Delayed 16 May 1996 (range availability)
	First (and only) God Crew
	First flight of three Block I SSMEs
	Carried Coca-Cola fountain dispenser (FGBA-2)

Wakeup Calls:

20 May	The Air Force Hymn – "Wild Blue Yonder"
21 May	"Up, Up and Away"
22 May	The Navy Hymn – "Anchors Aweigh"
23 May	"Milky Way" (children's song)
24 May	"Hold Me, Thrill Me, Kiss Me"
25 May	"Down Under"
26 May	"Up Down, and Touch the Ground" (*Winnie the Pooh*)
27 May	"Light My Fire"
28 May	"Start Me Up"
29 May	"I Can See Clearly Now"

Endeavour *and its reflection within the parabolic mirror of the inflatable antenna experiment. The center leg of the tripod also delineates the flat top of the Spacehab shape. The STS designation is featured as twin stylized chevrons adapted from the NASA logo. (NASA)*

NASA moved the desired 16 May 1996 launch date to the 19th because the Eastern Range was not available. There were no unscheduled holds in the count and *Endeavour* launched on time. This was the first mission completely controlled from the new White Flight Control Room at JSC, finally replacing the Apollo-era rooms used for the first 76 missions.

Much like the "Dog Crews" of STS-53 and STS-69, this became the "God Crew." Marc Garneau had been the first Canadian to fly in space on STS-17/41G in 1984. The other STS-77 crew members teased him that it was such a long time ago, that it must have been back in the Apollo days. The name "Apollo" stuck and slowly the entire crew received nicknames of mythological gods. John Casper became "Zeus," Curt Brown was "Saturn," Andy Thomas was "Thor," Mario Runco was "Neptune," and Dan Burch was "Pan" (later, "Baccus").

The Spacehab single module carried nearly 3,100 pounds of experiments and support equipment for 12 commercial space products in biotechnology, electronic materials, polymers, and agriculture as well as several experiments for other NASA organizations. One of these, CFZF (commercial float zone facility) was developed through international collaboration between America, Canada, and Germany. It heated various samples of semiconductor material through the float-zone technique. Spacehab also carried the SEF (space experiment facility) that grew crystals using vapor diffusion.

The crew deployed the SPARTAN-207 (Shuttle Pointed Autonomous Research Tool for Astronomy) free-flyer on flight day 2 to test the performance of a large inflatable antenna experiment (IAE) during a 90-minute deployment. The 132-pound IAE structure, mounted on three struts, inflated to its full 50-foot diameter. The potential benefits of inflatable antennas, and inflatable structures in general (such as the much later Bigelow ISS module) included their lower development costs and greater reliability, as well as the possible use of a smaller, lower-cost launch vehicle since they required less stowage space and volume. After the test, the crew jettisoned the antenna structure, then retrieved the free-flyer using the SRMS and berthed it in the payload bay.

The crew used an improved Coca-Cola dispenser (officially the fluids generic bioprocessing apparatus, FGBA-2) that held 1.65 liters each of Coca-Cola, Diet Coke, and Powerade. Earlier in the program, Coke had flown experiments on STS-26/51F and STS-63, while Pepsi had half-heartedly flown an experiment on STS-26/51F.

John Casper and Curt Brown performed the deorbit maneuver for the first opportunity at the Shuttle Landing Facility.

From the left, are Dan Bursch, Curtis Brown, Mario Runco, Marc Garneau, John Casper, and Andy Thomas. The Canadian flag at right is for Garneau, who is also wearing a Canadian flag on his left shoulder. Note Brown and Casper have silver Air Force wings. (NASA)

The History of the American Space Shuttle

An improved fluids generic bioprocessing apparatus (FGBA-2) (see page 185 for an earlier experiment). This time the drinks, including Powerade, were mixed from water and syrup and then carbonated rather than being dispensed from pre-mixed containers. (NASA)

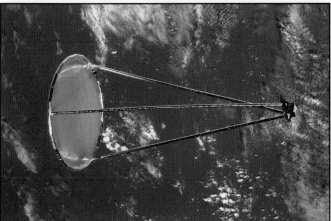

The 132-pound inflatable antenna experiment (IAE) antenna, mounted on three struts, inflated to its full 50-foot diameter, resulting in a structure about the size of a tennis court. The IAE laid the groundwork for future development in inflatable space structures. (NASA)

The SPARTAN-207 free-flyer being held by the SRMS in a low-hover mode above its berth in the payload bay. The Spacehab module can be seen in the foreground. Spacehab was a commercial venture, hence being able to "advertise" by having their name on the module. (NASA)

The Endeavour stack on its way to LC-39B. Note the black stripe around the top of the left solid rocket booster, a photo-reference mark that helped engineers and analysts identify it in photography after it separated from the external tank. (NASA)

STS-78

Mission:	78	NSSDC ID:	1996-036A
Vehicle:	OV-102 (20)	ET-79 (LWT)	SRB: BI081
Launch:	LC-39B	20 Jun 1996	14:49 UTC
Altitude:	154 nm	Inclination:	39.03 degrees
Landing:	KSC-33	07 Jul 1996	12:38 UTC
Landing Rev:	272	Mission Duration:	405 hrs 48 mins

Commander:	Terence T. "Tom" Henricks (4)
Pilot:	Kevin R. Kregel (2)
MS1:	Richard M. "Rick" Linnehan (1)
MS2:	Susan J. Helms (3)
MS3:	Charles E. "Chuck" Brady, Jr. (1)
PS1:	Jean-Jacques Favier (1)
PS2:	Robert B. "Bob" Thirsk (1)

Payloads:	Up: 31,854 lbs	Down: 0 lbs
	Spacelab LMS (27,642 lbs)	
	Extended duration orbiter pallet	
	No SRMS	

Notes:	Backup payload specialists were
	Pedro F. Duque Duque and Luca Urbani

Wakeup Calls:

21 Jun	"Free Fallin'"
22 Jun	"Bad to the Bone"
23 Jun	"Flight of the Bumblebee"
24 Jun	"Space Oddity"
25 Jun	"She Blinded Me With Science"
26 Jun	"Back on the Chain Gang"
27 Jun	"Every Breath You Take"
28 Jun	"Carolina in My Mind"
29 Jun	"Another Saturday Night" (Max-Q)
30 Jun	"Les Murs De Poussiere"
01 Jul	"Oh, Canada"
02 Jul	"Closer to Free"
03 Jul	"Wake Up Little Susie"
04 Jul	"God Bless the U.S.A."
05 Jul	"Birthday" (for Henricks)
06 Jul	"Don't Bring Me Down"
07 Jul	"Time For Me to Fly"

Columbia, *whose shape evokes the image of the eagle, an icon of power and prestige, and the national symbol of the United States. The eagle's feathers represent both peace and friendship. An orbit surrounding the STS designation recalls the red chevron in the NASA meatball.* (NASA)

The launch of *Columbia* on 20 June proceeded smoothly with no unscheduled holds. NASA had begun manifesting the Life and Microgravity Spacelab (LMS) in September 1994 and flew the mission in June 1996, meaning the planning and integration had taken only 21 months. This was the shortest planning period for any comparable effort, far shorter than the three to four years typical for other Spacelab missions. As a result, the costs were approximately half those of other Spacelab missions, answering one of the primary criticisms that had driven NASA to lease the (somewhat) competing Spacehab modules. There were 40 scientific investigations associated with LMS involving research sponsored by the NASA Office of Life and Microgravity Sciences and Applications (OLMSA). These included 24 microgravity science and 16 life sciences investigations, supported by data from three accelerometers.

Microgravity experiments included interfacial fluid physics studies of thermocapillary flow and electrohydrodynamics performed in the bubble, drop, and particle unit (BDPU). The mission also included an extensive set of microgravity experiments to study the morphology of semi-conductors and metal alloys in the advanced gradient heating facility (AGHF). These materials were formed under different processing conditions to achieve morphologies that included single crystals, equiaxed dendrites, and columnar dendrites. The final group of microgravity experiments was protein crystal growth investigations performed in the advanced protein crystallization facility (APCF), which included on-orbit observations of growth and post-flight crystallography studies.

The various life sciences investigations included studies on the role of corticosteroids in bone loss performed on rats in the animal enclosure module (AEM), lignin formation in the plant growth facility (PGF), and the development of medaka fish eggs performed in the space tissue loss module (STLM). In addition, a number of human physiology experiments involving musculoskeletal studies were performed.

Scientists on the ground remotely commanded the majority of the experiments, with the crew providing whatever hands-on attention was required. Teams from around the world monitored the LMS experiments. The American sites included JSC, KSC, MSFC, and NASA Lewis in Cleveland, Ohio. European sites in Brussels, Belgium, Milan and Naples, Italy, and Toulouse, France were also active during the flight.

Columbia returned to the Shuttle Landing Facility on the first opportunity on flight day 16, marking the end to the longest mission to-date and the second longest of the flight campaign.

Seated are Tom Henricks (left) and Kevin Kregel. Standing, from the left, are Jean-Jacques Favier, Rick Linnehan, Susan Helms, Chuck Brady, and Bob Thirsk. Favier was with the French Atomic Energy Commission (FAEC) and represented the French Space Agency (CNES). (NASA)

The round object with several holes at top center of the pressurized module was a vent and relief valve assembly. This protected the module from inadvertent overpressure and also provided a location to vent experiment gases overboard if needed. (NASA)

Chuck Brady working in Spacelab. The Europeans designed and built Spacelab and apparently did not like the dull gray and off-white used in American spacecraft or the green used by the Russians. Instead they opted for a pale yellow for most panels. (NASA)

Most launches looked essentially the same, although the distinctive black chines and SILTS pod made Columbia easy to identify. The GOX vent arm, and its characteristic "beanie cap" can be seen in the retracted position just under the back of the hammerhead crane. (NASA)

The crew departs the Operations and Checkout Building for their ride to the launch pad. At this point the crew had eaten breakfast (regardless of the time of day) and were dressed in their pressure suits. Gloves and helmets would not be donned until they were in the ship. (NASA)

STS-79 (S/MM-04)

Mission:	79	NSSDC ID:	1996-057A
Vehicle:	OV-104 (17)	ET-82 (LWT)	SRB: BI083
Launch: Altitude:	LC-39A 211 nm	16 Sep 1996 Inclination:	08:55 UTC 51.60 degrees
Landing: Landing Rev:	KSC-15 160	26 Sep 1996 Mission Duration:	12:15 UTC 243 hrs 18 mins

Commander:	William F. "Reads" Readdy (3)
Pilot:	Terrence W. "Terry" Wilcutt (2)
MS1:	Jerome "Jay" Apt III (4)
MS2:	Thomas D. "Tom" Akers (4)
MS3:	Carl E. Walz (3)
Up:	John E. Blaha (5)
Down:	Shannon W. Lucid (5)

Payloads:	Up: 27,812 lbs	Down: 2,126 lbs
	Spacehab LDM (15,555 lbs)	
	No SRMS	

Notes:
Rolled-back to VAB 10 Jul 1996 (SRB concerns)
Rolled-back to VAB 04 Sep 1996 (Hurricane Fran)
First flight of Spacehab double module
First crew rotation mission (Blaha/Lucid)
Backup Mir crew member was
  Jerry M. Linenger

Wakeup Calls:

16 Sep	"Duke of Earl"
17 Sep	"Rescue Me"
18 Sep	"Hold On (I'm Coming)"
19 Sep	"Whole Lotta Shakin' Goin' On"
20 Sep	"Cheeseburger in Paradise"
21 Sep	"Another Saturday Night" (Max-Q)
22 Sep	"Got Me Under Pressure"
23 Sep	"Please Don't Leave Me"
24 Sep	"Only Wanna Be With You"
25 Sep	"Danger Zone" (*Topgun*)

The direction of their names either up (Blaha) or down (Lucid) denotes transport up to the Mir space station or return to Earth on STS-79. The crew patch is in the shape of the orbiter airlock hatch, symbolizing the gateway to international cooperation in space. (NASA)

Atlantis was rolled-out to LC-39A early in the morning of 1 July 1996 for a planned 31 July launch. But concerns about sooting in the STS-78 solid rocket motor field joints caused NASA to roll-back the vehicle to the VAB on 10 July. An investigation into the seepage identified the most probable cause as being the use of a new water-based pressure-sensitive adhesive adopted to comply with Environmental Protection Agency (EPA) regulations to reduce ozone-depleting substances. Since the adhesive had already been used in several stacks, technicians played musical boosters to find a suitable set for this launch. *Atlantis* was rolled-out to the pad on 21 August with a launch date of 14 September. However, Hurricane Fran threatened to make landfall and NASA rolled the stack back to the VAB for the second time on 4 September. The storm passed overnight and the vehicle was again rolled-out to LC-39A the following day. The countdown on 16 September proceeded smoothly. It was a well-traveled vehicle before it ever lifted-off.

The fourth Mir docking proceeded without incident. This was the first space shuttle mission to rendezvous with a fully assembled Mir, following the arrival of its Priroda module on 26 April 1996. However, of most interest to the popular media was the return to Earth of Shannon Lucid after 188 days in space, a new American record, as well as then-record for a woman astronaut.

This was the first flight of the Spacehab logistics double module. The forward portion housed three experiments conducted by the crew: CPCG, (commercial protein crystal growth), MGM (mechanics of granular materials, and ETTF (extreme temperature translation furnace). The aft portion contained material to be transferred to Mir and would house trash on the way home. The crew transferred about 4,000 pounds of food, clothing, supplies, and spare equipment from Spacehab to Mir. In addition, they moved 20 contingency water containers with 2,025 pounds of water. The original plan was to provide the Mir with 15 containers of water, but during the mission the Russians requested 5 additional containers. Three experiments were also transferred including BTS (biotechnology system), CGBA (commercial generic bioprocessing apparatus), and MIDAS (material in devices as superconductors). About 2,000 pounds of experiment samples and equipment were transferred from Mir to *Atlantis*.

Near the end of the mission, Bill Readdy and Terry Wilcutt used the RCS vernier jets to lower the Mir orbit. A similar maneuver was made at end of STS-82 (SM2) to reboost the Hubble Space Telescope to a higher orbit.

The weather was good at KSC and *Atlantis* returned to the Shuttle Landing Facility on the first opportunity.

The STS-79 and EO-22 crews pose in the Mir Core Module. Front, from the left, are Aleksandr Kaleri, Jay Apt, John Blaha, Bill Readdy, and Shannon Lucid. In the back are Tom Akers, Carl Walz, Valeri Korzun, and Terry Wilcutt. Note that Blaha is wearing a Mir uniform. (NASA)

The Mir Spektr module as seen from Atlantis docked at the Docking Module. Spektr had begun as a module of a military space station before the collapse of the Soviet Union. It was launched aboard a Proton on 20 May 1995 and docked with Mir on 1 July 1995. (NASA)

This photograph of Atlantis was taken from approximately 170 feet away by Shannon Lucid in the Mir Base Block. The Spacehab double module is in the back of the payload bay; compare this to the single module carried in STS-76 (page 203). (NASA)

An unusual perspective of the Atlantis stack as it rolls-out of the Vehicle Assembly Building. This shows just how close the flight elements were to each other. At this point, the ET nose cone was still painted. Note the bipod attaching the front of the orbiter to the ET. (NASA)

The forward flight deck of Atlantis while docked to Mir. Note the two temporary work lights. The flight deck was not particularly well lit to work in (it was fine to fly in) with only a couple of small fluorescent lights and a couple of "map" lights over the front seats. (NASA)

The left side of the aft flight deck. The panel to the left controlled the orbiter docking system. The Russians painted their panels pale green and used decidedly different switches and indicators than the Americans or Canadians. Note the ever present roll of gray tape. (NASA)

STS-80

Mission:	80	NSSDC ID:	1996-065A
Vehicle:	OV-102 (21)	ET-80 (LWT)	SRB: BI084
Launch:	LC-39B	19 Nov 1996	19:56 UTC
Altitude:	203 nm	Inclination:	28.45 degrees
Landing:	KSC-33	07 Dec 1996	11:50 UTC
Landing Rev:	278	Mission Duration:	423 hrs 53 mins

Commander:	Kenneth D. "Taco" Cockrell (3)
Pilot:	Kent V. "Rommel" Rominger (2)
MS1:	F. Story Musgrave (6)
MS2:	Thomas D. "Tom" Jones (3)
MS3:	Tamara E. "Tammy" Jernigan (4)

Payloads:	Up: 43,635 lbs	Down: 0 lbs
	ORFEUS-SPAS-02 (7,784 lbs)	
	WSF-3 (4,644 lbs)	
	SEM-01	
	Extended duration orbiter pallet	
	SRMS s/n 202	

Notes:	Delayed 30 Oct 1996 (window replacements)
	Delayed 04 Nov 1996 (SRB concerns)
	Delayed 15 Nov 1996 (Eastern Range conflicts)
	Two EVAs canceled (airlock issues)
	Longest mission of the flight campaign

Wakeup Calls:

20 Nov	"I Can See For Miles"
21 Nov	"Theme from Fireball XL5"
22 Nov	"Roll With the Changes"
23 Nov	"Reelin' and Rockin'"
24 Nov	"Roll With It"
25 Nov	"Good Times Roll"
26 Nov	"Red Rubber Ball"
27 Nov	"Alice's Restaurant"
28 Nov	"Some Guys Have All the Luck"
29 Nov	"Changes"
30 Nov	"Break On Through"
01 Dec	"Shooting Star"
02 Dec	"Stay"
03 Dec	"Return to Sender"
04 Dec	"Should I Stay or Should I Go"
05 Dec	"Nobody Does It Better"
06 Dec	"Please Come Home For Christmas"

The crew patch depicts Columbia along with the two research satellites (ORFEUS-SPAS and WSF-3) it carried. Surrounding Columbia is a constellation of 16 stars, one for each day of the mission, and two bright blue stars representing the two planned EVAs. (NASA)

The launch of *Columbia* scheduled for 30 October 1996 was delayed three days to replace two forward windows after an engineering analysis suggested windows with a high number of flights could fracture more easily. One of the windows had flown eight times and the other seven times. On 4 November, a further seven-day delay was used to finish an analysis of unusual erosion of the SRB nozzles recovered from STS-79. NASA set a new launch date for 15 November, but two days beforehand rescheduled the mission for 19 November because of conflicts on the Eastern Range. The final count proceeded smoothly except for a short delay at T-31 seconds due to a minor hydrogen leak in the aft compartment, but engineers determined it did not represent a threat and continued. Story Musgrave became the only person to fly on all five orbiters.

Columbia carried two satellites that were released and retrieved. ORFEUS-SPAS (Orbiting and Retrievable Far and Extreme Ultraviolet Spectrograph-Shuttle Pallet Satellite) was a cooperative endeavor between NASA and the German Space Agency (DARA). The crew used the SRMS to deploy the ORFEUS-SPAS on flight day 1 and the satellite made 422 observations of almost 150 astronomical bodies, ranging from the moon to extra-galactic stars and a quasar. The crew retrieved the spacecraft on flight day 16.

On flight day 2 the crew used the SRMS to deploy the third Wake Shield Facility (WSF-3) and the satellite remained stable throughout its three-day free flight and completed all seven of its growth cycles. The Space Vacuum Epitaxy Center at the University of Houston designed and fabricated the spacecraft in conjunction with Space Industries, Inc.

The crew depressurized the cabin to 10.2 psia on flight day 7 in preparation for the first of two planned EVAs by Tammy Jernigan and Tom Jones. These would have been the first EVAs conducted from *Columbia*, despite having been in service for fifteen years. However, the astronauts reported they could not open the outer airlock hatch. After a great deal of troubleshooting without resolving the issue, the mission management team canceled the EVAs. Interestingly, officially this still counted as an EVA, although it has no time associated with it. The second EVA was also canceled.

During the mission, the flight was extended one day to allow the ORFEUS team more experiment time, but subsequent concerns about weather at KSC and Edwards forced mission management to cancel the extension. The landing was waved-off from all opportunities on 5 and 6 December due to weather, with *Columbia* finally returning to the Shuttle Landing Facility two days later than planned. This was the longest mission of the Space Shuttle Program.

During the TCDT, from the left, are Tom Jones, Story Musgrave, Kent Rominger, Tammy Jernigan, and Ken Cockrell. With this Columbia mission, Musgrave became the only person to fly on all five orbiters; this was his sixth and last spaceflight. (Courtesy of Tom Jones)

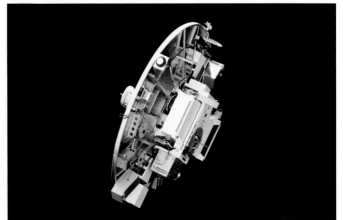

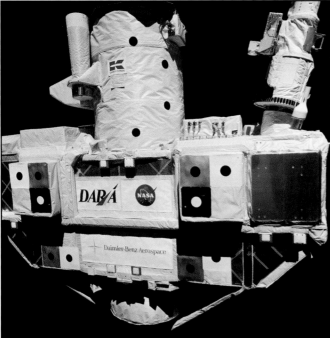

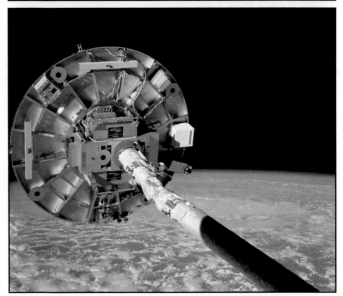

This was the third and last flight of the Wake Shield Facility. The forward edge of the WSF disk redirected atmospheric and other particles around the sides, leaving an "ultra-vacuum" in its wake. The resulting environment was used to study epitaxial film growth. (NASA)

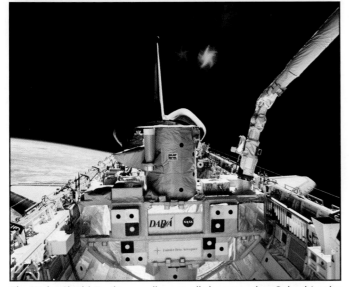

The Wake Shield Facility usually gets all the press, but Columbia also carried the Orbiting and Retrievable Far and Extreme Ultraviolet Spectrograph that used the Shuttle Pallet Satellite (ORFEUS-SPAS). This satellite carried a telescope with several spectrographs. (NASA)

The Columbia stack atop MLP-3 climbs the hill toward LC-39B. The crawler-transporter was capable of keeping the MLP absolutely level as it climbed the 5 percent grade to the launch pad. The crawlers dated from Apollo, having carried the Saturn V moon rockets. (NASA)

STS-81 (S/MM-05)

Mission:	81	NSSDC ID:	1997-001A
Vehicle:	OV-104 (18)	ET-83 (LWT)	SRB: BI082
Launch: Altitude:	LC-39B 213 nm	12 Jan 1997 Inclination:	09:27 UTC 51.60 degrees
Landing: Landing Rev:	KSC-33 160	22 Jan 1997 Mission Duration:	14:24 UTC 244 hrs 55 mins

Commander:	Michael A. "Mike" Baker (4)
Pilot:	Brent W. Jett, Jr. (2)
MS1:	Peter J. K. "Jeff" Wisoff (3)
MS2:	John M. Grunsfeld (2)
MS3:	Marsha S. Ivins (4)
Up:	Jerry M. Linenger (2)
Down:	John E. Blaha (5)

Payloads:	Up: 28,149 lbs	Down: 2,842 lbs
	Spacehab LDM (10,525 lbs)	
	No SRMS	

Notes:	Last mission where all crew members used S1032 launch-entry suits (LES) Backup Mir crew member was C. Michael Foale

Wakeup Calls:

13 Jan	"Free Ride"
14 Jan	"It Keeps You Runnin'"
15 Jan	"Hitchin' a Ride"
16 Jan	"Celebration"
17 Jan	"I Got You (I Feel Good)"
18 Jan	"Mack the Knife"
19 Jan	"Ticket to Ride"
20 Jan	"So Long, Farewell" (*The Sound of Music*)
21 Jan	"My Favorite Marcia"
22 Jan	"Day-O (The Banana Boat Song)"

The crew patch for the fifth Shuttle-Mir docking mission was shaped as the Roman numeral V. Atlantis is launching toward a rendezvous with the Russian space station Mir, silhouetted in the background. The American and Russian flags are depicted at the right. (NASA)

The 12 July 1996 decision to replace the boosters on STS-79 with the set intended for STS-81, then the series of frustrations experienced by STS-79, caused this *Atlantis* launch to be delayed several times. When the stars finally aligned, the countdown on 12 January 1997 went smoothly

The mission carried a Spacehab logistics double module of supplies to Mir, with Mike Baker and Brent Jett rendezvousing and docking with the Russian space station on flight day 4. While the vehicles were docked, the combined crews transferred 1,608 pounds of water, 1,138 pounds of equipment, 2,206 pounds of Russian logistics, and 268 pounds of miscellaneous material to Mir. *Atlantis* would bring 2,842 pounds of American science material, Russian logistics, and miscellaneous material back to Earth. This included the first plants to complete a life cycle in space, a crop of wheat grown from seed to seed. John Grunsfeld placed a call to the NPR show Car Talk, hosted by two fellow Massachusetts Institute of Technology (MIT) alumni, brothers Tom and Ray Magliozzi.

There were nine risk mitigation experiments related to the future space station. One of these was the treadmill vibration isolation and stabilization system (TVIS) designed for the Russian Zvezda Service Module. Preliminary analysis of the downlinked data showed less than 1-lbf was imparted during steady-state walking/running on the treadmill. Another activity related to a future space station involved firing the small reaction control system vernier jets during mated operations to gather engineering data.

Secondary payloads included CREAM (cosmic radiation effects and activation monitor), KidSat, and, as a payload of opportunity when the objectives could be met without impacting the flight, MSX (midcourse space experiment). CREAM included active and passive foil detectors that were placed around the middeck and airlock and recorded heavy ions that penetrate the structure as a function of location, time, and orbital attitude. Although the plan was to take 287 photographs as part of the KidSat effort, the crew eventually took 304 photographs prior to docking and 221 after docking. The MSX was unable to acquire any data on this mission.

As with all the Shuttle-Mir missions, STS-81 also provided American and Russian planners with data on how things would work on the eventual International Space Station.

On flight day 9, *Atlantis* undocked from Mir and executed the now-traditional fly-around. Mission control waved-off the first landing opportunity due to forecasted cloud cover at KSC. The weather cleared sufficiently for *Atlantis* to return to the Shuttle Landing Facility on the second opportunity.

Seated in the front are Brent Jett (left) and Mike Baker. In the back, from the left, are John Grunsfeld, John Blaha, Jeff Wisoff, Jerry Linenger, and Marsha Ivins. Blaha was already on Mir when Atlantis launched, being replaced by Linenger on-orbit. (NASA)

The History of the American Space Shuttle

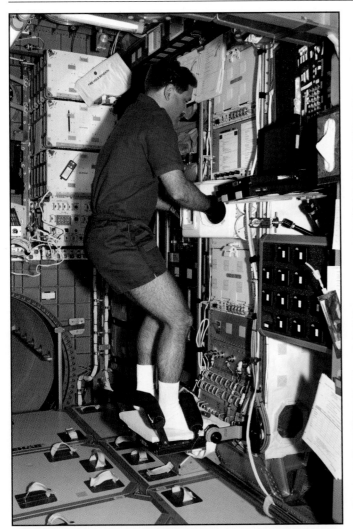

This flight carried a Spacehab double module. The forward module was configured as a laboratory space and John Grunsfeld (left) is shown working at a biorack glovebox. The aft module was configured for logistics (right), with a large number of bags stored on each side and on the floor. A small piece of a Mir gyrodyne can be seen on the right side. Each bag was the same size and shape as a middeck locker (the standard unit of measure was "middeck locker equivalent"). These modules are currently on display at the California Science Center. (NASA)

The LH$_2$ umbilical. In the lower left corner is where the orbiter was physically attached to the external tank. Along the left side are a flash and a camera to record ET separation. The electrical monoball connector is top center and the 17-inch LH$_2$ valve is below it. (NASA)

Lithium hydroxide (LiOH) canisters were used to remove carbon dioxide from the air. The canisters were generally changed once or twice daily through an access door in the middeck floor. Each canister was rated at 48 man-hours and the orbiter usually carried 30 canisters. (NASA)

STS-82 (HST-SM2)

Mission:	82	NSSDC ID:	1997-004A
Vehicle:	OV-103 (22)	ET-81 (LWT)	SRB: BI085
Launch: Altitude:	LC-39A 335 nm	11 Feb 1997 Inclination:	08:55 UTC 28.46 degrees
Landing: Landing Rev:	KSC-15 150	21 Feb 1997 Mission Duration:	08:33 UTC 239 hrs 37 mins

Commander:	Kenneth D. "Sox" Bowersox (4)
Pilot:	Scott J. "Doc" Horowitz (2)
MS1:	Joseph R. "Joe" Tanner (2)
MS2:	Steven A. Hawley (4)
MS3:	Gregory J. "Greg" Harbaugh (4)
MS4:	Mark C. Lee (4)
MS5:	Steven L. "Steve" Smith (2)

Payloads:	Up: 24,891 lbs	Down: 6,638 lbs
	Hubble Servicing Mission (16,735 lbs)	
	SRMS s/n 301	

Notes:	EVA1 13 Feb 1997 (Lee and Smith)
	EVA2 14 Feb 1997 (Harbaugh and Tanner)
	EVA3 15 Feb 1997 (Lee and Smith)
	EVA4 16 Feb 1997 (Harbaugh and Tanner)
	EVA5 17 Feb 1997 (Lee and Smith)

Wakeup Calls:

12 Feb	"Magic Carpet Ride"
13 Feb	"These Are Days"
14 Feb	"Two Princes"
15 Feb	"Higher Love"
16 Feb	"The Packerena"
17 Feb	"Shiny Happy People"
18 Feb	"Dreams"
19 Feb	"That Thing You Do!"
20 Feb	"Five Hundred Miles Away From Home"
21 Feb	(from crew) "Sloop John B."
22 Feb	"Born to Be Wild"

The crew patch features Hubble as the crew will see it as the orbiter approaches for rendezvous. To the right of the telescope is a cross-like structure known as a gravitational lens, one of the numerous fundamental discoveries made using Hubble imagery. (NASA)

NASA originally set this *Discovery* launch for 13 February 1997, but despite some minor glitches, the mission lifted-off two days early with no holds in the final countdown.

After *Discovery* rendezvoused with Hubble, Steven Hawley, who had deployed the telescope on STS-31R, used the SRMS to grapple and berth it in the payload bay.

Mark Lee and Steve Smith conducted the first EVA to replace the Goddard high-resolution spectrometer (GHRS) and the faint object spectrograph (FOS) with the space telescope imaging spectrograph (STIS) and the near infrared camera and multi-object spectrometer (NICMOS), respectively.

During the second EVA, Greg Harbaugh and Joe Tanner replaced a degraded fine guidance sensor (FGS-1R) and a failed engineering and science tape recorder (ESTR-2R) with new spares. They also installed the optical control electronics enhancement kit that further increased the capability of the fine guidance sensor. During this EVA, the astronauts noted cracking and wear on multi-layer insulation (MLI) on the side of Hubble facing the Sun and in the direction of travel. Engineers on the ground determined that at least some of this insulation needed to be replaced, although none was being carried aboard *Discovery*.

The third EVA featured Mark Lee and Steve Smith again, installing a new data interface unit (DIU-2R) and replacing a reel-to-reel engineering and science tape recorder with a new digital solid-state recorder (SSR-2). They also replaced one of four reaction wheel assemblies (RWA-1R) that provided attitude control for the telescope. During this time, mission managers decided to add a fifth EVA to repair the damaged MLI.

Greg Harbaugh and Joe Tanner conducted the fourth EVA, replacing a solar array drive electronics (SADE-2R) package that controlled the positioning of the solar arrays. The pair also replaced covers on the magnetometers and installed MLI blankets over two areas of degraded insulation just below the top of the observatory. Meanwhile, inside *Discovery*, Doc Horowitz and Mark Lee were fabricating additional insulation blankets using materials available on the orbiter. Mark Lee and Steve Smith used the unplanned fifth EVA top attach several of the hastily assembled insulation blankets to three equipment compartments at the top of the telescope support systems module that contained the data processing equipment.

The crew closed the payload bay doors on flight day 9, but the weather at KSC was unacceptable and mission control waved-off the first opportunity. The weather improved and *Discovery* returned to the Shuttle Landing Facility on the second opportunity.

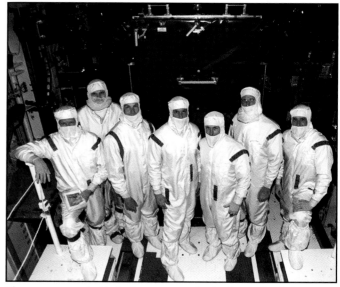

The crew poses in their clean room suits in the payload changeout room at LC-39A during a final inspection of the payload. From the left are Steven Hawley, Steve Smith, Mark Lee, Greg Harbaugh, Ken Bowersox, Joe Tanner, and Doc Horowitz. (NASA)

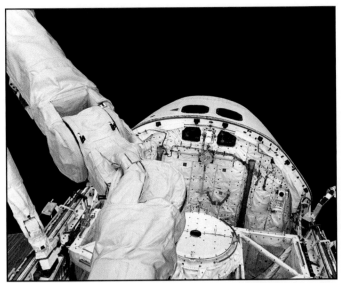

An unusual view of Discovery provided by either Greg Harbaugh or Joe Tanner on the SRMS arm. This mission marked the first flight for the "exit airlock" at lower center, which did not carry the normal orbiter docking system mechanism since the flight was not going to Mir. (NASA)

At the top of the payload bay is the external airlock, followed by the containers carrying the new instruments. At the bottom is the flight support system berthing and positioning system ring that will support the Hubble Space Telescope while it is in the payload bay. (NASA)

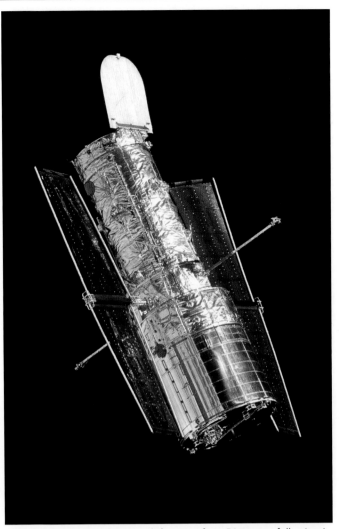

The Hubble Space Telescope drifts away from Discovery following its release from the payload bay. Ground controllers have already opened the large aperture door on the front of the telescope and have started checking out the various systems. (NASA)

Steve Smith waves at crewmates still in the orbiter while working near the SRMS foot restraint during EVA3. Note the extravehicular maneuvering unit suit has a crew patch on the chest. Mark Lee is out of frame. (NASA)

STS-83

Mission:	83	NSSDC ID:	1997-013A
Vehicle:	OV-102 (22)	ET-84 (LWT)	SRB: BI086
Launch:	LC-39A	04 Apr 1997	19:21 UTC
Altitude:	163 nm	Inclination:	28.46 degrees
Landing:	KSC-33	08 Apr 1997	18:34 UTC
Landing Rev:	64	Mission Duration:	95 hrs 13 mins

Commander:	James D. "Jim" Halsell, Jr. (3)
Pilot:	Susan L. Still (1)
MS1:	Janice E. Voss (3)
MS2:	Michael L. "Mike" Gernhardt (2)
MS3:	Donald A. "Don" Thomas (3)
PS1:	Roger K. Crouch (1)
PS2:	Gregory T. "Greg" Linteris (1)

Payloads:	Up: 34,373 lbs	Down: 0 lbs
	Spacelab MSL-1 (32,363 lbs)	
	CryoFD (hitchhiker)	
	Extended Duration orbiter pallet	
	No SRMS	

Notes:	Delayed 03 Apr 1997 (processing)
	Fuel cell failure cut mission 12 days short
	Mission remanifested as STS-83R (later, STS-94)
	Backup payload specialist was
	Paul D. Ronney

Wakeup Calls:

Since crew worked two shifts around the clock, mission control did not send any wakeup calls.

The center circle symbolized a free liquid in microgravity, representing various fluid and materials science experiments while the surrounding starburst of a blue flame represented the combustion experiments. The three-lobed shape symbolized the biotechnology experiments. (NASA)

Planners initially scheduled this *Columbia* launch for 3 April 1997, but the need to install thermal protection around a floodlight coldplate in the payload bay caused a slip to 4 April. On that day, the built-in cell performance monitor reported several anomalies with a fuel cell, but the unit appeared to be operating normally so engineers proceeded with the count.

The primary payload was the Microgravity Science Laboratory (MSL-1), a collection of experiments housed in a Spacelab long module. The experiments consisted of 19 materials science experiments in five major facilities that included the CSLM (coarsening in solid-liquid mixtures), DCECMF (droplet combustion experiment and combustion module facility), EXPRESS (expedite the processing of experiments to the space station), LIF (large isothermal furnace), and TEMPUS (electromagnetic containerless processing facility). Additional experiments were performed in the middeck glovebox (MGBX) and the crew used the HI-PAC DTV (high-packed digital television) system to provide multi-channel real-time analog video.

As was typical for science missions, the crew split into two teams to support around-the-clock operations. The blue team consisted of Jim Halsell, Susan Still, Don Thomas, and Greg Linteris while the red team included Janice Voss, Mike Gernhardt, and Roger Crouch. The crew completed activating the Spacelab about five hours after launch. However, shortly after on-orbit operations began, the differential voltage in a fuel cell began trending upward, repeating the pre-launch anomaly. The crew attempted to purge the fuel cell while engineers in Houston evaluated the condition. Despite several purges, the fuel cell continued to behave anomalously, so mission control directed the crew to shut it down. The crew managed to accomplish some science before the mission management team declared a minimum duration flight and directed the crew to return some 11 days earlier than planned.

The post-flight failure analysis of the fuel cell did not determine a root cause for the anomaly, but revealed 20 adjacent cells had some amount of degradation, thus showing the built-in cell performance monitor was operating properly. As a result of the analysis and investigation, NASA revised the launch commit criteria to not launch with a fuel cell showing similar pre-launch readings.

This was the fourth shortest mission of the flight campaign, following STS-1, STS-2, and STS-20/51C. Since the space station assembly schedule was encountering significant delays, leaving space shuttle with little to do, NASA decided to refly the MSL payload on STS-94 (known as STS-83R during planning) using the same crew.

In the front, from the left, are Janice Voss, Jim Halsell, Susan Still, and Don Thomas. In the back are Roger Crouch, Greg Linteris, and Mike Gernhardt. Unintentionally, this became the second shortest operational mission, after the DoD-dedicated STS-20/51C. (NASA)

Susan Still appears excited about the chore of vacuuming in the Spacelab as Mike Gernhardt approaches with the vacuum cleaner, still inside its middeck locker tray. (NASA)

Janice Voss displays a successful results from the CM-1 (combustion module) experiment, designed to study the SOFBALL (structures of flame balls at low Lewis) numbers. (NASA)

Don Thomas prepares the IFFD (internal flows in free drops) experiment, designed to study the flows within drops of several fluids under varying acoustic pressure. (NASA)

For most of the flight campaign, deployable payloads, such as satellites or ISS modules, were installed while the orbiter was vertical at the launch pad. Non-deployable payloads, such as Spacelab and Spacehab, were installed while the orbiter was horizontal in the Orbiter Processing Facility. Here the Microgravity Science Laboratory (MSL-1) Spacelab module is installed into a payload canister (left) in the Operations and Checkout Building prior to being transported to OPF-1 where it will be integrated (right) into the payload bay of Columbia. (NASA)

A series of artistic photos of Columbia as she rolled to LC-39A and of liftoff. Rollouts from the Vehicle Assembly Building almost always happened during the very early morning since that is when the weather on the central coast of Florida was most forgiving (few thunderstorms and the resulting lightning). This resulted in some gorgeous sunrise photography (left). In the center, the stack is passing by the turn basin where the barges delivered the external tanks. At right, an old tree frames launch. (NASA)

STS-84 (S/MM-06)

Mission:	84	NSSDC ID:	1997-023A
Vehicle:	OV-104 (19)	ET-85 (LWT)	SRB: BI087
Launch:	LC-39A	15 May 1997	08:08 UTC
Altitude:	215 nm	Inclination:	51.60 degrees
Landing:	KSC-33	24 May 1997	13:29 UTC
Landing Rev:	145	Mission Duration:	221 hrs 20 mins

Commander:	Charles J. "Charlie" Precourt (3)
Pilot:	Eileen M. Collins (2)
MS1:	Jean-François A. Clervoy (2)
MS2:	Carlos I. Noriega (1)
MS3:	Edward T. "Ed" Lu (1)
MS4:	Elena VladiMirovna Kondakova (2)
Up:	C. Michael "Mike" Foale (4)
Down:	Jerry M. Linenger (2)

Payloads:	Up: 28,497 lbs	Down: 2,576 lbs
	Spacehab LDM (9,231 lbs)	
	No SRMS	

Notes: First mission where all crew members wore ACES suits

Backup Mir crew member was
    James S. Voss

Wakeup Calls:

16 May	"Those Magnificent Men in Their Flying Machines"
17 May	"Hold On, I'm Coming"
18 May	British National Anthem – "God Save the Queen"
19 May	French National Anthem – "La Marseillaise"
20 May	Peruvian National Anthem – "Himno Nacional del Perú"
21 May	Russian National Anthem – "State Anthem"
22 May	"Triste et Bleu" ("Sad and Blue")
23 May	Medley of Armed Forces Hymns
24 May	American National Anthem – "The Star-Spangled Banner"

The shape of the crew patch reflects the flat-top shape of the Spacehab module. The Phase One program is represented by the rising Sun and the Greek letter Phi followed by one star. This sixth docking mission is symbolized by the six stars surrounding the Cyrillic Mir. (NASA)

As with all Mir missions, the exact time of launch was determined only 90 minutes before liftoff based on the location of Russian space station. A smooth countdown on 15 May 1997 resulted in *Atlantis* being launched on time.

The primary science facility in the Spacehab double module was the Biorack, a large multi-purpose unit that provided temperature-controlled environments, centrifuges for simulating gravity, and a protected workspace for specimen handling. This was the sixth flight of the Biorack facility and it carried 11 American, French, and German experiments. Spacehab also carried CVDA (commercial vapor diffusion apparatus), EORF (enhanced orbiter refrigerator/freezer), LME (liquid motion experiment), the LSLE (life science laboratory equipment) refrigerator/freezer, and SSD/MOMO (self-standing drawer/morphological transition and model substances). Experiments on the middeck included CREAM (cosmic radiation effect and activation monitor), EPICS (electrolysis performance improvement concept study), PCG-STES (protein crystal growth-single thermal enclosure system), and RME-III (radiation monitoring experiment). Secondary payloads included MSX (midcourse space experiment) and SIMPLEX (shuttle ionospheric modification with pulsed local exhaust).

Charlie Precourt and Eileen Collins docked with Mir on flight day 2 as they flew above the Adriatic Sea and the crews opened the hatches between the two spacecraft about two hours later. Jerry Linenger and Mike Foale officially traded places on flight day 3.

The crews transferred 7,315 pounds of water and logistics to and from Mir. During the docked phase, 1,025 pounds of water, 845 pounds of American science equipment, 2,576 pounds of Russian logistics, and 393 pounds of miscellaneous material were transferred to Mir. Returning to Earth aboard *Atlantis* were 898 pounds of American science material, 1,171 pounds of Russian logistics, 31 pounds of ESA material and 376 pounds of miscellaneous material.

Unlike previous Shuttle-Mir missions, *Atlantis* did not conduct a fly-around of the space station, although the orbiter stopped three times while backing away to collect data from an ESA proximity operations sensor (EPOS) designed to assist future rendezvous of the proposed ATV resupply vehicle. The EPOS receiver maintained lock with the GPS satellites through docking, much better than expected because of the pre-launch concerns about multi-path effects close to the Russian space station.

The mission management team waved-off the first landing opportunity because of cloud cover at KSC, but the weather was acceptable for the second opportunity.

In the front, from the left, are Jerry Linenger, Charlie Precourt, and Mike Foale. In the back are Jean-Francois Clervoy, Eileen Collins, Ed Lu, Elena Kondakova, and Carlos Noriega. Foale replaced Linenger when he joined the EO-23 crew once on-orbit. (NASA)

American, European, and Russian flags in the back of the Spacehab double module. At left is the glovebox used for various experiments along with the ever present roll of gray tape (NASA did not use brand names, hence, it was not Duct Tape or Duck Tape). (NASA)

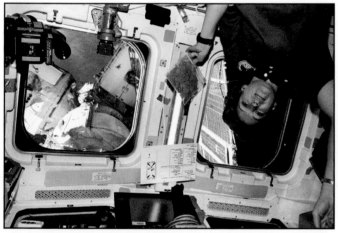

Eileen Collins shows a bag of snacks on the aft flight deck. Of particular interest is the view of Mir through the overhead windows behind her since is how the crew saw a space station as they were rendezvousing and docking with it. (NASA)

The view of the Mir Docking Module as seen by Atlantis just prior to docking. Two of the four Mir environmental effects payload (MEEP) experiment panels can be seen above and below the square panels with black dots on them. (NASA)

Veteran cosmonaut Valery Victorovich Ryumin greets his wife, Elena Kondakova, with some flowers after Atlantis returned to Runway 33 at the Shuttle Landing Facility. This was Kondakova's second space flight after spending 169 days as flight engineer on the EO-17 crew. (NASA)

A head-on shot of Atlantis and her drag chute as the orbiter decelerates on the concrete runway at KSC. Note the open speedbrake. The rudder/speedbrake consisted of panels that could operate together or separately as needed for directional and velocity control. (NASA)

STS-94 (STS-83R)

Mission:	85	NSSDC ID:	1997-032A
Vehicle:	OV-102 (23)	ET-86 (LWT)	SRB: BI088
Launch: Altitude:	LC-39A 163 nm	01 Jul 1997 Inclination:	18:02 UTC 28.45 degrees
Landing: Landing Rev:	KSC-33 251	17 Jul 1997 Mission Duration:	10:47 UTC 376 hrs 45 mins

Commander:	James D. "Jim" Halsell, Jr. (4)
Pilot:	Susan L. Still (2)
MS1:	Janice E. Voss (4)
MS2:	Michael L. "Mike" Gernhardt (3)
MS3:	Donald A. "Don" Thomas (4)
PS1:	Roger K. Crouch (2)
PS2:	Gregory T. "Greg" Linteris (2)

Payloads:	Up: 34,359 lbs Down: 0 lbs Spacelab MSL-1R (32,327 lbs) CryoFD-R (hitchhiker) Extended Duration orbiter pallet No SRMS

Notes:	Reflight of STS-83
	Called STS-83R (reflight) during planning
	Only time the same crew flew two missions
	Payload was not removed from *Columbia* during
	  ground processing between flights
	Backup payload specialist was
	  Paul D. Ronney

Wakeup Calls:

Since crew worked two shifts around the clock, mission control did not send any wakeup calls.

The crew updated their STS-83 patch for the reflight, changing the outer border from red to blue and the flight number from 83 to 94. Otherwise, the patch was the same as for the original flight, hardly surprising given the limited amount of time available. (NASA)

This was a reflight of the STS-83 Microgravity Science Laboratory (MSL) mission that had been launched on 4 April 1997. This was the only true "reflight" of the Space Shuttle Program where the same payload and the same crew flew twice. The first mission had been intended to be on-orbit for 15 days, 16 hours; but the failure of a fuel cell shortened it to 3 days, 23 hours. During the extremely quick planning phase (by far the shortest for any space shuttle flight) this mission was called STS-83R, but was subsequently assigned a number in the normal mission sequence.

Hoping to escape before the expected afternoon thunderstorms, the mission management team moved-up this *Columbia* launch by 47 minutes. The decision to launch early removed one end-of-mission daylight landing opportunity at Edwards, but allowed two daylight landing opportunities at the Shuttle Landing Facility.

The primary payload was identical to STS-83 and included the Microgravity Science Laboratory (MSL-1R) collection of experiments housed in a Spacelab long module. The experiments consisted of 19 materials science investigations in five major facilities that included the CSLM (coarsening in solid-liquid mixtures), DCECMF (droplet combustion experiment and combustion module facility), EXPRESS (expedite the processing of experiments to the space station), LIF (large isothermal furnace), and TEMPUS (electromagnetic containerless processing facility). Additional experiments were performed in the middeck glovebox (MGBX) and the crew used the high-packed digital television (HI-PAC DTV) system to provide multi-channel real-time analog video.

Experiments that measured microgravity included MMA (microgravity measurement assembly), OARE (orbital acceleration research experiment), QSAM (quasi-steady acceleration measurement), and SAMS (space acceleration measurement system). This was the third flight of the QSAM system that had flown previously on the Russian Foton-11 free-flying capsule in October 1997 and STS-83. The experiment flew at least one additional time, on the Foton-12 mission in September 1999.

The major secondary payloads included CRYOFD (cryogenic flexible diode experiment), MSX (midcourse space experiment), and SAREX-II (shuttle amateur radio experiment). The CRYOFD hitchhiker payload consisted ALPHA (American loop heat pipe with ammonia) and CFDHP (cryogenic flexible diode heat pipe).

This mission proved much more successful than the first attempt and the crew accomplished all of the preflight objectives. They closed the payload bay doors and made the deorbit burn on the first KSC landing opportunity.

In the front are Susan Still and Janice Voss. The middle row includes Mike Gernhardt, Jim Halsell, and Greg Linteris. In the back are Don Thomas and Roger Crouch. This is the only mission that did not have an official crew portrait, reusing the original STS-83 photo instead. (NASA)

Greg Linteris floating in the pressurized Spacelab long module. Each Spacelab module was 13.5 feet in diameter and 9.0 feet long plus a pair of 2.5-foot endcaps; combining two modules created a long module that was 23 feet long (including the endcaps). Ultimately the pressurized modules flew 16 times although other Spacelab components, mostly pallets, flew on 58 flights from STS-2 through STS-125. (NASA)

Normally, the exercise equipment was located on the middeck where there was more room. For whatever reason, on this mission the crew decided the aft flight deck offered a better solution as shown by Don Thomas on the ergometer. (NASA)

The view into the payload bay. This was the only time the same orbiter carried exactly the same payload twice. In fact, the payload was not removed from the payload bay between missions. (NASA)

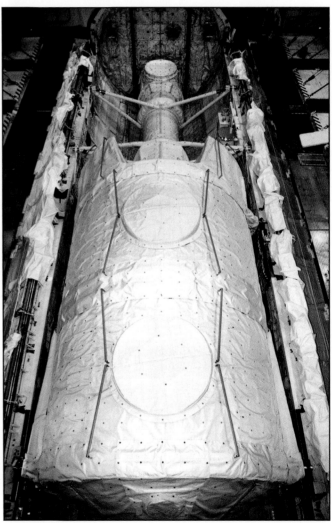

This is what a Spacelab long module looked from the back of the payload bay looking forward. The round areas on top of the module could carry various windows or retractable airlocks, depending on the mission requirements. Note the yellow handrails. (NASA)

STS-85

Mission:	86	NSSDC ID:	1997-039A
Vehicle:	OV-103 (23)	ET-87 (LWT)	SRB: BI089
Launch:	LC-39A	07 Aug 1997	14:41 UTC
Altitude:	160 nm	Inclination:	57.00 degrees
Landing:	KSC-33	19 Aug 1997	11:09 UTC
Landing Rev:	189	Mission Duration:	284 hrs 27 mins

Commander:	Curtis L. "Curt" Brown, Jr. (4)
Pilot:	Kent V. "Rommel" Rominger (3)
MS1:	N. Jan Davis (3)
MS2:	Robert L. "Beamer" Curbeam, Jr. (1)
MS3:	Stephen K. "Steve" Robinson (1)
MS4:	Bjarni V. Tryggvason (1)

Payloads:	Up: 39,321 lbs	Down: 0 lbs
	CRISTA-SPAS-02 (7,154 lbs)	
	SEM-02	
	IEH-2 (hitchhiker)	
	TAS-1 (hitchhiker)	
	GAS (2)	
	SRMS s/n 301	

Notes: Jeffrey S. "Bones" Ashby was originally assigned as the pilot but was relieved to care for his wife, who was being treated for cancer

Wakeup Calls:

08 Aug	"To the Moon and Back"
09 Aug	"Don't Look Down"
10 Aug	"My Home's In Alabama"
11 Aug	"Chances Are"
12 Aug	"The House is Rockin'"
13 Aug	"Good Vibrations"
14 Aug	"You Will Go to the Moon"
15 Aug	"Stay"
16 Aug	"Mighty Iron Arm Atom" (Japanese cartoon theme)
17 Aug	"You're Not From Texas"
18 Aug	"So Far Away"
19 Aug	"Running On Empty"

CRISTA was depicted on the right side of the crew patch pointing its trio of telescopes at the atmosphere. The high-inclination orbit was shown as a yellow band over the northern latitudes. Comet Hale-Bopp, visible from Earth during the mission, was depicted at upper right. (NASA)

The 7 August 1997 countdown proceeded smoothly and resulted in an on-time launch. The primary payload was CRISTA-SPAS-II, an improved version of a payload first carried on STS-66. CRISTA (cryogenic infrared spectrometers and telescopes for the atmosphere) consisted of three telescopes and four spectrometers that measured trace gases and the dynamics of the middle atmosphere. The science instruments were mounted on the Shuttle Pallet Satellite (SPAS) that provided power, command, and communication with *Discovery* during free flight.

Jan Davis deployed CRISTA-SPAS about eight hours after launch. Davis used the SRMS to retrieve the free-flyer at the beginning of flight day 9 and berthed it about an hour later. CRISTA-SPAS had measured atmospheric spectra for 183 hours, obtaining 44,000 altitude profiles of emission of up to 17 trace gases. The measurements covered an altitude range of 4 to 115 miles by using different pointing orientation modes of the SPAS platform. Davis again used the SRMS to grappled CRISTA-SPAS later during flight day 9 and maneuvered it over the payload bay for about four hours to support ACVS (autotrac computer vision system) and SVS (space vision system) operations. The middle atmosphere high resolution spectrograph investigation (MAHRSI), located in the payload bay, focused on obtaining new vertical profile data on the distribution of hydroxyl (OH) in the mesosphere and upper stratosphere under different conditions (both seasonal and diurnal) from the previous flight on STS-66. There were 38 different groups of researchers associated with the CRISTA/MAHRSI science team.

Bjarni Tryggvason operated the MIM (microgravity vibration isolation mount) experiment, a double-locker size device designed to isolate ISS payloads and experiments from disturbances created by thruster firings or crew activity. The surface effects sample monitor (SESAM) was a passive carrier for optical surfaces to study the impact of the atomic oxygen on materials. The crew also operated the southwest ultraviolet imaging system (SWUIS-01) from the Southwest Research Institute (SwRI) along with scientific collaborators from Applied Physics Laboratory, Jet Propulsion Laboratory, and the University of Maryland. SWUIS (pronounced "swiss") was a wide-field UV imager used to observe comet Hale-Bopp based around a 7-inch Maksutov UV telescope and a UV-sensitive, xybion image-intensified CCD camera. Each SWUIS observation period lasted approximately three hours.

Mission management extended the mission one day because of ground fog at the Shuttle Landing Facility. The deorbit, entry, and landing went smoothly.

An impromptu on-orbit crew portrait was snapped while the crew members were setting up for a more formal portrait on the middeck. From the left are Kent Rominger, Robert Curbeam, Steve Robinson, Curt Brown, Jan Davis, and Bjarni Tryggvason. (NASA)

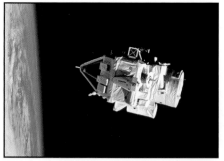
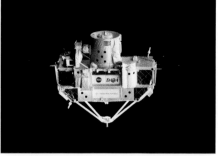

At left, the Cryogenic Infrared Spectrometers and Telescopes for the Atmosphere on the Shuttle Pallet Satellite (CRISTA-SPAS-02) had just been deployed over northwest Canada. The clouds were a storm that tracked over Canada during the flight, limiting the topographic photo opportunities for several days. At right, the satellite just prior to being retrieved. (NASA)

The crew used a 35 mm camera with a time exposure to record this image of the southern lights (Aurora Australis) along with the vertical stabilizer of Discovery. (NASA)

Steve Robinson wears a training version of the extravehicular mobility unit suit as he is lowered into the pool at the Weightless Environment Test Facility (WET-F). Two crew members on every mission were trained for contingency EVAs, even if no spacewalks were scheduled. (NASA)

Discovery flying over the Sea of Japan. In center foreground is the manipulator flight demonstration (MFD) that evaluated the use of the small fine arm (SFA) as part of the future Japanese Experiment Module remote manipulator system on the station. (NASA)

Retrieving the boosters was always a major undertaking, although at least the seas were calm for this mission. Each booster had a dedicated retrieval ship. (NASA)

STS-86 (S/MM-07)

Mission:	87	NSSDC ID:	1997-055A
Vehicle:	OV-104 (20)	ET-88 (LWT)	SRB: BI090
Launch:	LC-39A	26 Sep 1997	02:34 UTC
Altitude:	212 nm	Inclination:	51.60 degrees
Landing:	KSC-15	06 Oct 1997	21:57 UTC
Landing Rev:	170	Mission Duration:	259 hrs 21 mins

Commander:	James D. "WxB" Wetherbee (4)
Pilot:	Michael J. "Bloomer" Bloomfield (1)
MS1:	VladiMir Georgiyevich Titov (4)
MS2:	Scott E. Parazynski (2)
MS3:	Jean-Loup Chrétien (3)
MS4:	Wendy B. Lawrence (2)
Up:	David A. Wolf (2)
Down:	C. Michael "Mike" Foale (4)

Payloads:	Up: 29,728 lbs	Down: 2,859 lbs
	Spacehab LDM (14,447 lbs)	
	No SRMS	

Notes:
Lawrence was scheduled to replace Foale on Mir, but concerns about the minimum size of the Russian Orlan spacesuit resulted in Wolf being selected instead
First joint American-Russian EVA
EVA1 01 Oct 1997 (Parazynski and Titov)

Wakeup Calls:

26 Sep	"Roll on Down the Highway"
27 Sep	"Dancing in the Dark"
28 Sep	"Takin' Care of Business"
29 Sep	"What I Like About You"
30 Sep	"Grand Ol' Flag"
01 Oct	"What a Wonderful World"
02 Oct	"Impression That I Get"
03 Oct	"Fanfare For the Common Man"
04 Oct	"Let It Ride"
05 Oct	"Shake, Rattle and Roll"
06 Oct	"Homeward Bound"

The American, French, and Russian flags are incorporated into the astronaut symbol. The rays streaking across the sky depict the orbital tracks of the spacecraft as they prepare to dock. The mercator projection of Earth illustrates the global nature of the flight. (NASA)

On 25 June 1997, a Progress supply vehicle collided with the Mir Spektr module, damaging a radiator, one of four solar arrays, and depressurizing the station. The mishap occurred while EO-23 commander Vasily Vasiliyevich Tsibliyev was guiding the Progress to a manual docking. The crew sealed the hatch to the leaking module and repressurized the rest of the station. Following their arrival aboard Soyuz TM-26 on 7 August, EO-24 commander Anatoly Yakovlevich Solovyev and flight engineer Pavel VladiMirovich Vinogradov conducted an intravehicular activity inside the Spektr module. During the 22 August effort, the pair re-routed 11 power cables from the three still-functioning solar arrays through a new Spektr hatch that had a cable pass-through. On 5 September, Mike Foale and Solovyev surveyed the damage outside Spektr and attempted, unsuccessfully, to locate the breach in the hull.

NASA had planned for Wendy Lawrence to succeed Mike Foale onboard Mir. However, on 30 July, NASA announced David Wolf was replacing Lawrence to provide a backup crew member for the EVAs necessary to repair the damaged Spektr module. Lawrence was too small to safely wear the Russian Orlan space suit and had not undergone EVA training. David Wolf was originally scheduled to fly on the STS-89 mission to Mir. The STS-86 countdown on 26 September 1997 proceeded with no unplanned holds.

This mission featured the first joint American-Russian EVA when VladiMir Titov and Scott Parazynski installed a solar array cap on the Mir Docking Module for future use by Mir crew members to seal off the suspected leak in Spektr. The pair also retrieved four MEEP (Mir cnvironmental effects payload) panels.

While docked, Jim Wetherbee and Mike Bloomfield fired the RCS vernier jets to provide data for MiSDE (Mir structural dynamics experiment) that measured disturbances to Mir components and its solar arrays. Other experiments included CPCG (commercial protein crystal growth, CCM-A (cell culture module, CREAM (cosmic radiation effects and activation monitor (CREAM), and RME-III (radiation monitoring experiment). On flight day 8, Jim Wetherbee and Mike Bloomfield undocked from Mir and backed *Atlantis* off about 600 feet while Anatoly Solovyev and Pavel Vinogradov opened a pressure valve to allow air into the Spektr module to see if the STS-89 crew could locate the leak. The *Atlantis* crew identified a possible hull breach location.

The mission management team waved-off both landing attempts on flight day 10 because of bad weather at the Shuttle Landing Facility. The weather cleared sufficiently for *Atlantis* to return on the first KSC opportunity on flight day 11.

Wearing their partial pressure launch-entry suits are, from the left, Jean-Loup Chretien, David Wolf, Mike Bloomfield, Jim Wetherbee, Wendy Lawrence, and Mike Foale. Wearing the extravehicular mobility unit suits in the back are Scott Parazynski (left) and VladiMir Titov. (NASA)

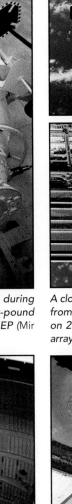

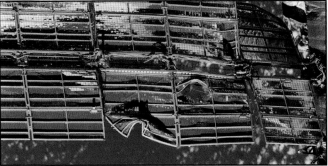

This mission featured the first joint American-Russian EVA during a space shuttle mission. Titov and Parazynski affixed a 120-pound solar array cap to the Docking Module and retrieved four MEEP (Mir environmental effects payload) panels. (NASA)

A close-up of the solar array panel on the Spektr module shows damage from an by an unmanned Progress that collided with the space station on 25 June 1997, causing Spektr to depressurize. The damaged solar array is the second from the right at the bottom of the top photo. (NASA)

A view into the middeck while Atlantis was on the launch pad. Sitting facing upward, from the left, are Jean-Loup Chretien, David Wolf, and Wendy Lawrence during the terminal countdown demonstration test (TCDT) on 10 September 1997. (NASA)

Carried atop the orbiter transporter system, Atlantis exits the Orbiter Processing Facility (OPF-3) during the rollover to the Vehicle Assembly Building. The photo was taken from the roof the VAB. It wasn't a long trip since the facilities were just across the street from each other. (NASA)

STS-87

Mission:	88	NSSDC ID:	1997-073A
Vehicle:	OV-102 (24)	ET-89 (LWT)	SRB: BI092
Launch:	LC-39B	19 Nov 1997	19:46 UTC
Altitude:	155 nm	Inclination:	28.45 degrees
Landing:	KSC-33	05 Dec 1997	12:21 UTC
Landing Rev:	251	Mission Duration:	376 hrs 34 mins

Commander:	Kevin R. Kregel (3)
Pilot:	Steven W. "Steve" Lindsey (1)
MS1:	Kalpana Chawla (1)
MS2:	Winston E. Scott (2)
MS3:	Takao Doi (1)
PS1:	Leonid Kostyantynovych Kadenyuk (1)

Payloads:	Up: 37,392 lbs	Down: 0 lbs
	USMP-4 (31,397 lbs)	
	SPARTAN-201-04 (2,980 lbs)	
	LHP/NaSBE (hitchhiker)	
	SOLSE-1 (hitchhiker)	
	GAS (1)	
	Extended duration orbiter pallet	
	SRMS s/n 301	

Notes:	First roll-to-heads-up maneuver during ascent
	First EVA from *Columbia*
	EVA1 25 Nov 1997 (Scott and Doi)
	EVA2 03 Dec 1997 (Scott and Doi)
	Backup payload specialist was
	Yaroslav Igoryevich Pustovyi

Wakeup Calls:

20 Nov	"Hitchin' a Ride"
21 Nov	"Theme from New York, New York"
22 Nov	"Ginga Shounen Tai" (Japanese puppet theme)
23 Nov	The Air Force Hymn – "Wild Blue Yonder"
24 Nov	"Walk of Life"
25 Nov	"Mishra Piloo"
26 Nov	Ukrainian National Anthem –
	"Shche ne vmerla Ukrayiny"
27 Nov	"America the Beautiful"
28 Nov	Florida State University Seminoles Fight Song
29 Nov	"California Dreamin'"
30 Nov	"This Island Earth"
01 Dec	"Ultraman"
02 Dec	"Centerfield"
03 Dec	"Flight of the
	Bumble Bee"
03 Dec	"Should I Stay
	or Should I Go"

The helmet-shaped crew patch symbolize the EVA that evaluated ISS assembly tools. The three red lines represent the astronaut symbol as well as the SRMS used to handle SPARTAN. The gold flames were the corona of the Sun studied by SPARTAN. (NASA)

The *Columbia* countdown on 19 November 1997 proceeded with no unplanned holds. This was the first mission to perform a roll-to-heads-up (RTHU) maneuver that allowed the orbiter to communicate with the TDRS satellites during the climb uphill and eliminated the need for the Bermuda tracking station. All future low-inclination missions used this procedure.

The primary payload was the fourth United States Microgravity Payload (USMP-4) carried on two mission-peculiar experiment support structures (MPESS) in the payload bay. Secondary payloads included the AERCam/Sprint (autonomous EVA robotic camera/sprint), EDFT-05 (EVA demonstration flight test), G-744 (turbulent GAS jet diffusion), LHP (loop heat pipe), NaSBE (sodium sulfur battery experiment), OARE (orbital acceleration research experiment), SOLSE (shuttle ozone limb sounding experiment), and SPARTAN-201-04 (Shuttle Pointed Autonomous Research Tool for Astronomy). SPARTAN used the ultraviolet coronal spectrometer (UCS) and the white light coronagraph (WLC) to investigate heating and the acceleration of the solar wind that originates in the outer layers of the solar corona. AERCam/Sprint was a small free-flying camera platform, about the size of a soccer ball, that could fly around a spacecraft to check for damage (or anything else).

Mission management delayed the deployment of SPARTAN one day to provide time for the Solar Heliospheric Observatory (SOHO) to reconfigure from a shutdown that occurred on 19 November. SOHO was in solar orbit and would provide simultaneous observations with SPARTAN. On flight day 3 the crew used the SRMS to deploy the spacecraft. However, SPARTAN failed to perform an expected pirouette maneuver and the crew attempted to re-grapple it before it moved out of the reach of the SRMS. Unfortunately, they missed and, unintentionally, caused the satellite to spin as it drifted away from the orbiter. *Columbia* began running low on RCS propellant before it could catch-up with the spacecraft.

After a day of evaluating alternatives, the mission management team allowed Winston Scott and Takao Doi to retrieve SPARTAN during the first scheduled EVA. Interestingly, this was the first EVA (if one does not count the zero-time EVA on STS-80) from *Columbia* despite having been in service for 16 years; in fact, the only other *Columbia* EVAs were during the fourth Hubble servicing mission (SM3B) on STS-109. Scott and Doi managed to stop SPARTAN from spinning and the crew used the SRMS to grapple the satellite and berth it in the payload bay.

The weather was good at KSC and *Columbia* returned to the Shuttle Landing Facility on the first opportunity.

The traditional on-orbit crew portrait, posed in other-than-traditional attire on the middeck. In the front row, from the left, are Steve Lindsey, Takao Doi, and Winston Scott. In the back are Kevin Kregel, Kalpana Chawla, and Leonid Kadenyuk. (NASA)

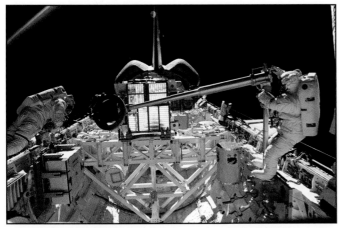

Winston Scott (red stripes) watches as Takao Doi works with a crane during the second EVA. Doi is using the 156-pound crane to grasp an orbital replacement unit near Scott to determine if a similar crane could be used to move ORUs on the ISS. (NASA)

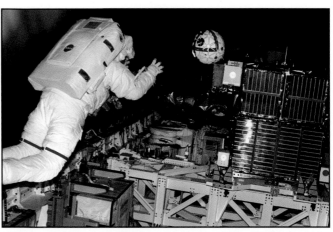

Winston Scott lets loose the prototype free-flying AERCam Sprint (autonomous extravehicular activity robotic camera sprint). This was a small, self-contained, spherically shaped television camera that could be used for remote inspections of the exterior of the ISS. (NASA)

Leonid Kadenyuk, of the National Space Agency of Ukraine (NSAU), works with the brassica rapa (Wisconsin fast plants) seedlings being grown for the collaborative Ukrainian experiment (CUE). This was part of a collection of ten plant space biology experiments. (NASA)

The STS-87 crew participates in the crew equipment integration test (CEIT) with the SPARTAN-201 in the Vertical Processing Facility. From the left are Steve Lindsey, Takao Doi, Kalpana Chawla, Kevin Kregel, and Leonid Kadenyuk. (NASA)

Two members of the STS-87 crew, Kalpana Chawla (second right, top) and Takao Doi (right top), are seen with the 1995 astronaut candidate class getting a taste of microgravity aboard the aptly named "Vomit Comet," a modified Boeing KC-135 operated by JSC. (NASA)

After the retrieval ships brought the expended solid rocket boosters back to Hangar AF on Cape Canaveral AFS, large movable cranes picked them out of the water and took them to the hangar for disassembly. The name of the crane was purely coincidental. (NASA)

STS-89 (S/MM-08)

Mission:	89	NSSDC ID:	1998-003A	
Vehicle:	OV-105 (12)	ET-90 (LWT)	SRB: BI093	
Launch:	LC-39A	23 Jan 1998	02:48 UTC	
Altitude:	215 nm	Inclination:	51.60 degrees	
Landing:	KSC-15	31 Jan 1998	22:36 UTC	
Landing Rev:	138	Mission Duration:	211 hrs 47 mins	

Commander:	Terrence W. "Terry" Wilcutt (3)
Pilot:	Joe F. Edwards, Jr. (1)
MS1:	James F. "JR" Reilly II (1)
MS2:	Michael P. Anderson (1)
MS3:	Bonnie J. Dunbar (5)
MS4:	Salizhan Shakirovich Sharipov (1)
Up:	Andrew S. W. "Andy" Thomas (2)
Down:	David A. Wolf (2)

Payloads:	Up: 28,040 lbs	Down: 3,508 lbs
	Spacehab LDM (12,917 lbs)	
	GAS (4)	
	No SRMS	

Notes:
Delayed 15 Jan 1998 (Russian request)
Delayed 20 Jan 1998 (Russian request)
First flight of a Block IIA space shuttle main engine
First Endeavour flight with an external airlock
Was originally scheduled to return
   Wendy B. Lawrence but returned David A. Wolf
   and left Andrew Thomas on Mir;
   Thomas returned on STS-91
Backup Mir crew member was
   James S. Voss

Wakeup Calls:

23 Jan	"It's Not Unusual"
24 Jan	"Calypso"
25 Jan	"Friends, We Are Migrant Birds" (Russian pilots' song)
26 Jan	"Singer From Down Under"
27 Jan	"Clap For the Wolfman"
28 Jan	"Hide Away"
29 Jan	"Here We Go Loopty-Loo" ("Lupe de Lue")
30 Jan	"Bad To the Bone"
31 Jan	"Breakfast Blues"

The link between America and Russia is represented by Endeavour and Mir orbiting above the Bering Strait between Alaska and Siberia. The white inside line in the shape of the number eight and the nine stars symbolize the STS designation. (NASA)

On 22 May 1997, NASA announced *Endeavour* would fly STS-89 instead of *Discovery* and become the first orbiter other than *Atlantis* to dock with Mir. Flight planners originally targeted launch for 15 January 1998, but the Russians requested a delay to 20 January and, subsequently, to 23 January.

There were a record number of human beings in space during this flight: one American and two Russians aboard Mir, one Frenchman and two Russians on Soyuz, and six Americans and one Russian aboard *Endeavour*.

This was the first flight of an external airlock and orbiter docking system (ODS) configured for the ISS, although it was still carried farther aft in the payload bay than would become normal for ISS missions and continued to use the APAS-89 (androgynous peripheral attach system) mechanism instead of the later APAS-95. David Wolf was returning after spending 119 days aboard Mir, having arrived on STS-86 in September 1997, but there was an initial problem. His replacement, Andy Thomas, could not get his Russian Sokol pressure suit to fit correctly and mission rules would not allow him to stay on Mir without a properly fitting garment to wear should the crew need to return in a Soyuz. Fortunately, the crew managed to adjust Wolf's Sokol suit to fit Thomas and mission management allowed the crew exchange to proceed. Once on Mir, Thomas was able to make adequate adjustments to his own suit. Thomas spent approximately four months on Mir before returning on *Discovery* in late May during STS-91.

Spacehab payloads included ADV-XDT (advanced X-ray detector), ADV-CGBA (advanced commercial generic bioprocessing apparatus), EORF (enhanced orbiter refrigerator/freezer), MGM (mechanics of granular materials), RME-1312 (radiation monitor experiment), SAMS (space acceleration measurement system), VOA (volatile organic analyzer), and the volatile removal assembly (VRA) prototype for the ISS water recovery system.

Middeck payloads included BIO3D (biochemistry of 3D tissue engineering), CEBAS (closed equilibrated biological aquatic system), COCULT (co-culture experiment), EarthKAM, MPNE (microgravity plant nutrient experiment), OSVS (orbiter space vision system), SIMPLEX (shuttle ionospheric modification with pulsed local exhaust), TEHM (thermo-electric holding module), and TMIP (telemedicine instrumentation pack). There were also four get-away special cans in the payload bay carrying one American two German, and one Chinese experiment.

Unusually, the weather forecast for KSC was so good that NASA did not even activate the alternate landing site at Edwards.

In the front row, from the left, are Joe Edwards, Terry Wilcutt, and Bonnie Dunbar. In the back are David Wolf, Salizhan Sharipov, James Reilly, Andy Thomas, and Michael Anderson. Wolf was aboard Mir as a guest researcher when the mission launched. (NASA)

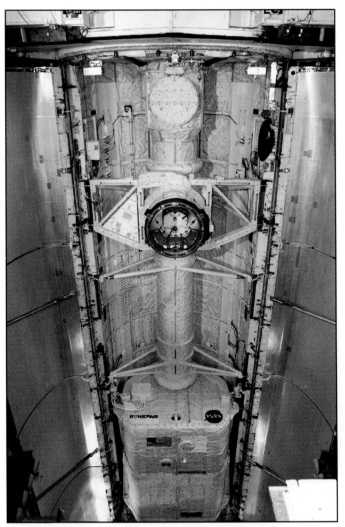

A seal on the left payload bay door failed at the pad and was replaced prior to the doors being closed. Note the position of the external airlock compared to the photo below. The Spacehab double module was in the back of the payload bay connected by a long tunnel. (NASA)

The Endeavour crew recorded a series of 35 mm and 70 mm fly-around survey photos of Mir after undocking from the Russian space station but before heading home. The orbiters mated to the orange Docking Module at the bottom of the station. (NASA)

This is the ISS configuration with the external airlock at the front of the payload bay. Compare this location to the top photo. The forward location was used for the last Mir mission (STS-91) and all ISS missions except the first one (STS-88), which used the Mir configuration. (NASA)

The cavity in the nose of the orbiter that normally held the forward reaction control system module while Endeavour was in OPF-1 being processed. The module itself was at the Hypergolic Maintenance Facility for servicing. (NASA via Mike McClure)

STS-90

Mission:	90		NSSDC ID:		1998-022A
Vehicle:	OV-102 (25)		ET-91 (LWT)		SRB: BI094
Launch:	LC-39B		17 Apr 1998		18:19 UTC
Altitude:	154 nm		Inclination:		39.00 degrees
Landing:	KSC-33		03 May 1998		16:10 UTC
Landing Rev:	256		Mission Duration:		381 hrs 50 mins

Commander:	Richard A. "Rick" Searfoss (3)
Pilot:	Scott D. "Scooter" Altman (1)
MS1:	Dafydd R. "Dave" Williams (1)
MS2:	Kathryn P. "Kay" Hire (1)
MS3:	Richard M. "Rick" Linnehan (2)
PS1:	Jay C. Buckey, Jr. (1)
PS2:	James A. "Jim" Pawelczyk (1)

Payloads:	Up: 36,049 lbs	Down: 0 lbs
	Spacelab Neurolab (26,530 lbs)	
	GAS (3)	
	Extended duration orbiter pallet	
	No SRMS	

Notes:	Delayed 16 Apr 1998 (NSP-2 failure)
	First OMS-Assist burn during ascent
	Last flight of a Spacelab pressurized module
	Last daytime landing for *Columbia*
	Backup payload specialists were
	Alexander W. Dunlap and Chiaki Mukai

Wakeup Calls:	
18 Apr	"Think"
19 Apr	"Take Me Out to the Ball Game"
20 Apr	"Doctor My Eyes"
21 Apr	"Bad To the Bone"
22 Apr	"Bad Case of Loving You"
23 Apr	"I Got You (I Feel Good)" /
	"This Land Is Your Land"
24 Apr	"She Drives Me Crazy"
25 Apr	"Every Breath You Take"
26 Apr	"Fight On, State" (Penn State fight song)
27 Apr	"Turn, Turn, Turn"
28 Apr	"Take a Chance on Me"
29 Apr	"Round and Round"
30 Apr	"Cruise Control"
01 May	"If I Only Had a Brain" (*The Wizard of Oz*)
02 May	"Stir It Up"

The crew patch reflected the mission's dedication to neuroscience as part of the Decade of the Brain. The nine stars of the constellation Cetu (the whale) represent the seven crew members and two backup payload specialists as part of the International Year of the Ocean. (NASA)

The launch team scrubbed the 16 April 1998 attempt before the crew had boarded *Columbia* due to a faulty orbiter network signal processor. The launch attempt the following day proceeded without incident. This was 16th and last flight of the ESA-developed Spacelab pressurized module although Spacelab pallets and other equipment continued to be used through the last Hubble servicing mission (STS-125).

During second stage ascent, *Columbia* performed a 102.4-second OMS-assist maneuver for the first time. The use of the OMS engines during ascent provided approximately 250 pounds of additional payload capability for each 4,000 pounds of propellant used. This was different than the normal OMS-1 or OMS-2 burns, which adjusted orbital altitude. The OMS-assist burn soon became routine.

This mission carried Neurolab, a Spacelab mission focusing on the effects of microgravity on the human nervous system. During the flight, the Neurolab crew served both as experiment subjects and investigators. Eleven experiments involving the autonomic nervous system, sensory motor and performance, vestibular, and sleep used humans as subjects, while fifteen experiments in neuronal plasticity, mammalian development, aquatic, and neurobiology used rats, mice, crickets, snails, and two kinds of fish. Research proceeded as planned, with the exception of the mammalian development team, which had to reprioritize science activities because of the unexpected high mortality rate among the neonatal rats.

Researchers wanted to increase their fundamental understanding of neurological and behavioral changes during spaceflight. Specifically, experiments studied the adaptation of the vestibular system and space adaptation syndrome, the adaptation of the central nervous system and the pathways that control the ability to sense location in the absence of gravity, and the effect of microgravity on a developing nervous system. Agencies participating in this mission included six of the National Institutes of Health, the National Science Foundation, and the Office of Naval Research, as well as the national space agencies of Canada (CSA), France (CNES), Germany (DARA), and Japan (NASDA), and the European Space Agency (ESA). Secondary payloads included BDS-04 (bioreactor demonstration system), SVF (shuttle vibration forces), and three GAS cans.

The mission management team decided against a one-day extension because the science community indicated approximately 96 percent of the desired data had already been collected and the KSC weather was deteriorating. The crew performed a deorbit burn on the first opportunity and landed successfully at the Shuttle Landing Facility. STS-90 marked the last daytime landing of *Columbia*.

The crew during the crew equipment interface test (CEIT) in OPF-3. From the left are Scott Altman, Jim Pawelczyk, Rick Searfoss, Dave Williams, Kay Hire, Jay Buckey, and Rick Linnehan. Hire was the first former KSC employee to fly as a crew member. (NASA)

Columbia lifts-off. Despite a variety of modifications over the years, the thermal protection system configuration on Columbia always differed from the other orbiters, making her easy to identify. Note the water pouring out of the sound suppression water system. (NASA)

Ultimately, the two pressurized modules flew 16 missions, nine with the first module and seven with the second, but Spacelab components, mostly pallets, were used on 58 flights from STS-2 to STS-125. The first pressurized module (seen here departing KSC) is on display Steven F. Udvar-Hazy Center in Chantilly, Virginia. The second module was initially displayed in the Bremenhalle exhibition at the Bremen airport, but was later moved to Building 4C at the EADS (Airbus) complex. It is somehow ironic that the module paid for by the Europeans is in an American museum and the one bought by NASA is in Europe. (NASA)

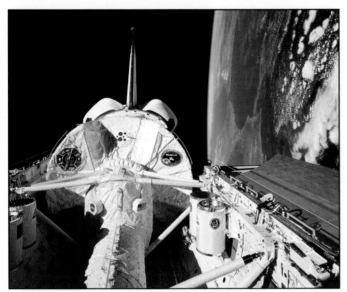

The predecessors to the European Space Agency (ESA) designed and manufactured Spacelab as their contribution to the Space Shuttle Program. The Europeans paid for the first set of hardware and NASA paid for the second set, as well as the infrastructure at KSC. (NASA)

The Spacelab processing area, which once processed Apollo spacecraft, in the Operations and Checkout Building at KSC. Spacelab was a grand enterprise with a lot of infrastructure and a large standing army, something that became too expensive for NASA to sustain. (NASA)

STS-91 (S/MM-09)

Mission:	91		NSSDC ID:		1998-034A
Vehicle:	OV-103 (24)		ET-96 (SLWT)		SRB: BI091
Launch:	LC-39A		02 Jun 1998		22:06 UTC
Altitude:	204 nm		Inclination:		51.60 degrees
Landing:	KSC-15		12 Jun 1998		18:01 UTC
Landing Rev:	154		Mission Duration:		235 hrs 54 mins

Commander: Charles J. "Charlie" Precourt (4)
Pilot: Dominic L. P. "Dom" Gorie (1)
MS1: Franklin R. Chang-Diaz (6)
MS2: Wendy B. Lawrence (3)
MS3: Janet L. Kavandi (1)
MS4: Valery Victorovich "Victor" Ryumin (5)
Down: Andrew S. W. "Andy" Thomas (2)

Payloads: Up: 35,549 lbs   Down: 2,964 lbs
Spacehab LSM (22,251 lbs)
AMS-01 (9,196 lbs)
SEM-03
SEM-05
GAS (4)
SRMS s/n 201

Notes: SLWT tanking test 18 May 1998
First flight of super lightweight tank (SLWT)
Last flight of Shuttle/Mir Program

Wakeup Calls:

03 Jun "Shake, Rattle, and Roll"
04 Jun "Come Go With Me"
05 Jun "South Australia"
06 Jun "You Really Got Me"
07 Jun "Travelin' Band"
08 Jun "Manic Monday"
09 Jun "How Bizarre"
10 Jun "Theme from Superman" (TV show)
11 Jun "Interplanet Janet" (Schoolhouse Rock)
12 Jun "Homeward Bound"

The crew patch depicts the rendezvous of Discovery with Mir. The American and Russian flags are displayed at the top of the patch and both countries are visible on Earth behind the two spacecraft. Ryumin's name is in Cyrillic at the lower right. (NASA)

Because this *Discovery* mission was the first to use a super lightweight tank (SLWT), engineers performed a tanking test on 18 May 1998 to evaluate predicted environments and operational procedures. The test was successful, although two weeks prior to launch the Ice Team found three loose pieces of insulating foam on the ET. All of the debonded sites were typical of an ET detanking and engineers considered the damage acceptable for flight. The tank performed as expected during ascent, with reentry and breakup occurring within the predicted footprint.

This was the first mission to use an external airlock in its final position all the way forward in the payload bay. Both *Atlantis* and *Endeavour* had used an airlock approximately 6 feet farther aft than would be the norm for the International Space Station. In addition, this was the first docking to use the androgynous peripheral attach system (APAS-95) mechanism intended for the ISS that was later renamed the androgynous peripheral docking system (APDS). All previous Mir dockings had used the interim APAS-89. In reality, NASA called it all an orbiter docking system (ODS).

This mission brought the Shuttle-Mir Program to a successful conclusion with the completion of the logistics transfer and the retrieval of the seventh and final astronaut (Andy Thomas) after almost five months on Mir. *Discovery* also carried the prototype Alpha Magnetic Spectrometer (AMS-01) to search for anti-matter and study astrophysics. Issues with the Ku-band system on the orbiter precluded high-data-rate and television transmissions throughout the mission. Unfortunately, a communications problem between a ground station and the Russian mission control center outside Moscow prevented television broadcasts from Mir, limiting communications to audio only on NASA television.

While *Discovery* was docked to Mir, the crew moved several long-term American experiments from Mir onto *Discovery*. These included the tissue engineering co-culture (COCULT) and space acceleration measurement system (SAMS) investigations. Secondary payloads included CPCG (commercial protein crystal growth), CREAM (cosmic radiation effects and active monitor), and SSCE (solid surface combustion experiment). There were also four GAS cans and two space experiment modules (SEM-03 and SEM-05) in the payload bay. Other experiments conducted by the *Discovery* crew included evaluating new electronics and software in the SRMS and testing the orbiter space vision system (OSVS) for use during ISS assembly missions.

The weather at the Shuttle Landing Facility was good and *Discovery* came back to KSC on the first opportunity.

Six plus one equals seven, as the STS-91 crew is joined by Andy Thomas for its on-orbit crew portrait. On the bottom, from the left, are Charlie Precourt, Janet Kavandi, and Franklin Chang-Diaz. At the top are Dom Gorie, Wendy Lawrence, Andy Thomas, and Valery Ryumin. (NASA)

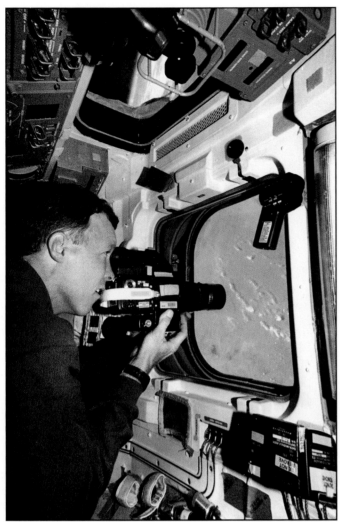

Dom Gorie uses a handheld 70 mm camera to record images of Africa through one of the overhead windows on the aft flight deck. The four windows on the aft flight deck were the primary locations of Earth observation photography during most space shuttle missions. (NASA)

Dom Gorie makes a meal on the middeck. The galley is behind his arms while a wall of middeck lockers is in front of him. The original middeck accommodations rack (MAR) is the blue object behind him. (NASA)

Nikolai Budarin, the Mir flight engineer, shows the lived-in appearance of the Russian space station. This was the Mir Core Module that included the main living quarters and the environmental control systems. Note the colorful Russian uniform. (NASA/RSA)

Charlie Precourt talks with somebody from the inside Discovery during the crew equipment interface test (CEIT) in OPF-2 at KSC. During CEIT, the crew had an opportunity to get a hands-on look at the orbiter and payloads they would work with while on-orbit. (NASA)

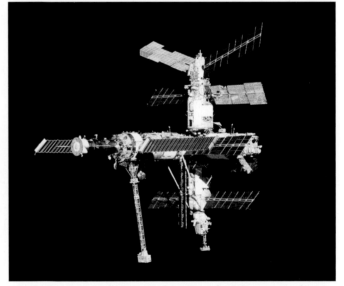

Charlie Precourt and Dom Gorie made a final fly-around of Mir as Discovery departed. This would be the last time a space shuttle visited Mir. The station, which weighed some 285,940 pounds, would destructively reenter the atmosphere on 23 March 2001. (NASA)

STS-95

Mission:	92	NSSDC ID:		1998-064A
Vehicle:	OV-103 (25)	ET-98 (SLWT)		SRB: BI096
Launch:	LC-39B	29 Oct 1998		19:20 UTC
Altitude:	303 nm	Inclination:		28.45 degrees
Landing:	KSC-33	07 Nov 1998		17:05 UTC
Landing Rev:	134	Mission Duration:		213 hrs 44 mins

Commander:	Curtis L. "Curt" Brown, Jr. (5)
Pilot:	Steven W. "Steve" Lindsey (2)
MS1:	Pedro F. Duque Duque (1)
MS2:	Scott E. Parazynski (3)
MS3:	Stephen K. "Steve" Robinson (2)
PS1:	Chiaki Mukai (2)
PS2:	John H. Glenn, Jr. (2)

Payloads:	Up: 41,591 lbs	Down: 0 lbs
	Spacehab SM (34,247 lbs)	
	SPARTAN-201-05 (2,978 lbs)	
	HOST/TAS-02 (2,800 lbs)	
	PANSAT (127 lbs)	
	SEM-04	
	GAS (4)	
	IEH-3 (hitchhiker)	
	CryoTSU (hitchhiker)	
	SRMS s/n 201	

Notes:	HOST platform validated components for HST-SM3
	Oldest space shuttle flyer
	(Glenn, 77 years, 3 months, 11 days)

Wakeup Calls:

30 Oct	"What a Wonderful World"
31 Oct	"Cachito"
01 Nov	"This Pretty Planet"
02 Nov	"Moon River"
03 Nov	"The House is Rockin'"
04 Nov	"Wakaki Chi" (Keio University cheering song)
05 Nov	"I Know You're Out There Somewhere"
06 Nov	"Voyage Into Space"
07 Nov	"La Cucaracha"

The small Mercury capsule and red streamer extending up toward the center of the orbiter to form a "7" are references to Glenn's association with the Mercury Seven and their spacecraft. All of the manned Mercury spacecraft had seven as part of their name. (NASA)

The 29 October 1998 countdown for this *Discovery* mission encountered two unplanned holds that lasted about 20 minutes. The launch was rare in that the official weather forecast provided by the 45th Weather Squadron was 100 percent favorable, something that seldom happens in central Florida.

During the SSME ignition sequence, pad cameras showed the drag chute compartment door falling away from the orbiter. A review of imagery showed the door detached three seconds before liftoff and struck the center main engine nozzle as it fell. Technicians found the remains of the door in the launch pad area; no other vehicle hardware was found. Engineers were somewhat concerned the drag chute might deploy during entry or on approach, but there was little they could do to alter the eventual course of events other than brief the crew on the possibilities.

Bill Clinton became the second (and so far, last) American president to witness a manned space launch, joined by Hillary on the roof of the VAB. The first such viewing was when Richard Nixon witnessed the launch of Apollo 12 on 14 November 1969.

The primary objectives included conducting experiments in the Spacehab single module, deploying the SPARTAN-201-05 (Shuttle Pointed Autonomous Research Tool for Astronomy) free-flyer, operating the IEH (international extreme ultraviolet hitchhiker), and evaluating the HOST (Hubble orbiting systems test) equipment in the payload bay. Despite its science objectives, most of the publicity surrounding the flight came from one payload specialist. At 77 years, 3 months, 11 days, Glenn was the oldest person and the third politician to fly in space, preceded by Jake Garn (STS-23/51D) and Bill Nelson (STS-32/61C). Since researchers believed that the aging process and spaceflight experience share several physiological responses, NASA and the National Institute on Aging sponsored a number of experiments using Glenn as a test subject. The citizens of Perth and Rockingham in Australia greeted Glenn by leaving their house lights on while *Discovery* passed overhead, repeating a tribute they had paid Friendship 7 in February 1962.

Because the drag-chute door had fallen off during launch, mission management directed the crew not to use the drag chute during landing since it might have been damaged during ascent or entry. The post-flight inspection showed the chute remained in its compartment with little evidence of discoloration or damage.

The Space Foundation awarded the crew of STS-95 its Douglas S. Morrow Public Outreach Award for 1999. The award has been given annually since 1995 to an individual or organization making significant contributions to public awareness of space programs.

Clockwise from right center are Curt Brown, Steve Lindsey, Steve Robinson, Pedro Duque, Chiaki Mukai, Scott Parazynski, and John Glenn. Flying in Discovery, particularly being able to move about, was very different than the five hours Glenn had spent in Friendship 7. (NASA)

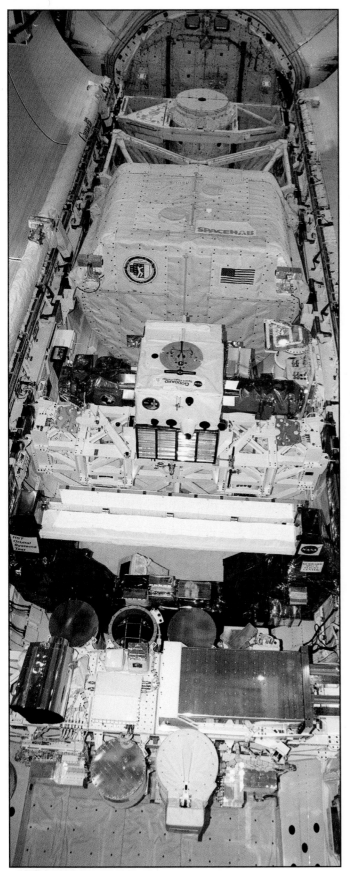

Discovery had a full payload bay on this mission. From the top is the external airlock, sans ODS, and the tunnel adapter that connected it to the Spacehab single module. This was followed by IEH-3, SPARTAN-201-05, and HOST/TAS-02 at the bottom. (NASA)

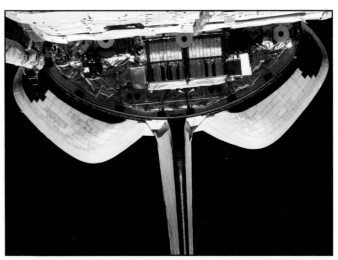

A good view of the final "eyeball" pattern on the OMS pods. The rest of the leading edge continued to be covered with white LRSI tiles until the end of the flight campaign, although most of the side areas used AFRSI (FIB) blankets. (NASA)

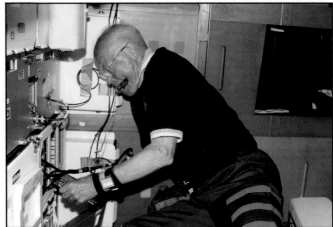

In addition to being the oldest person to fly in space, Glenn was also the third sitting member of Congress to do so. The experiments conducted on Glenn during the mission were sponsored by NASA and the National Institute on Aging based on the theory that the aging process and spaceflight experience share a number of similar physiological responses. Data obtained during this mission were compared to data obtained from the Friendship 7 flight in 1962. (NASA)

ET-98 was the first tank towed by a SRB retrieval ship instead of a commercial tug company. Freedom Star left Michoud towing the Poseidon barge on 12 June 1998. United Space Alliance estimated that using the otherwise idle ships saved about $50,000 per trip. (NASA)

STS-88 (ISS-2A)

Mission:	93	NSSDC ID:	1998-069A
Vehicle:	OV-105 (13)	ET-97 (SLWT)	SRB: BI095
Launch:	LC-39A	04 Dec 1998	08:36 UTC
Altitude:	213 nm	Inclination:	51.60 degrees
Landing:	KSC-15	16 Dec 1998	03:54 UTC
Landing Rev:	185	Mission Duration:	283 hrs 18 mins

Commander:	Robert D. "Bob" Cabana (4)
Pilot:	Frederick W. "Rick" Sturckow (1)
MS1:	Jerry L. Ross (6)
MS2:	Nancy J. Currie (3)
MS3:	James H. "Jim" Newman (3)
MS4:	Sergei Konstantinovich Krikalev (4)

Payloads:	Up: 37,731 lbs	Down: 335 lbs
	Node 1 (Unity) (8,890 lbs)	
	MightySat/SAC-A	
	SEM-07	
	GAS (1)	
	SRMS s/n 202	

Notes:	Scrubbed 03 Dec 1998 (orbiter master alarm)
	First international space station assembly flight
	Third Dog Crew
	EVA1 07 Dec 1998 (Ross and Newman)
	EVA2 09 Dec 1998 (Ross and Newman)
	EVA3 12 Dec 1998 (Ross and Newman)

Wakeup Calls:

04 Dec	"Get Ready"
05 Dec	The Navy Hymn – "Anchors Aweigh"
06 Dec	"Somewhere Over the Rainbow"
07 Dec	"Jerry the Rigger"
08 Dec	"Streets of Bakersfield"
09 Dec	"Floating in the Bathtub"
10 Dec	"God Bless the U.S.A."
11 Dec	"Trepak" (Russian dance)
12 Dec	"Hound Dog"
13 Dec	"Goodnight Sweetheart, Goodnight"
14 Dec	"I Got You (I Feel Good)"
15 Dec	"Ride of the Valkyries"

The rising Sun symbolizes the dawning of a new era of international cooperation in space. The Earth scene outlines the International Space Station partners: America, Canada, Japan, Russia, and the members of the European Space Agency. (NASA)

An improperly positioned switch on *Endeavour* resulted in scrubbing the 3 December 1998 countdown at T-31 seconds. The countdown on the following day for the first American ISS assembly mission proceeded smoothly.

This mission featured yet another "Dog Crew"—this time "Stealth Dog Crew III" (stealth because NASA management did not necessarily share the spirit of fun). The crew featured Jim "Pluto" Newman, Bob "Mighty Dog" Cabana (a veteran of the original STS-53 litter), Rick "Devil Dog" Sturckow, Jerry "Hooch" Ross, Nancy "Laika" Currie, and Sergei "Spotnik" Krikalev.

The mating of Node 1 (Unity) to the Functional Cargo Block (FGB, Zarya), and three spacewalks to connect cables between the modules, highlighted this seven-day mission. Unity had been scheduled for launch in December 1997 but was rescheduled after repeated delays on the Russian module. Zarya, built by Boeing for the Russian Space Agency, was launched on a Proton-K from the Baikonur Cosmodrome in Kazakhstan on 20 November 1998. Unity had a pressurized mating adapter (PMA) at either end; one PMA was permanently mated to the FGB and the other was used for orbiter dockings. *Endeavour* carried her external airlock in the Mir position, farther aft in the payload bay, to provide the STS-88 crew with a better view of Zarya while they were performing the initial assembly. Engineers feared Node 1 would completely block the view from the overhead windows had the airlock been in its final position, and everybody was being cautious.

To begin the assembly sequence, the crew conducted a series of rendezvous maneuvers to reach the orbiting FGB. On the way, Nancy Currie used the SRMS to place Node 1 within 4 inches of the orbiter docking system. At that time, the SRMS was in a "limp" configuration and the flight crew used the RCS primary jets to bump PMA-2 into mating with the ODS. Bob Cabana completed the rendezvous by flying *Endeavour* to within 33 feet of the FGB, allowing Currie to capture the FGB with the SRMS and mate it to PMA-1. Reflecting the international cooperation involved in assembling the ISS, on 11 December Cabana and Sergei Krikalev opened the hatch to Unity and floated into the new station together. About an hour later, they opened the hatch to Zarya, which was the control center for the embryonic station.

The mission was originally scheduled to land on Runway 33 at KSC but, two hours before landing, the mission management team switched to Runway 15 (north to south) because of changing winds. As a result of the unresolved drag chute anomaly on STS-95, the drag chute was not used on this flight.

The crew pose at LC-39A after completing the pre-launch terminal countdown demonstration test (TCDT). From left are Jim Newman, Rick Sturckow, Bob Cabana, Nancy Currie, Jerry Ross, and Sergei Krikalev. Note the Russian flag worn by Krikalev. (NASA)

Nancy Currie watches the two SRMS television monitors in the background as she maneuvers the Zarya Functional Cargo Block and mates it to the pressurized mating adapter attached to Unity (Node 1). This marked the beginning of the ISS assembly sequence. (NASA)

Bob Cabana looks out the overhead windows as he maneuvers Endeavour within 33 feet of the Functional Cargo Block, allowing Nancy Currie to capture it with the SRMS. Zarya had been launched from Baikonur on 20 November 1998 using a Proton-K booster. (NASA)

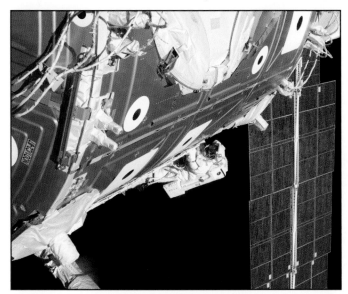

Jim Newman works at the edge of the Unity module during the third EVA. One of the solar array panels on the Russian Zarya Functional Cargo Block is at right. Note the "Nod1" label at the extreme left, telling astronauts exactly where they are on the station. (NASA)

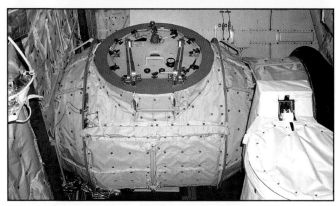

Unity in the payload bay of Endeavour (top). This was the only ISS mission where the external airlock was carried in this position with a connecting tunnel between the middeck and the airlock (bottom photo). All other missions carried the airlock six feet further forward. (NASA)

STS-96 (ISS-2A.1)

Mission:	94	NSSDC ID:	1999-030A
Vehicle:	OV-103 (26)	ET-100 (SLWT)	SRB: BI098
Launch:	LC-39B	27 May 1999	10:50 UTC
Altitude:	214 nm	Inclination:	51.60 degrees
Landing:	KSC-15	06 Jun 1999	06:04 UTC
Landing Rev:	153	Mission Duration:	235 hrs 13 mins

Commander:	Kent V. "Rommel" Rominger (4)
Pilot:	Rick D. Husband (1)
MS1:	Daniel T. "Dan" Barry (2)
MS2:	Ellen Ochoa (3)
MS3:	Tamara E. "Tammy" Jernigan (5)
MS4:	Julie Payette (1)
MS5:	Valeri Ivanovich Tokarev (1)

Payloads:	Up: 33,808 lbs	Down: 213 lbs
	Spacehab LDM (16,072 lbs)	
	ICC-G/UCP (4,960 lbs)	
	STARSHINE (hitchhiker)	
	SRMS s/n 303	

Notes:	Rolled-back to VAB 16 May 1999 (ET hail damage)
	EVA1 30 May 1999 (Jernigan and Barry)

Wakeup Calls:

27 May	"California Dreamin'"
28 May	"Danger Zone" (*Topgun*)
29 May	"Theme from Star Wars"
30 May	"Morning Colors"
31 May	"Amarillo by Morning"
01 Jun	"Exultate Jubilate"
02 Jun	"Vasha Blagarodye" (Russian song)
03 Jun	"Free Bird"
04 Jun	"Good Morning Starshine"
05 Jun	"Theme from The Longest Day"

ROMINGER HUSBAND PAYETTE TOKAPEB JERNIGAN OCHOA BARRY

The triangular shape represents building on the experience of earlier missions while the three vertical bars point toward future human endeavors in space. The five-pointed star is symbolic of the five space agencies participating in the development of ISS. (NASA)

Hail damage to spray-on foam insulation on ET-100 forced a roll-back to the VAB on 8 May 1999 for repairs. The *Discovery* stack returned to LC-39B on 20 May, resulting in a launch delay of seven days. The countdown on 27 May proceeded with no unplanned holds.

This was the second flight for the cameras carried on each SRB forward skirt to support the investigation of foam loss from the external tank thrust panel area. The cameras revealed a technique where technicians had punched small vent holes in foam seemed to reduce the potential for debris, but did not completely eliminate it.

Following rendezvous with the ISS on flight day 3, Kent Rominger and Rick Husband made the first direct docking with PMA-2. The following day, the crew entered Node 1 (Unity) and the Functional Cargo Block (FGB, Zarya) and began installing equipment brought up in the Spacehab logistics double module. Ultimately, the crew transferred 98 items weighing 2,881 pounds as well as 686 pounds of water from the orbiter to the ISS. The crew also transferred 18 items weighing 197 pounds from the ISS to *Discovery*.

Tammy Jernigan and Dan Barry conducted the only EVA of the mission to transfer the orbital replacement unit (ORU) transfer device (OTD) and Russian Strela crane as well as pre-positioning hardware for subsequent ISS assembly missions. The crane was carried on an integrated cargo carrier (ICC-G) with an unpressurized cargo pallet (UCP) mounted in the payload bay. During the Strela crane transfer, the EVA crew members were unable to break the torque on two of the bolts using the EVA power grip tool. Undeterred, they retrieved the manual ratchet wrench and cheater bar from the starboard tool stowage assembly and successfully removed the bolts. Strela was then moved to its location on PMA-2.

Secondary payloads included the multi-mirrored STARSHINE (student tracked atmospheric research satellite for heuristic international networking experiment) and the SVF (shuttle vibration forces) experiment. The development of STARSHINE, an inert 19-inch-diameter sphere covered by 1-inch mirrors, involved thousands of students. Scientists used the satellite to train volunteers to visually track the optically reflective spacecraft for several months, calculate its orbit from shared observations, and derive atmospheric density from its orbital decay. On flight day 9, the crew deployed SUNSHINE from the payload bay, an event witnessed by the developers on the ground outside the Payload Operations Control Center (POCC) at NASA Goddard.

The weather at KSC was good enough to allow mission management to not activate Edwards as an alternate landing site.

Julie Payette is in the middle of this shot of the crew in the hatch leading to the Zarya Functional Cargo Block (FGB). Valeri Tokarev is holding the 28000 km/h sign. Clockwise from him are Tammy Jernigan, Kent Rominger, Rick Husband, Ellen Ochoa, and Dan Barry. (NASA)

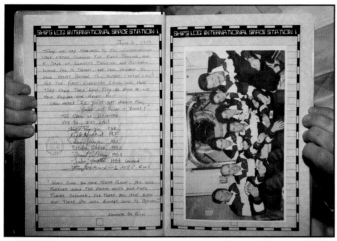

The STS-96 entry in the ISS logbook. Every flight that has visited and every Expedition that has served on the ISS has maintained the logbook. Not shown here, but each mission also places a crew patch decal inside the station in honor of their visit. (NASA)

Spacehab did not make the ideal logistics carrier since everything had to be moved from Spacehab through a narrow tunnel to the external airlock and then through another narrow tunnel to the station. This is Ellen Ochoa in the Spacehab near the end of the move effort. (NASA)

Kent Rominger at the controls on the aft flight deck. The docking mechanism of the approaching station is just a few feet away on the other side of the overhead window. (NASA)

Discovery and her mobile launch platform atop a crawler-transporter during the trip to LC-39B. The VAB had four high bays but space shuttle only used the two eastern bays for stacking. (NASA)

The nascent space station as seen from Discovery as the orbiter approached for docking. At this point the station consisted of the Zarya Functional Cargo Block (right) and Node 1 (Unity, on the left) connected by the pressurized mating adapter in the middle. (NASA)

Dan Barry wears a training version of the extravehicular mobility unit suit during an underwater simulation in the Neutral Buoyancy Laboratory at JSC. The new NBL had replaced the Weightless Environment Training Facility (WET-F) during 1997. (NASA)

STS-93

Mission:	95	NSSDC ID:	1999-040A
Vehicle:	OV-102 (26)	ET-99 (SLWT)	SRB: BI097
Launch:	LC-39B	23 Jul 1999	04:31 UTC
Altitude:	154 nm	Inclination:	28.45 degrees
Landing:	KSC-33	28 Jul 1999	03:21 UTC
Landing Rev:	79	Mission Duration:	118 hrs 50 mins

Commander:	Eileen M. Collins (3)
Pilot:	Jeffrey S. "Bones" Ashby (1)
MS1:	Michel Tognini (2)
MS2:	Steven A. Hawley (5)
MS3:	Catherine G. "Cady" Coleman (2)

Payloads:	Up: 52,382 lbs	Down: 0 lbs
	Chandra X-ray Observatory/IUS (50,162 lbs)	
	No SRMS	

Notes:
Scrubbed 20 Jul 1999 (hydrogen readings)
Scrubbed 22 Jul 1999 (KSC weather)
First female space shuttle commander (Collins)
Heaviest single payload carried by space shuttle
Hydrogen leak during ascent
Premature SSME shutdown resulted in
15 fps underspeed

Wakeup Calls:

23 Jul	"Beep Beep"
24 Jul	"Brave New Girls"
25 Jul	"Someday Soon"
26 Jul	"The Sounds of Silence"
27 Jul	"A Little Traveling Music"

The launch for this *Columbia* mission was scheduled for 20 July 1999, but engineers detected a hydrogen buildup in the aft compartment and scrubbed prior to main engine ignition at T-6 seconds. A subsequent reading at T-8 seconds (after the scrub) registered normal levels, meaning the earlier indication was likely false, but by then it was too late. The mission management team rescheduled the launch for 22 July, but a thunderstorm caused the launch director to cancel the attempt at T-5 minutes. This gave NASA just one more opportunity to launch before the Air Force closed the Eastern Range for modifications. The countdown on 23 July was delayed seven minutes by a ground communications problem but was otherwise uneventful.

About five seconds after SSME start, but prior to liftoff, one main engine experienced a small hydrogen leak in the engine nozzle caused by a small pin routinely used during repairs departing from the engine, impacting the nozzle, and rupturing three hydrogen cooling-tubes. The hydrogen leak caused an improper mixture ratio that resulted in low engine performance, which in turn caused an increase in oxygen flow. This resulted in an early low-level oxygen cutoff and a 16-fps underspeed. Despite the premature main engine shutdown, the OMS burns made up the performance shortfall and the mission proceeded as planned. Contrary to many reports, this was not an abort-to-orbit. As if this was not enough excitement, an electrical short affected two of the main engine controllers. Both controllers automatically switched to backup computers, but the change resulted in the loss of all data from one engine for the duration of the mission and eliminated redline protection for the other engine. Fortunately, the engines continued to operate well.

The crew successfully deployed the $1,550-million Chandra X-ray Observatory a little more than seven hours after launch. This was the largest and heaviest payload carried during the flight campaign and was the last significant spacecraft deployed by space shuttle. The third of the Great Observatories was the most sophisticated X-ray telescope of its time and was designed to observe high-energy regions of the universe, such as hot gas in the remnants of exploded stars. This was originally called the Advanced X-ray Astrophysics Facility (AXAF), but was renamed in honor of Subrahmanyan Chandrasekhar, the 1983 Nobel Laureate in Physics. "Chandra" also means "moon" or "luminous" in Sanskrit.

The landing marked the twelfth night landing of the flight campaign, including five at Edwards. To date, there had been 19 consecutive landings at the Shuttle Landing Facility, and 25 of the last landings 26 had been there.

The crew patch depicted Chandra separating from Columbia after a successful deployment, with a spiral galaxy shown in the background as a possible target for observations. The two flags represent the American and French crew members. (NASA)

In front are Eileen Collins and Michel Tognini, who represented France's Centre National d'Etudes Spatiales (CNES). Behind them, from the left, are Steven Hawley, Jeff Ashby, and Cady Coleman. Note the flag from the Chandra X-Ray Center in the background. (NASA)

A diver untangles parachute lines from an SRB aft skirt more than 100 feet underwater. It was dangerous work, often in high seas off the coast of Florida, but the program had a remarkably good safety record. (NASA)

The most worrisome issue during ascent was when a main engine ejected a small pin used for a repair. The pin ruptured three of the liquid hydrogen cooling tubes that make up the nozzle, resulting in a hydrogen leak that caused an improper mixture ratio and a corresponding increase in oxygen flow. Eventually, the engines shut down because they had used all the liquid oxygen in the ET. Fortunately, the OMS engines made up the difference in performance. (NASA)

The Chandra X-ray Observatory and its inertial upper stage as captured during separation using an HDTV camcorder from inside the crew module. NASA designed the observatory to last five years; it is still functioning in late 2018, almost 20 years after it was deployed. (NASA)

It was quite a debut for the first female commander, piloting what was perhaps the most harrowing ascent of the flight campaign with multiple main engine issues going uphill. Here, Eileen Collins looks over a procedures checklist on the forward flight deck. (NASA)

The Chandra X-ray Observatory occupied the entire payload bay, to the point of requiring some minor equipment be removed from the forward and aft bulkheads to provide sufficient clearance. It was the longest and heaviest payload carried by space shuttle. (NASA)

STS-103 (HST-SM3A)

Mission:	96	NSSDC ID:	1999-069A
Vehicle:	OV-103 (27)	ET-101 (SLWT)	SRB: BI099
Launch:	LC-39B	20 Dec 1999	00:50 UTC
Altitude:	330 nm	Inclination:	28.45 degrees
Landing:	KSC-33	28 Dec 1999	00:02 UTC
Landing Rev:	119	Mission Duration:	191 hrs 11 mins

Commander:	Curtis L. "Curt" Brown, Jr. (6)
Pilot:	Scott J. Kelly (1)
MS1:	John M. Grunsfeld (3)
MS2:	Jean-François Clervoy (3)
MS3:	C. Michael "Mike" Foale (5)
MS4:	Steven L. "Steve" Smith (3)
MS5:	Claude Nicollier (4)

Payloads:	Up: 20,276 lbs	Down: 5,351 lbs
	Hubble Servicing Mission (11,927 lbs)	
	SRMS s/n 301	

Notes:	Delayed 06 Dec 1999 (wiring inspections)
	Delayed 16 Dec 1999 (MPS welds)
	Scrubbed 18 Dec 1999 (KSC weather)
	EVA1 22 Dec 1999 (Smith and Grunsfeld)
	EVA2 23 Dec 1999 (Foale and Nicollier)
	EVA3 24 Dec 1999 (Smith and Grunsfeld)

Wakeup Calls:

20 Dec	"Taking Care of Business"
21 Dec	"Rendezvous"
22 Dec	"Hucklebuck"
23 Dec	"Traditional Swiss music" /
	"Only When I Sleep
24 Dec	"Magic Carpet Ride" /
	"Skinnamarink" (children's song)
25 Dec	"I'll Be Home for Christmas"
26 Dec	"We're So Good Together"
27 Dec	"The Cup of Life"

The failure of three of the six rate sensor units on the Hubble Space Telescope caused NASA to advance part of the third servicing mission scheduled for June 2000. Having fewer than three working gyros would have precluded science observations, although the telescope would have remained safely on-orbit. The flight rules dictated a "call-up" mission before a fourth gyro failed.

This resulted in the only launch-on-need mission (STS-LON-3) of the flight campaign and caused the planned third servicing mission to be split in two; SM3A would conduct emergency repairs plus a few modifications, while SM-3B would perform the upgrades originally planned since the new instruments were not yet ready for installation. Four new gyros had been installed during the first servicing mission (STS-61) in December 1993 and all six gyros were working during the second servicing mission (STS-82) in February 1997. Then a single gyro failed in each of the following three years. The Hubble team believed they understood the cause of the failures, although they could not be certain until the gyros were returned from space.

The agency delayed an early planning date of 6 December 1999 to perform inspections of the Kapton wiring on the entire orbiter fleet, including *Discovery* and postponed a 16 December date to inspect welds in the main propulsion system. The launch team scrubbed the first real attempt on 18 December due to bad weather at the launch site. The count on 20 December proceeded without incident, but by this time NASA had cut the planned ten-day mission to eight days to ensure *Discovery* was back on the ground before the Year-2000 (Y2K) rollover, which engineers feared might trigger unexpected problems in the various computer systems.

The primary objectives were to replace three rate sensor units (RSU), open a cooling valve on NICMOS (near infrared camera and multi-object spectrometer) and install six voltage improvement kits, a new computer, and a fine guidance sensor. The mission would also replace the S-band transmitter, a solid-state recorder, and several areas of worn multilayer insulation (MLI). Steve Smith and John Grunsfeld conducted the first EVA, installing the three RSUs, opening the NICMOS valve, and fitting the six voltage improvement kits. Mike Foale and Claude Nicollier conducted the second EVA the following day, replacing the computer and fine guidance sensor. During the third EVA, Smith and Grunsfeld replaced the S-band transmitter and solid-state recorder, and installed the new multilayer insulation.

Mission control waved-off the first KSC opportunity because of excessive crosswinds. The weather cleared for the second opportunity. This was the last time *Discovery* would be alone in space; all of her future missions would be to the ISS.

The crew patch depicted Discovery approaching Hubble prior to its capture and berthing. The horizontal and vertical lines centered on the telescope symbolized the ability to reach and maintain a desired attitude in space, essential to its scientific operation. (NASA)

The crew pose following emergency egress training with an M113 armored personnel carrier behind them. From the left are Scott Kelly, Steve Smith, John Grunsfeld, Claude Nicollier, their trainer George Hoggard, Curt Brown, Mike Foale, and Jean-Francois Clervoy. (NASA)

Steve Smith (red stripes) and John Grunsfeld appear as small figures in this wide scene photographed during the first EVA. On this spacewalk they replaced three rate sensing units, opened the NICMOS valve, and installed six voltage improvement kits. (NASA)

Wearing Santa hats, John Grunsfeld (left) and Steve Smith share a brief celebration on the middeck before the final EVA. Both are wearing the liquid cooling garments that go under the EMU suits. (NASA)

Claude Nicollier, from the European Space Agency (ESA), is carrying one of the Hubble power tools as he works at a storage container during the second spacewalk. A small piece of the telescope, including an antenna, can be seen in the upper right corner. (NASA)

Mike Foale uses some early virtual reality hardware to rehearse his extravehicular activity duties. Although somewhat heavy and cumbersome, the VR setups proved useful for training, despite what today would be called extremely low-resolution graphics. (NASA)

Mike Foale, and the rest of the EVA crew members, also trained in more traditional ways, such as this rehearsal using the Hubble Space Telescope mockup in the new Neutral Buoyancy Laboratory at JSC. Note the safety divers in the background, just in case. (NASA)

STS-99

Mission:	97	NSSDC ID:	2000-010A
Vehicle:	OV-105 (14)	ET-92 (LWT)	SRB: BI100
Launch:	LC-39A	11 Feb 2000	17:44 UTC
Altitude:	130 nm	Inclination:	57.00 degrees
Landing:	KSC-33	22 Feb 2000	23:23 UTC
Landing Rev:	182	Mission Duration:	269 hrs 39 mins

Commander:	Kevin R. Kregel (4)
Pilot:	Dominic L. P. "Dom" Gorie (2)
MS1:	Gerhard P. J. "Hoss" Thiele (1)
MS2:	Janet L. Kavandi (2)
MS3:	Janice E. Voss (5)
MS4:	Mamoru "Mark" Mohri (2)

Payloads:	Up: 35,410 lbs	Down: 0 lbs
	SRL-3/SRTM (26,835 lbs)	
	No SRMS	

Notes:	Delayed 16 Sep 1999 (wiring concerns)
	Delayed 19 Nov 1999 (manifest)
	Delayed 13 Jan 2000 (manifest)
	Scrubbed 31 Jan 2000 (KSC weather)

Wakeup Calls (Blue Team / Red Team):

11 Feb	"Time for Me to Fly" (Blue Team)
12 Feb	"Eye in the Sky" / "Some Guys Have All The Luck"
13 Feb	"Linus and Lucy" / "Jumpin' Jive"
14 Feb	"Journey to the Stars" / "Radar Love"
15 Feb	"Canon in D" / "New York, New York"
16 Feb	"We Saw the Sea" / "Smack Dab in the Middle"
17 Feb	"Take a Little Less" / "Die Moldau"
18 Feb	"Rawhide" / "Magic Carpet Ride"
19 Feb	"One After 909" /
	"Catch the Moments As They Fly By"
20 Feb	"Stay" / "Walk Don't Run"

The clear portion of Earth illustrates the radar beams penetrating its cloudy atmosphere and the grid on reflects the mapping character of the SRTM mission. The rainbow along the horizon resembles an orbital sunrise, symbolic of the bright future in space. (NASA)

Flight planners originally scheduled this *Endeavour* mission to fly on 16 September 1999, but in mid-August, the agency postponed the launch date until October because of Kapton wiring concerns throughout the orbiter fleet. Given the amount of wiring requiring inspection, NASA shifted the target date to no earlier than 19 November. Mission management subsequently decided to launch STS-103, the urgent third Hubble servicing mission, and set the launch of STS-99 for 13 January 2000. In December, NASA postponed the launch again, this time until 31 January. The launch team scrubbed that attempt because of unacceptable weather. The launch on 11 February went smoothly.

The Shuttle Radar Topography Mission (SRTM) was an international effort led by the National Imagery and Mapping Agency (NIMA, now the National Geospatial-Intelligence Agency, NGA) and NASA, with participation of the German Aerospace Center (DLR). The primary objective was to create a complete high-resolution digital topographic database using an C-band and X-band interferometric synthetic aperture radar (IFSAR) to gather data that produced 3D images of the surface of the Earth.

The crew split into two teams to work around-the-clock. The blue team consisted of Dom Gorie, Janice Voss, and Mamoru Mohri while the red team included Kevin Kregel, Gerhard Thiele, and Janet Kavandi. They successfully deployed the 197-foot-long radar mast and began mapping an area from 60 degrees north to 56 degrees south less than 12 hours after launch. Data was sent to the Jet Propulsion Laboratory (JPL) for analysis and early indications showed the data to be of excellent quality. During 222 hours, 23 minutes of mapping, they covered 99.96 percent of the planned mapping area at least once and 94.6 percent of it twice. Only about 80,000 square miles in scattered areas remained unimaged, most of them in North America and most already well mapped by other methods. Also aboard was the EarthKAM student experiment that took 2,715 digital photos through an overhead window, a significant improvement over the combined total of 2,018 images from four previous missions. The pictures were used in classroom projects on Earth science, geography, mathematics, and space science. More than 75 middle schools around the world participated in the project.

Mission control waved-off the first landing opportunity due to weather at KSC. On the next opportunity, *Endeavour* flew into a direct 13 knot crosswind, enabling the pilots to fulfill a crosswind development test objective that had been planned for 57 previous missions. This was the last time *Endeavour* would be alone in space; all of her future missions would be to the ISS

Despite working opposite shifts, the crew took time to pose for an on-orbit portrait. In the rear, from the left, are blue team members Mamoru Mohri, Dom Gorie, and Janice Voss. In front are red team members Janet Kavandi, Kevin Kregel, and Gerhard Thiele. (NASA)

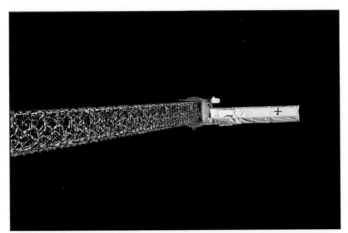

The 197-foot-long mast was the longest rigid structure deployed in space at the time and supported C-band and X-band antennas at its end. Coupled with antennas in the payload bay, this allowed the radars to collect 3D data, greatly increasing its effectiveness. (NASA)

Mamoru Mohri (foreground) and Janice Voss during a simulation in one of the trainers in the Space Vehicle Mockup Facility. After the Challenger accident, one crew member had to occupy seat 5 (where Voss is sitting) to operate the hatch jettison handle if it had ever been needed. (NASA)

Flight controllers in the Mission Control Center at JSC follow pre-launch activity at KSC. Houston had little involvement until the vehicle cleared the launch tower. This attempt on 31 January was eventually scrubbed due to the adverse weather shown on the large monitor at right. (NASA)

Endeavour sitting on LC-39A. The crew cabin access arm is in place at the crew hatch and the two large tail service masts (TSM) can be seen beside the aft fuselage. The TSMs transfered propellants, purge gases, and power to the orbiter on the pad. (NASA)

STS-101 (ISS-2A.2a)

Mission:	98	NSSDC ID:	2000-027A
Vehicle:	OV-104 (21)	ET-102 (SLWT)	SRB: BI101
Launch:	LC-39A	19 May 2000	10:11 UTC
Altitude:	207 nm	Inclination:	51.60 degrees
Landing:	KSC-15	29 May 2000	06:21 UTC
Landing Rev:	156	Mission Duration:	236 hrs 09 mins

Commander:	James D. "Jim" Halsell, Jr. (5)
Pilot:	Scott J. "Doc" Horowitz (3)
MS1:	Mary Ellen Weber (2)
MS2:	Jeffrey N. "Jeff" Williams (1)
MS3:	James S. "Jim" Voss (4)
MS4:	Susan J. Helms (4)
MS5:	Yury VladiMirovich Usachov (3)

Payloads:	Up: 35,604 lbs	Down: 1,391 lbs
	Spacehab LDM (18,893 lbs)	
	ICC-G/UCP (3,970 lbs)	
	SEM-06	
	SRMS s/n 202	

Notes:	Scrubbed 24 Apr 2000 (KSC weather)
	Scrubbed 25 Apr 2000 (KSC weather)
	Scrubbed 26 Apr 2000 (TAL weather)
	Delayed 18 May 2000 (range conflict)
	First flight of the "glass cockpit" (MEDS)
	EVA1 22 May 2000 (Voss and Williams)

Wakeup Calls:

19 May	"Free Fallin'"
20 May	"Shining Brightly"
21 May	"Lookin' Out the Window"
22 May	"Haunted House"
23 May	"I Only Have Eyes for You"
24 May	"I'm Gonna Fly"
25 May	"Don't It Make You Wanna Dance"
26 May	Untitled Russian song
27 May	"25 or 6 to 4"
28 May	"El Capitan"

The three large stars represent the third ISS assembly mission. The crew patch depicts the ISS consisting of the Unity and Zarya modules as Atlantis approaches. The elements and colors of the border reflect the American and Russian flags. (NASA)

On 18 February 2000, NASA confirmed plans to fly an additional space shuttle mission to the ISS. The plan distributed the original STS-101 mission objectives between two flights, STS-101 and STS-106, both using *Atlantis*.

The agency ultimately set the STS-101 launch date as 24 April, but scrubbed the attempt at T-5 minutes because of excessive crosswinds at the Shuttle Landing Facility and high winds at Edwards. An attempt the following day was also scrubbed for high winds. On 26 April everything looked good at KSC, but excessive crosswinds at the Ben Guerir TAL site resulted in another scrub. A crowded Eastern Range schedule delayed the next attempt until 18 May, but the scrub of an Atlas III on 16 May and its need to recycle caused STS-101 to be delayed another day to 19 May. This time all went well, and *Atlantis* was launched at the beginning of a five-minute window.

This was the first mission to use the new multifunction electronic display system (MEDS) "glass cockpit." In reality, the orbiter had always used a set of monochrome (green) CRTs to display most mission-specific information, but MEDS took it to a new level with 11 full color LCD flat panel displays based on the units developed for the Boeing 777. Unfortunately, for the most part the new screens presented the same information as the earlier system and efforts to provide better situational awareness for the flight crew were canceled after the decision to retire space shuttle.

The payload bay contained a pressurized Spacehab logistics double module. This was the second flight of the integrated cargo carrier (ICC-G) that included an unpressurized cargo pallet (UCP) used for carrying parts of the Russian Strela crane, the SHOSS (Spacehab-Oceaneering space system) box that carried EVA tools and flight equipment, and the space integrated global positioning system/inertial navigation system (SIGI) operational attitude readiness (SOAR) hardware. Secondary payloads included AST (astroculture), BioTube (BioTube precursor experiment), CPCG (commercial protein crystal growth), MARS (mission to America's remarkable schools), PCG-BAG (protein crystal growth-biotechnology ambient generic), and SEM-06 (space experiment module).

On flight day 2, *Atlantis* rendezvoused with the ISS and docked at PMA-2. After the docking, but before opening the hatches, Jim Voss and Jeff Williams conducted the only EVA of the mission on flight day 3. The crew members repositioned the orbital replacement unit (ORU) transfer device (OTD), assembled the Russian Strela crane, and replaced the early communications (ECOMM) antenna.

The weather was good and *Atlantis* returned to the Shuttle Landing Facility on the first opportunity.

Seated in front are Doc Horowitz (left) and Jim Halsell. The others, from the left, are Mary Ellen Weber, Jeff Williams, Yury Usachov, James Voss, and Susan Helms. Voss and Williams were wearing EMU suits while the others are wearing launch-entry suits. (NASA)

Susan Helms carries part of a treadmill into Node 1 (Unity). All of the equipment and supplies had to be moved from the Spacehab module, through the transfer tunnel and orbiter, and then into the station. (NASA)

A fish-eye lens on a 35 mm camera captured this view of Atlantis as she orbited a cloud covered Earth. The ODS is in the foreground with the Spacehab single module in the background. (NASA)

Jim Voss maneuvers himself into the Atlantis airlock following the 6-hour, 44-minute EVA he shared with Jeff Williams. (NASA)

This was the first mission to use the multifunction electronic display system (MEDS) glass cockpit. This photo, often cited as being of Atlantis on STS-101, was actually one of the fixed-base simulators in the Mission Simulation and Training Facility at JSC. (NASA)

The exhaust from the solid rocket boosters light up the clouds of smoke and steam trailing behind Atlantis as it climbs uphill into the pre-dawn sky on 19 May 2000. This was the fifth launch attempt for STS-101 and the liftoff was at the beginning of the five-minute window. (NASA)

STS-106 (ISS-2A.2b)

Mission:	99	NSSDC ID:	2000-053A
Vehicle:	OV-104 (22)	ET-103 (SLWT)	SRB: BI102
Launch:	LC-39B	08 Sep 2000	12:46 UTC
Altitude:	206 nm	Inclination:	51.60 degrees
Landing:	KSC-15	20 Sep 2000	07:58 UTC
Landing Rev:	186	Mission Duration:	283 hrs 11 mins

Commander:	Terrence W. "Terry" Wilcutt (4)
Pilot:	Scott D. "Scooter" Altman (2)
MS1:	Edward T. "Ed" Lu (2)
MS2:	Richard A. "Rick" Mastracchio (1)
MS3:	Daniel C. "Dan" Burbank (1)
MS4:	Yuri Ivanovich Malenchenko (2)
MS5:	Boris VladiMirovich Morukov (1)

Payloads:	Up: 34,991 lbs		Down: 948 lbs
	Spacehab LDM (18,508 lbs)		
	ICC-G (4,528 lbs)		
	SEM-08		
	GAS (1)		
	SRMS s/n 202		

Notes: EVA1 11 Sep 2000 (Lu and Malenchenko)

Wakeup Calls:

08 Sep	"I'll Be"
09 Sep	"I Say a Little Prayer"
10 Sep	"All Star"
11 Sep	"The Hukilau Song"
12 Sep	"Brown-Eyed Girl"
13 Sep	"Kombaht" (Russian song)
14 Sep	"Haze Has Melted Away"
15 Sep	University of Connecticut Fight Song
16 Sep	The Coast Guard Hymn – "Semper Paratus"
17 Sep	"YMCA"
18 Sep	"Home in the Islands"
19 Sep	"Houston"

The newly arrived Zvezda Service Module was depicted on the crew patch mated with the orbiting Unity and Zarya modules. The astronaut symbol provided a connection between Atlantis and the ISS. Stylized versions of the American and Russian flags meet at the station. (NASA)

Out for a short scenic tour, *Atlantis* visited VAB High Bay 2 for a fit check of the "safe haven" modifications after being rolled out of High Bay 1. The check revealed nothing of significance and the Space Shuttle Program continued to maintain the safe haven for the duration of the flight campaign. In this case, safe haven meant a place to store a stack during a hurricane, not the on-orbit CSCS (contingency shuttle crew support) safe haven developed after the *Columbia* accident. The countdown on 8 September went smoothly the launch occurred at the opening of the five-minute window with no unplanned holds.

The Ku-band radar acquired the ISS at 145,000 feet and tracked the station to within 320 feet when the crew placed the system in communications mode to provide television coverage of the docking. The primary objective was to change the Zvezda Service Module from the launch to the flight configuration. Zvezda had joined the ISS configuration on 26 July as the early living quarters for the station and was one of the primary Russian contributions to the ISS. In addition, the crew removed the docking unit and aft docking probe from the Functional Cargo Block (FGB, Zarya). The crew also replaced two batteries on Zarya and installed three batteries on Zvezda, which had been launched with only five of its eight batteries because of performance limits of the Proton-K launch vehicle.

On flight day 3, Ed Lu and Yuri Malenchenko conducted the only EVA of the mission, connecting cables between Zvezda and Zarya. They also installed a 6-foot-long magnetometer that provided Earth orientation information to minimize Zvezda propellant usage during attitude control. This marked the sixth spacewalk in support of the station assembly and the 50th of the Space Shuttle Program (if one does not count the zero time EVA on STS-80).

The crew entered the ISS on flight day 5 and began transferring 4,619 pounds of materiel from the Spacehab logistics double module and 3,000 pounds of hardware from a waiting Progress supply vehicle to the ISS. The crew also transferred 780 pounds of water. Some 948 pounds of trash, mostly used packing material, was loaded into the Spacehab for the return to Earth. Since the crew was conserving sufficient consumables, mission management approved a one-day extension. Originally, the crew was only expected to complete only 30 of 48 tasks approved for the STS-106 mission during an 11-day mission. The additional day on-orbit allowed the crew to complete all of the original 48 tasks as well as 22 additional tasks.

The deorbit burn occurred on the first KSC landing opportunity, marking the 30th landing at the Shuttle Landing Facility in the last 31 missions.

The crew poses in front of the Spacehab module during the crew equipment interface test. In the front, from the left, are Dan Burbank, Scott Altman, Terry Wilcutt, and Rick Mastracchio; in the background are Yuri Malenchenko, Boris Morukov, and Ed Lu. (NASA)

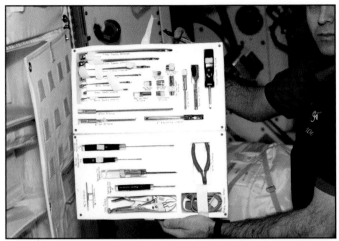

One of the items that STS-106 brought to the ISS was a rather normal looking set of tools, being held by Rick Mastracchio, partially out of frame. There were also a lot of specialized tools. (NASA)

Yuri Malenchenko (left) drinks a beverage as Boris Morukov prepares to exercise using an ergometer on the middeck of Atlantis. Malenchenko and Morukov represented Rosaviakosmos. (NASA)

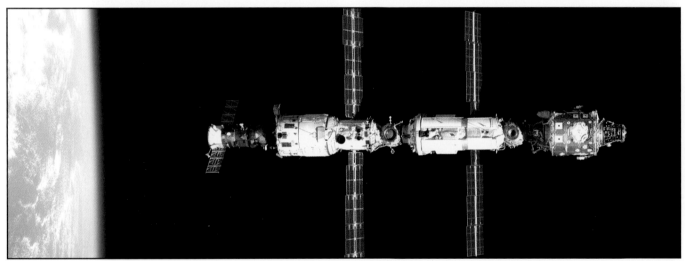

A view of the ISS from Atlantis after undocking over northeastern Ukraine. A Progress supply vehicle is at left. At this point, the ISS was still unmanned, but that would change with the arrival of the Expedition 1 crew aboard Soyuz TM-31 on 2 November 2000. (NASA)

The crew examine the orbiter docking system during the crew equipment interface test while Atlantis was in OPF-3. Note everybody is wearing bunny suits. (NASA)

From the bottom are the Spacehab double module, integrated cargo carrier, transfer tunnel, and external airlock with ODS. A SHOSS toolbox is on top of the ICC. (NASA)

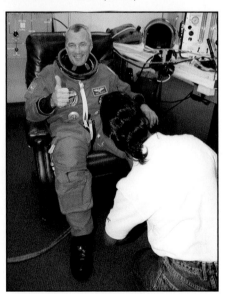

Terry Wilcutt suiting up in the Operations and Checkout Building prior to launch. Despite the help in the O&C Building, the crew could self-don their suits while on-orbit. (NASA)

STS-92 (ISS-3A)

Mission:	100	NSSDC ID:		2000-062A
Vehicle:	OV-103 (28)	ET-104 (SLWT)		SRB: BI104
Launch:	LC-39A	11 Oct 2000		23:17 UTC
Altitude:	213 nm	Inclination:		51.60 degrees
Landing:	EDW-22	24 Oct 2000		21:01 UTC
Landing Rev:	203	Mission Duration:		309 hrs 43 mins

Commander:	Brian Duffy (4)
Pilot:	Pamela A. "Pam" Melroy (1)
MS1:	Koichi Wakata (2)
MS2:	William S. "Bill" McArthur, Jr. (3)
MS3:	Peter J. K. "Jeff" Wisoff (4)
MS4:	Michael E. "LA" López-Alegría (2)
MS5:	Leroy Chiao (3)

Payloads:	Up: 35,250 lbs	Down: 293 lbs
	Z1 Truss (21,998 lbs)	
	SLP/PMA-3 (6,257 lbs)	
	SRMS s/n 301	

Notes:	Scrubbed 05 Oct 2000 (attach bolt concerns)
	Scrubbed 09 Oct 2000 (KSC weather)
	Scrubbed 10 Oct 2000 (17-inch disconnect)
	EVA1 15 Oct 2000 (Chiao and McArthur)
	EVA2 16 Oct 2000 (Wisoff and López-Alegría)
	EVA3 17 Oct 2000 (Chiao and McArthur)
	EVA4 18 Oct 2000 (Wisoff and López-Alegría)
	Orbiter returned to KSC on 03 Nov 2000 (N905NA)

Wakeup Calls:

12 Oct	"Incense and Peppermints" (*Austin Powers*)
13 Oct	"Girls Just Want to Have Fun"
14 Oct	"Eikan wa kimi mi Kagayku" (Japanese marching song)
15 Oct	"Camelot" (*Monty Python and the Holy Grail*)
16 Oct	"Je t'aimais, Je t'aime, J'taimerai" (French song)
17 Oct	The Army Hymn – "Caissons Go Rolling Along"
18 Oct	"Theme from Mission Impossible"
19 Oct	The Navy Hymn – "Anchors Aweigh"
20 Oct	"The River"
21 Oct	"Saturday Night"
22 Oct	The Air Force Hymn – "Wild Blue Yonder"
23 Oct	"Bad Bad Leroy Brown"
24 Oct	"Deja Vu"

The black silhouette of Discovery stands out against the deep blue background of space. In the foreground, in gray, is a profile of the ISS configuration as it appears when the crew arrives; the red elements are the Z1 truss and PMA-3 delivered on this mission. (NASA)

The 100th space shuttle flight; in 1979 this had been expected to occur before 1983, not in the year 2000. The *Discovery* launch attempt on 5 October 2000 was scrubbed early because of concerns over the forward orbiter-ET attach bolt. Imagery of the same bolt during STS-106 showed it did not retract as completely as expected, leading to concerns the orbiter could "hang" on the exposed part of the bolt and not separate from the ET correctly. Analysis eventually determined the attach bolt issue was acceptable to fly and NASA rescheduled the launch for 9 October. Weather on that day caused a 24-hour scrub. The launch team canceled an attempt on 10 October at T-3-hours when a metal pin, typically used to secure removable handrails to work platforms, was found on a strut connecting the orbiter and ET. NASA briefly considered the consequences of launching without removing the pin, but concluded it might damage the thermal protection system and elected to retrieve it instead. The pin could not be removed with propellants loaded on the vehicle, causing another 24-hour scrub. *Discovery* was finally launched without further delay on 11 October.

The primary objectives of this mission were to bring the Z1 truss (mounted on a Spacelab pallet), control moment gyros (CMG), pressurized mating adapter (PMA-3), and two DC-to-DC converter units (DDCU) to the ISS. The Z1 truss was the first exterior framework installed on the station and would allow the first set of American solar arrays to be temporarily installed on Node 1 (Unity) during STS-97 (ISS-4A) and the early Ku-band communication system during STS-100 (ISS-6A).

Just prior to beginning the rendezvous with the ISS, the Ku-band system failed, taking the rendezvous radar with it. This forced the crew to accomplish the docking maneuver, a task they were trained for and completed successfully. Perhaps most disappointing to those on the ground, the lack of Ku-band communications meant there was no live video. Following the successful docking on flight day 3, the crew entered PMA-2 and began transferring supplies from *Discovery* to the station. Following the successful checkout of the common berthing mechanism (CBM), the crew mated the Z1 truss to the ISS.

The crew closed the payload bay doors only to have both KSC landing opportunities waved-off because of high winds. The following day, mission control again waved-off both KSC opportunities because of winds and both Edwards opportunities because of rain. Persistent high winds at KSC forced NASA to divert landing to Edwards, for the first time in four years (STS-76), on the third day. The return ferry flight aboard N905NA was delayed one day at Edwards because one of the tailcone bolts broke.

In front are Pam Melroy (left) and Brian Duffy. In the rear, from the left, are Leroy Chiao, Mike López-Alegría, Bill McArthur, Jeff Wisoff, and Koichi Wakata. Adding an international flavor, Wakata represented the National Space Development Agency of Japan (NASDA). (NASA)

Landing operations continue as the Sun sets over Edwards AFB. Discovery would soon be towed to the Dryden Flight Research Center. (NASA)

A shot taken from Discovery just after undocking shows the two new ISS elements. PMA-3 is the black object at the top, while the Z1 truss structure is visible at the bottom. The orbiter had been docked to the port facing the camera (PMA-2 attached to Node 1 forward). (NASA)

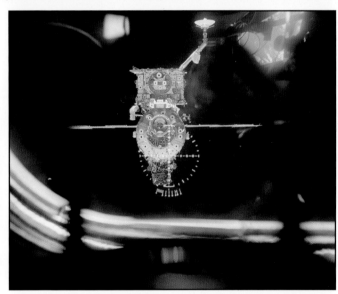

A farewell shot of the ISS was captured through the crew optical alignment system (COAS) during the separation maneuver. The discoloration of the station is associated with the COAS. (NASA)

Flight director Chuck Shaw (lower right) monitors one of the EVAs from his console in the White Flight Control Room in the Mission Control Center in Houston. The White FCR monitored, partly or completely, all space shuttle missions beginning with STS-77. (NASA)

STS-97 (ISS-4A)

Mission:	101	NSSDC ID:	2000-078A
Vehicle:	OV-105 (15)	ET-105 (SLWT)	SRB: BI103
Launch:	LC-39B	01 Dec 2000	03:06 UTC
Altitude:	206 nm	Inclination:	51.60 degrees
Landing:	KSC-15	11 Dec 2000	23:04 UTC
Landing Rev:	170	Mission Duration:	259 hrs 57 mins

Commander:	Brent W. Jett, Jr. (3)
Pilot:	Michael J. "Bloomer" Bloomfield (2)
MS1:	Joseph R. "Joe" Tanner (3)
MS2:	J. Marc Garneau (3)
MS3:	Carlos I. Noriega (2)

Payloads:	Up: 42,804 lbs	Down: 227 lbs
	P6 Truss (34,762 lbs)	
	SRMS s/n 303	

Notes: EVA1 03 Dec 2000 (Tanner and Noriega)
EVA2 05 Dec 2000 (Tanner and Noriega)
EVA3 07 Dec 2000 (Tanner and Noriega)

Wakeup Calls:

01 Dec	"Stardust"
02 Dec	"I Believe I Can Fly"
03 Dec	"Sunshine of Your Love"
04 Dec	"Lovin' You Lots & Lots"
05 Dec	"Fight On" (USC fight song)
06 Dec	"O Mio Babbino Caro" (opera)
07 Dec	"Here Comes the Sun"
08 Dec	"Rattled"
09 Dec	"Back in the Saddle Again"
10 Dec	"Beyond the Sea"
11 Dec	"I'll be Home for Christmas"

On 30 November 2000, Brent Jett exercised the push-to-talk switch on the rotational hand controller only to find it inoperative. However, he was able to talk by engaging an alternate push-to-talk button on his S1035 pressure suit. Because of the availability of the alternate function, the apparent switch failure did not impact the mission. *Endeavour* was launched as planned with no unscheduled holds during a "flawless" countdown.

On flight day 3, Brent Jett and Mike Bloomfield rendezvoused with the ISS some 200 nm above Kazakhstan. Marc Garneau used the SRMS to remove the P6 truss from the payload bay, maneuvering it into an overnight park position to warm its components. Joe Tanner and Carlos Noriega opened the hatch to the ISS docking port to leave supplies and computer hardware on the doorstep of the station. The pair also completed three EVAs, during which they prepared a docking port for the arrival of the Destiny laboratory on STS-98 (ISS-5A), installed probes to measure electrical potential surrounding the station, and installed a camera cable outside Node 1 (Unity). They were so efficient that a planned fourth EVA was not required.

On Friday, 8 December, the orbiter crew paid the first visit to the Expedition 1 crew residing in the station. Until then the orbiter and ISS had kept one hatch closed to maintain the respective atmospheric pressures, allowing the *Endeavour* crew to conduct their EVAs from the orbiter airlock. After a welcome ceremony and safety briefing, the eight crew members conducted structural tests of the station and its solar arrays, transferred equipment, supplies, and trash back and forth between the spacecraft, and checked out the television camera cable installed by Joe Tanner and Carlos Noriega in preparation for the upcoming mission.

The solar array wings were the largest ever-deployed in space at the time, weighing more than 2,400 pounds. The crew successfully mated the P6 truss to the Z1 truss during the first EVA and deployed the solar arrays. However, an anomaly during deployment prevented one solar wing from being fully tensioned. Nevertheless, the crew completed all of the necessary connections and the array provided electrical power to the ISS. Joe Tanner and Carlos Noriega managed to fully extend and tension the wing during the third EVA.

On 9 December, the two crews completed final transfers of supplies to the station and other items being returned to Earth. The *Endeavour* crew bade farewell to the Expedition 1 crew and closed the hatches between the spacecraft and *Endeavour* undocked. Mike Bloomfield then made an hour-long, tail-first circle of the ISS and the final separation burn took place near the coast of South America. *Endeavour* performed a deorbit maneuver for the first landing opportunity at the Shuttle Landing Facility.

The crew patch depicted Endeavour docked to ISS after the activation of the P6 electrical power system. Gold and silver were used to highlight the portion of ISS that was installed by the STS-97 crew. The Sun, central to the patch design, is the source of energy for ISS. (NASA)

The crew waves for the camera as they gather outside LC-39B. From the left are Carlos Noriega, Mike Bloomfield, Brent Jett, Joe Tanner, and Marc Garneau. The crew is wearing the ubiquitous blue flight suits that seem to be part of the modern astronaut persona. (NASA)

This view of Endeavour approaching the ISS was taken by one of the Expedition 1 crew members onboard the station. Although of only passing interest at the time, similar photography became a mandatory part of the TPS inspection after the Columbia accident. (NASA)

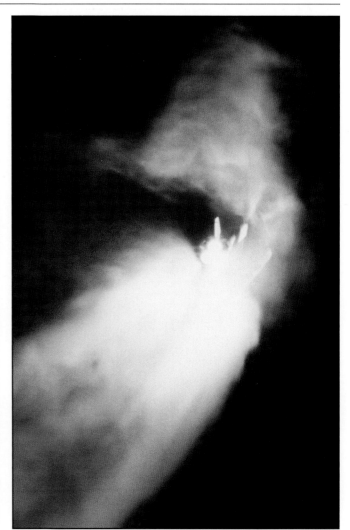

Night launches provided a spectacular sight. Here is solid rocket booster separation, something that occurred about 123 seconds after launch, some 141 nm downrange. This was the 25th night launch of the flight campaign and the sixth for Endeavour. (NASA)

Unusually, the crawler-transporter that moved the stack from the VAB to the launch pad suffered a broken cleat as it started up the incline at the pad. Technicians replaced the cleat on the crawlerway and the move continued, albeit a bit behind schedule. (NASA)

Endeavour about to touch down on Runway 15 at the Shuttle Landing Facility at KSC. This was the 16th night landing of the flight campaign and the fourth for Endeavour. The large xenon lights positioned at the end of the runway provided a fair amount of illumination. (NASA)

STS-98 (ISS-5A)

Mission:	102	NSSDC ID:	2001-006A
Vehicle:	OV-104 (23)	ET-106 (SLWT)	SRB: BI105
Launch: Altitude:	LC-39A 211 nm	07 Feb 2001 Inclination:	23:13 UTC 51.60 degrees
Landing: Landing Rev:	EDW-22 202	20 Feb 2001 Mission Duration:	20:34 UTC 309 hrs 20 mins

Commander:	Kenneth D. "Taco" Cockrell (4)
Pilot:	Mark L. "Roman" Polansky (1)
MS1:	Robert L. "Beamer" Curbeam, Jr. (2)
MS2:	Marsha S. Ivins (5)
MS3:	Thomas D. "Tom" Jones (4)

Payloads:	Up: 39,162 lbs		Down: 872 lbs
	US Laboratory (Destiny) (30,452 lbs)		
	SRMS s/n 202		

Notes:
MS1 was originally Mark C. Lee
Rolled-back to VAB 19 Jan 2001 (SRB concerns)
Last mission to use LES suit (Ivins)
EVA1 10 Feb 2001 (Jones and Curbeam)
EVA2 12 Feb 2001 (Jones and Curbeam)
EVA3 14 Feb 2001 (Jones and Curbeam)
Orbiter returned to KSC on 05 Mar 2001 (N911NA)

Wakeup Calls:

08 Feb	"Where You At"
09 Feb	"Who Let the Dogs Out"
10 Feb	"Girl's Breakdown"
11 Feb	"Blue Danube Waltz"
12 Feb	"Fly Me to the Moon"
13 Feb	"For Those About to Rock"
14 Feb	"To the Moon and Back"
15 Feb	"Sally Ann"
16 Feb	"The Trail We Blaze"
17 Feb	"Blue"
18 Feb	"Fly Away"
19 Feb	"Bad To the Bone"
20 Feb	"Should I Stay or Should I Go"

The crew patch depicted Atlantis with Destiny held high above the payload bay just before being attached to the ISS. Red and white stripes, with a deep blue field of white stars symbolized the continuing American contribution to the ISS. (NASA)

The *Atlantis* stack was rolled-out to LC-39A on 3 January 2001, but the need to investigate a cable anomaly noted on the retrieved STS-97 solid rocket boosters forced a roll-back to the VAB on 19 January. The vehicle rolled-back to LC-39A on 26 January and the countdown on 7 February was briefly delayed by some troubleshooting, but the flight was ultimately launched with no particular issues. Mark Lee was scheduled to fly on this mission but was replaced by Robert Curbeam for undisclosed reasons.

This was the first flight of revised software that automatically fired the forward reaction control system jets for about 2.02 seconds during SRB separation. This changed the flow pattern from the booster separation motors and eliminated the hazing that occurred when their exhaust impacted the forward-facing orbiter windows.

During the first EVA, Tom Jones and Robert Curbeam ventured into the payload bay to prepare the Destiny laboratory. After Destiny had been mated to the forward port on Node 1 (Unity), the pair began connecting the power and data cables. During the EVA, Curbeam was exposed to ammonia as he disconnected an umbilical on Destiny. In accordance with the flight rule concerning EMU decontamination during an EVA, Jones used a brush to remove contaminants from the EMU and Curbeam remained in sunlight for a 30-minute bake-out period while Jones completed the remaining tasks. Crew members inside *Atlantis* put on their "quick-don" masks before the airlock hatch opened but, since nobody noticed any odors, everything reverted to normal. All of this delayed activating Destiny and crew sleep by 2.5 hours.

Marsha Ivins operated the SRMS to remove PMA-2 from the Z1 truss and install it on the forward common berthing adapter of Destiny while Tom Jones and Robert Curbeam stood by to assist during the second EVA. Once that task was completed, the EVA crew members installed a power data and grapple fixture on Destiny to be used with the space station remote manipulator system (SSRMS). During the third EVA, Jones and Curbeam attached a spare antenna, checked connections between Destiny and its docking port, released a cooling radiator, and tested the ability of a spacewalker to carry an immobile crew member back to the orbiter airlock.

Atlantis undocked on flight day 9 and made a one-half revolution fly-around of the station. After the crew had closed the payload bay doors, mission management waved-off both landing opportunities on the planned landing day because of excessive crosswinds at the Shuttle Landing Facility. Mission management again waved-off both opportunities on the first extension day and decided to return to Edwards on the first opportunity the following day.

The photo that formed the basis of their "Think Safety" poster that was distributed to all of the space shuttle contractors prior to launch. From the left are Marsha Ivins, Robert Curbeam, Ken Cockrell, Mark Polansky, and Tom Jones. (Courtesy of Tom Jones and Marsha Ivins)

Atlantis fires her aft thrusters to maneuver toward the ISS. Part of Earth's limb can barely be seen on either side of the OMS pods. The orbiter docking system, located atop the external airlock, is in the foreground, while Destiny is partially visible behind it. (NASA)

Tom Jones and Mark Polansky change a lithium hydroxide canister on the middeck. Carbon dioxide reacted with the LiOH to produce lithium carbonate. Activated charcoal in the canisters removed odors and trace contaminants. Each canister was rated at 48 man-hours. (NASA)

The Destiny laboratory after it was berthed to the forward port of Node 1 (Unity). Boeing began fabricating the laboratory in 1995 at MSFC and shipped it to KSC in 1998 for ground checkout and integration. Destiny was 14 feet in diameter and 28 feet long. (NASA)

Reusing the solid rocket boosters was an involved undertaking. After they were retrieved and brought back to Hangar AF on the Cape Canaveral AFS, workers began the process of disassembling and inspecting them. At left is a shot showing the diver operated plug that was inserted into the nozzle before air was injected to float the booster. The center photo depicts the general arrangement inside Hangar AF, At right, workers remove the aft exit cone from a booster. Ultimately, the motor segments returned to Utah to be refilled with propellant. (NASA)

STS-102 (ISS-5A.1)

Mission:	103		NSSDC ID:		2001-010A
Vehicle:	OV-103 (29)		ET-107 (SLWT)		SRB: BI106
Launch:	LC-39B	08 Mar 2001			11:42 UTC
Altitude:	206 nm	Inclination:			51.60 degrees
Landing:	KSC-15	21 Mar 2001			07:33 UTC
Landing Rev:	201	Mission Duration:			307 hrs 50 mins

Commander:	James D. "WxB" Wetherbee (5)
Pilot:	James M. "Vegas" Kelly (1)
MS1:	Andrew S. W. "Andy" Thomas (3)
MS2:	Paul W. Richards (1)
Up:	Yury VladiMirovich Usachov (4)
Up:	James S. "Jim" Voss (5)
Up:	Susan J. Helms (5)
Down:	William M. "Shep" Shepherd (4)
Down:	Yuri Pavlovich Gidzenko (2)
Down:	Sergei Konstantinovich Krikalev (5)

Payloads:	Up: 37,328 lbs	Down: 1,086 lbs
	MPLM (Leonardo) (23,864 lbs)	
	ICC-G/ESP-1 (2,590 lbs)	
	SEM-09	
	GAS (1)	
	SRMS s/n 301	

Notes:	First ISS crew rotation
	EVA1 10 Mar 2001 (Voss and Helms)
	EVA2 12 Mar 2001 (Thomas and Richards)

Wakeup Calls:

08 Mar	"Living the Life"
09 Mar	"Vashe Blagorodiye" (*White Sun of the Desert*)
10 Mar	"Nothing's Gonna Stop Us Now"
11 Mar	"Blast Off" (*Scooby Doo and the Alien Invaders*)
12 Mar	"From A Distance"
13 Mar	"Free Fallin'"
14 Mar	"Should I Stay or Should I Go"
15 Mar	"She Blinded Me With Science"
16 Mar	"The Rising of the Moon"
17 Mar	Notre Dame Victory March
18 Mar	"Moscow Windows"
19 Mar	"Just What I Needed"

The ISS was shown along the direction of the flight as seen by the Discovery crew during docking. The names of the orbiter crew were depicted in gold around the top of the patch and surnames of the rotating Expedition crews were in the lower banner. (NASA)

This *Discovery* mission was launched on 8 March 2001 with no unplanned holds. The mission, the eighth to the ISS, carried the Expedition 2 crew to the ISS and returned the Expedition 1 crew to Earth after 140 days on-orbit, marking the first crew rotation by space shuttle.

This was the first use of the Italian-built multi-purpose logistics module (MPLM, named Leonardo) to bring supplies to the station. In addition, an integrated cargo carrier (ICC-G) carried an ammonia servicer, pump flow control subsystem, laboratory cradle assembly, rigid umbilical, and the external stowage platform (ESP-1). There was also a get-away special student experiment (G-783) and another GAS can that carried the WSVFM (wide-band shuttle vibration forces measurement) experiment.

Rendezvous with the ISS was completed satisfactorily, although the docking was about an hour late because of a delay in securing a latch on the ISS P6 solar array. Andy Thomas, assisted by James Kelly, used the SRMS to grapple PMA-3, unberth it from the Node 1 nadir active common berthing mechanism, and move it to the port ACBM. This allowed the MPLM to be docked to the nadir port on this and future missions. Jim Voss and Susan Helms used the first EVA to support the relocation of PMA-3 as well as installing the laboratory cradle assembly and rigid umbilical on Destiny. The official EVA time was 8 hours, 55 minutes, 56 seconds but only 6 hours, 42 minutes of this was outside the airlock since the crew members were connected to the servicing and cooling umbilical (SCU) in the external airlock for 2 hours 14 minutes while awaiting the completion of PMA-3 activities. Following the relocation of PMA-3, the crew docked Leonardo and its 9,649 pounds of material at the Node 1 nadir location using the SRMS. However, the crews faced a few challenges regarding the return stowage configuration for the MPLM, so mission management approved an additional day to complete the packing operations, extending the overall mission to 13 days. The crew completed all planned tasks between flight days 5 and 13.

During the second EVA, Andy Thomas and Paul Richards connected the rigid umbilical cables, installed ESP-1 on the port side of Destiny as a storage location for orbital replacement units and them moved the pump flow control subsystem and ammonia servicer to ESP-1 as spares. In addition, the pair photographed various portions of the ISS for engineering documentation. Mission management waved-off the first Florida landing opportunity because of rain and crosswinds, but the weather cleared sufficiently for the second opportunity and *Discovery* returned to the Shuttle Landing Facility at KSC.

The STS-102, Expedition 1, and Expedition 2 crews. In the first row, from the left, are Yuri Gidzenko, Sergei Krikalev, and Jim Voss; in the second row are Bill Shepherd, Susan Helms, and Yury Usachov; in the back are James Kelly, Paul Richards, James Wetherbee, and Andy Thomas. (NASA)

Sergei Krikalev in the Leonardo multi-purpose logistics module. The MPLMs were large logistic carriers and were an improvement over the Spacehab modules since they were removed from the payload bay and berthed to the ISS, simplifying moving the material. (NASA)

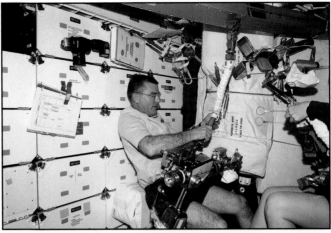

Jim Voss and Susan Helms (partially visible at right) check the tools they will use during their EVA. This image was taken shortly after Discovery reached orbit, still a day away from docking with the ISS. (NASA)

Jim Voss and Susan Helms hold their EMU suits. The pair spent most of the day before docking preparing for their EVA. Note the crew escape pole has been moved out of the way and positioned along the top of the middeck. The ever present roll of gray tape is above Voss. (NASA)

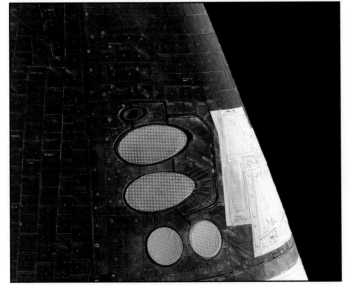

A close-up of the left forward reaction control system thrusters as Discovery sat on LC-39B. To prevent rain from entering the thruster nozzles, workers covered them with butcher paper (early in the flight campaign) or Tyvek cloth (later) that blew off during ascent. (NASA)

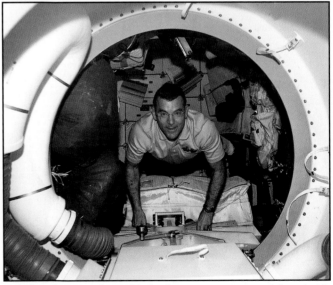

Jim Voss floats on the middeck of Discovery, as viewed from inside the external airlock. He is holding onto the hatch that opened into the middeck. The various white and brown tubes provided comfort air to from the orbiter systems to the airlock. (NASA)

STS-100 (ISS-6A)

Mission:	104	NSSDC ID:	2001-016A
Vehicle:	OV-105 (16)	ET-108 (SLWT)	SRB: BI107
Launch:	LC-39A	19 Apr 2001	18:41 UTC
Altitude:	219 nm	Inclination:	51.60 degrees
Landing:	EDW-22	01 May 2001	16:12 UTC
Landing Rev:	186	Mission Duration:	285 hrs 30 mins

Commander:	Kent V. "Rommel" Rominger (5)
Pilot:	Jeffrey S. "Bones" Ashby (2)
MS1:	Chris A. Hadfield (2)
MS2:	John L. Phillips (1)
MS3:	Scott E. Parazynski (4)
MS4:	Umberto Guidoni (2)
MS5:	Yury Valentinovich Lonchakov (1)

Payloads:	Up: 38,330 lbs	Down: 1,608 lbs
	MPLM (Raffaello) (19,501 lbs)	
	Spacelab LP/SSRMS (6,753 lbs)	
	SRMS s/n 303	

Notes:	EVA1 22 Apr 2001 (Hadfield and Parazynski)
	EVA2 24 Apr 2001 (Hadfield and Parazynski)
	Orbiter returned to KSC on 09 May 2001 (N905NA)

Wakeup Calls:

20 Apr	"Then the Morning Comes"
21 Apr	"Danger Zone" (Topgun)
22 Apr	"Take It From Day to Day"
23 Apr	"Both Sides Now"
24 Apr	"What A Wonderful World"
25 Apr	"Con te Partiro" ("With You I Will Go")
26 Apr	"Behind the Fog" (Russian song)
27 Apr	"Buckaroo"
28 Apr	"Dangerous"
29 Apr	"Miles from Nowhere"
30 Apr	Soundtrack to Gladiator
01 May	"True"

An EMU helmet frames the crew patch, with the Canadian-built SSRMS shown below the visor. Endeavour is reflected in the visor with the ISS rising above the horizon at orbital sunrise. Ten stars adorn the sky, representing the children of the STS-100 crew. (NASA)

There was a minor scare during the *Endeavour* main engine start sequence on 19 April 2001, when engineers at KSC observed structural flexing on the left OMS pod. Fortunately, photographic analysis did not reveal any visible damage or lost tiles, and the subsequent post-flight inspection did not find any damage.

The ISS-6A launch package consisted of the Raffaello multi-purpose logistics module (MPLM), a Spacelab logistics pallet (SLP), utilization experiments, and the direct current switching unit. The SLP carried the space station remote manipulator system (SSRMS, Canadarm2) and the ultrahigh frequency antenna, both of which were installed on the ISS during the two scheduled EVAs. The SSRMS was necessary to attach a new airlock during the STS-104 (ISS-7A) mission. A final component of the Canadarm2 would be the mobile base system (MBS) installed on the station during STS-111 (ISS-UF2).

Endeavour docked with the station on flight day 3. The crew attached the Canadarm2 cradle to the outside of the Destiny laboratory and later directed the arm to "walk off" the cradle and grab onto an electrical grapple fixture that provided data and power to the arm. Days later, the crew used the arm to hand off the cradle to the SRMS. The exchange of the cradle from station arm to shuttle arm marked the first robotic-to-robotic transfer in space.

During the first EVA on flight day 4, Chris Hadfield and Scott Parazynski connected cables between the ISS and the Canadarm2, then installed the UHF antenna. On flight day 5, the crew opened the hatches between *Endeavour* and the ISS and began transferring 6,000 pounds of cargo from the Raffaello MPLM, including two scientific experiment racks for Destiny and the first three American commercial payloads. Later, 1,600 pounds of material were transferred from the ISS to Raffaello for return to Earth.

On flight day six, Chris Hadfield and Scott Parazynski conducted the second EVA, connecting yet more cables and otherwise completing the installation of the SSRMS. In addition, Hadfield removed the early communications system antenna from the Unity module and stowed it in the payload bay. A planned contingency EVA on flight day 8 was not needed and was canceled.

Endeavour undocked on flight day 10 and Jeff Ashby performed a three-quarter-circle fly-around of the ISS. The first use of the space-to-space-orbiter radio (SSOR) and space-to-space-station radio (SSSR) during the undocking/separation maneuvers was successful. Mission management waved-off the first and second landing opportunities at KSC because of weather. The following day, an unfavorable forecast at the Shuttle Landing Facility for the rest of the week resulted in the landing being diverted to Edwards.

Seated are Kent Rominger (left) and Jeff Ashby. Standing, from the left, are Yuri Lonchakov, Scott Parazynski, Umberto Guidoni, Chris Hadfield, and John Phillips. As usual for the official portraits, the crew members scheduled to EVA wore EMU suits while the others wore ACES. (NASA)

Scott Parazynski (red stripes) and Chris Hadfield work with the Spacelab pallet that carried the space station remote manipulator arm (SSRMS, Canadarm2). The final component of the Canadarm2 was the Mobile Base System (MBS) that would be installed during STS-111. (NASA)

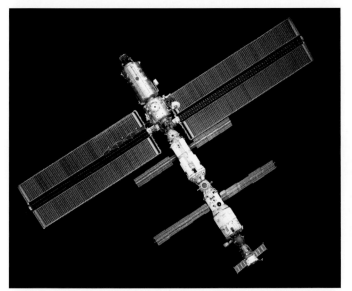

The ISS at the time. From the top were the Destiny laboratory, Node 1 (Unity), the Z1 truss, Zarya Functional Cargo Block, Zvezda Service Module, and a Soyuz. There was always a Soyuz docked at the station in case the crew needed to return in a hurry. (NASA)

Kent Rominger (left) and Jeff Ashby on the flight deck of Endeavour during deorbit preparations. Note the printed checklists and procedures scattered around the area. The pair have not yet donned their advanced crew escape suits. (NASA)

Endeavour punches through a low cloud after lifting-off from LC-39A. This was the 16th flight for Endeavour and the 13th flight of a super lightweight tank. The components were standard for ISS assembly missions, which used most of the available performance. (NASA)

STS-104 (ISS-7A)

Mission:	105	NSSDC ID:		2001-028A	
Vehicle:	OV-104 (24)	ET-109 (SLWT)		SRB: BI108	
Launch: Altitude:	LC-39B 211 nm	12 Jul 2001 Inclination:		09:04 UTC 51.60 degrees	
Landing: Landing Rev:	KSC-15 200	25 Jul 2001 Mission Duration:		03:41 UTC 306 hrs 35 mins	

Commander:	Steven W. "Steve" Lindsey (3)
Pilot:	Charles O. "Scorch" Hobaugh (1)
MS1:	Michael L. "Mike" Gernhardt (4)
MS2:	Janet L. Kavandi (3)
MS3:	James F. "JR" Reilly II (2)

Payloads:	Up: 35,135 lbs	Down: 626 lbs
	US Airlock (Quest) (13,408 lbs)	
	Spacelab LDP/HPGA (10,553 lbs)	
	SRMS s/n 202	

Notes:	First flight of a Block II space shuttle main engine
	EVA1 14 Jul 2001 (Gernhardt and Reilly)
	EVA2 17 Jul 2001 (Gernhardt and Reilly)
	EVA3 20 Jul 2001 (Gernhardt and Reilly)

Wakeup Calls:

12 Jul	"Wallace Courts Murron" (*Braveheart*)
13 Jul	"God of Wonders"
14 Jul	"Space Cowboy"
15 Jul	"No Woman, No Cry"
16 Jul	"Nobody Does it Better" (*The Spy Who Loved Me*)
17 Jul	"Happy Birthday, Darlin'"
18 Jul	"All I Wanna Do"
19 Jul	"A Time to Dance"
20 Jul	"I Could Write a Book" (*When Harry Met Sally*)
21 Jul	"Who Let the Dogs Out?"
22 Jul	"Orinoco Flow"
23 Jul	"Honey, I'm Home"
24 Jul	"Hold Back the Rain"

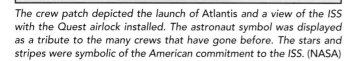

The crew patch depicted the launch of Atlantis and a view of the ISS with the Quest airlock installed. The astronaut symbol was displayed as a tribute to the many crews that have gone before. The stars and stripes were symbolic of the American commitment to the ISS. (NASA)

The countdown on 12 July 2001 resulted in an on-time *Atlantis* launch with no unplanned holds. This was the first flight of a Block II space shuttle main engine.

The primary objective was to install the Quest airlock that consisted of two cylindrical chambers attached end-to-end by a connecting bulkhead and hatch. Interestingly, although built by Boeing at MSFC and officially named the U.S. Airlock, Quest used a Russian depressurization pump and pressure equalization valves. The airlock has two main components: a crew airlock and a section for storing EVA gear and conducting preflight preparations. The airlock largely recycles its air during pressurization/depressurization, eliminating the loss of environmental consumables typical of most airlock operations. After it was installed, Quest became the primary path for EVA egress and ingress from the ISS using both American EMU and Russian Orlan spacesuits. The mission also carried a Spacelab logistics double pallet with four high-pressure gas assembly tanks that were later attached to the exterior of the airlock.

Mike Gernhardt and James Reilly conducted the first EVA on flight day 3 to assist the SRMS with installing Quest. They also removed the covers from the passive common berthing mechanism and installed various antennas and cables. In addition, they attempted four get-ahead tasks, three of which were successful. These included reattaching loose velcro on PMA-2 and opening the nadir CBM window cover.

During the second EVA, Mike Gernhardt and James Reilly installed two of the four high-pressure gas tanks that had been carried on the Spacelab pallet. Since adequate time remained, they installed a third high-pressure gas tank, something originally planned for the third EVA. The crew used the SRMS to move the tanks from the payload bay to the installation area, and the spacewalkers manually moved them into place and secured them. The third EVA was the first conducted from the new Quest airlock. With the aid of the SRMS, Mike Gernhardt and James Reilly installed the last of the four high-pressure gas tanks on the outside of Quest. Portions of the entire mission, including the three EVAs, were filmed using an IMAX 3D camera in the crew module and payload bay.

Mission management waved-off both KSC landing opportunities because of unacceptable weather, which cleared sufficiently to allow *Atlantis* to return to the Shuttle Landing Facility on the first opportunity the following day.

This was the last mission of the flight campaign to carry a five-man crew. All succeeding missions would have six or seven crew members (except the final mission STS-135, which carried four).

In front are Steve Lindsey (left) and Charles Hobaugh. In back, from the left, are Mike Gernhardt, Janet Kavandi, and JR Reilly. Note the various embroideries on their Lands' End shirts, ranging from stylized versions of the crew patch to American flags and simple STS-104 script. (NASA)

The History of the American Space Shuttle

The Atlantis payload bay shows the Quest airlock at the bottom, followed by a Spacelab logistics double pallet carrying four oxygen and nitrogen storage tanks that made up the high-pressure gas assembly. The tanks would be installed around the perimeter of Quest. (NASA)

Scattered clouds cast shadows as Atlantis crawls back inside VAB High Bay 1. After earlier starting its trek to LC-39B, Atlantis returned to the VAB due to lightning in the area. To the left of the VAB is the Launch Control Center that housed the firing rooms. (NASA)

Most of the crews had a welcome-home ceremony in Houston, usually at Ellington Field near JSC. These were popular events and a good opportunity to seek autographs, with NASA handing out free lithographs of the crew and individual crew member portraits. (NASA)

This was the first flight of the Block II main engine, shown being installed in OPF-3. The Block II configuration included a new Pratt & Whitney high-pressure fuel turbopump and was the final development of the SSME during the Space Shuttle Program. (NASA)

STS-105 (ISS-7A.1)

Mission:	106		NSSDC ID:		2001-035A
Vehicle:	OV-103 (30)		ET-110 (SLWT)		SRB: BI109
Launch:	LC-39A		10 Aug 2001		21:10 UTC
Altitude:	219 nm		Inclination:		51.60 degrees
Landing:	KSC-15		22 Aug 2001		18:23 UTC
Landing Rev:	186		Mission Duration:		285 hrs 13 mins

Commander:	Scott J. "Doc" Horowitz (4)
Pilot:	Frederick W. "Rick" Sturckow (2)
MS1:	Patrick G. "Pat" Forrester (1)
MS2:	Daniel T. "Dan" Barry (3)
Up:	Frank L. Culbertson, Jr. (3)
Up:	Mikhail Vladislavovich Tyurin (1)
Up:	VladiMir Nikolayevich Dezhurov (2)
Down:	Yury VladiMirovich Usachov (4)
Down:	James S. "Jim" Voss (5)
Down:	Susan J. Helms (5)

Payloads:	Up: 37,107 lbs	Down: 3,802 lbs
	MPLM (Leonardo) (19,306 lbs)	
	ICC-G (4,756 lbs)	
	SEM-10	
	HEAT (hitchhiker)	
	GAS (2)	
	SRMS s/n 301	

Notes:	Scrubbed 09 Aug 2001 (KSC weather)
	EVA1 16 Aug 2001 (Barry and Forrester)
	EVA2 18 Aug 2001 (Barry and Forrester)

Wakeup Calls:

11 Aug	"Back in the Saddle Again"
12 Aug	"The White Eagle" (Russian song)
13 Aug	"Overture from The Barber of Seville"
14 Aug	"Theme from Arthur (Best That You Can Do)"
15 Aug	"Big Boy Toys"
16 Aug	"The Marvelous Toy"
17 Aug	"Time Bomb"
18 Aug	"Hotel California"
19 Aug	"Under the Boardwalk"
20 Aug	"Brand New Day"
21 Aug	"East Bound and Down"
22 Aug	"Again"

The three gold stars near the ascending orbiter represent the American-commanded Expedition 3 crew as they journey into space, while the two gold stars near the descending orbiter represent the Russian-commanded Expedition 2 crew and their return to Earth. (NASA)

During the T-9-minute hold on 9 August 2001, the launch director called a 24-hour scrub because of lightning and rain around the pad. The following day, *Discovery* launched on time with no unplanned holds. This was the eleventh space shuttle mission to the ISS, with three Expedition 3 crew members trading places with the Expedition 2 crew. This was the first time the same orbiter had ferried a crew (Expedition 2) in both directions.

Discovery carried the Leonardo multi-purpose logistics module (MPLM) with five resupply return stowage racks (RSR), four resupply/return stowage platforms (RSP), and two EXPRESS (expedite the processing of experiments to the space station) racks. In addition, the payload bay carried an integrated cargo carrier (ICC-G) carrying the early ammonia servicer and two MISSE (materials international space station experiment) rack for exposure tests of materials and components planned for future spacecraft. The experiments were packaged in four passive experiment containers (PEC) initially used during 1996–97 aboard Mir. These suitcase-like containers clamped to the host spacecraft and opened to expose experiments to the space environment. Middeck payloads included the APCF (advanced protein crystallization facility), BCSS-4 (biotechnology cell science stowage), BTR (biotechnology refrigerator), DCPCG-V (dynamically controlled protein crystal growth-vapor), and H-Reflex (human research facility Hoffman-reflex experiment) hardware.

The Expedition 2 crew aboard the ISS used the DREAMTIME camcorder to capture a number of on the orbiter and the station including the orbiter approaching the ISS, the crew hatch opening, the MPLM and Node 1 hatch closure, the MPLM unberthing, and a general educational tour of the ISS. The *Discovery* crew unberthed Leonardo from the payload bay and installed it the nadir port of Node 1 (Unity).

During the first EVA, Dan Barry and Pat Forrester installed the early ammonia system and the two MISSE experiments, relocated the articulating portable foot restraint, and retrieved the rectangular scoop tool in preparation for a future mission. For the second EVA, the SRMS translated Barry and Forrester to the Destiny laboratory where they stowed handrails and cables that would be used on the S0 truss when it was delivered by STS-110.

The crew moved Leonardo back to the payload bay and undocked from the ISS. Mission management waved-off the first Florida landing opportunity because of rain near the end of the runway, but the weather cleared sufficiently for *Discovery* to come back to the Shuttle Landing Facility on the second opportunity.

In the center are the STS-105 crew of Rick Sturckow, Pat Forrester, Dan Barry, and Doc Horowitz. The Expedition 2 crew of Jim Voss, Yury Usachov, and Susan Helms is at left and the Expedition 3 crew of Mikhail Tyurin, VladiMir Dezhurov, and Frank Culbertson is at right. (NASA)

Pat Forrester works with one of the MISSE suitcases on the outside of the Quest airlock during the second EVA. (NASA)

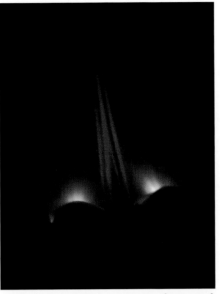

A two-engine OMS burn as seen from an aft flight deck window. The mission used six two-engine and two left-engine burns. (NASA)

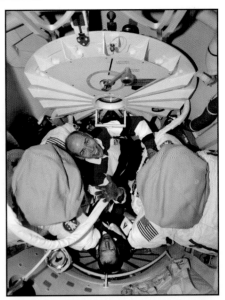

Dan Barry and Pat Forrester check-out their EMU suits in the airlock. The bags protected the helmets and their fragile visors. (NASA)

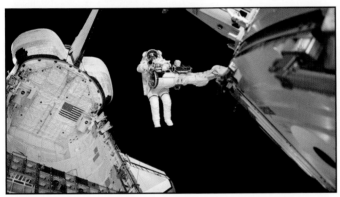

Dan Barry near the end of the SRMS during early stages of the second EVA. Note that the MPLM has been removed from the payload bay so that its contents could be transferred to the ISS. (NASA)

Dan Barry works on the integrated cargo carrier (ICC) in the payload bay at the end of EVA2. The ICC was a simple frame that carried whatever materiel was destined for the ISS. (NASA)

A close-up of the space experiment module (SEM-10) and one of the get-away special canisters (G-780) on the left side of the payload bay. Both of these programs offered schools, universities, and some companies inexpensive ways to conduct on-orbit experiments. (NASA)

The STS-105 stack sitting on LC-39A. Note the gaseous oxygen vent arm ("beanie cap") on top of the external tank is partly retracted. When sealed against the ET vents, the beanie cap ducted gaseous oxygen away from the tank to prevent it from causing ice. (NASA)

STS-108 (ISS-UF1)

Mission:	107		NSSDC ID:		2001-054A
Vehicle:	OV-105 (17)		ET-111 (SLWT)		SRB: BI110
Launch:	LC-39B		05 Dec 2001		22:19 UTC
Altitude:	204 nm		Inclination:		51.60 degrees
Landing:	KSC-15		17 Dec 2001		17:56 UTC
Landing Rev:	185		Mission Duration:		283 hrs 36 mins

Commander:	Dominic L. P. "Dom" Gorie (3)
Pilot:	Mark E. Kelly (1)
MS1:	Linda M. Godwin (4)
MS2:	Daniel M. "Dan" Tani (1)
Up :	Yuri Ivanovich Onufriyenko (2)
Up :	Carl E. Walz (4)
Up :	Daniel W. "Dan" Bursch (4)
Down :	Frank L. Culbertson, Jr. (3)
Down:	Mikhail Vladislavovich Tyurin (1)
Down:	VladiMir Nikolayevich Dezhurov (2)

Payloads:	Up: 38,177 lbs	Down: 4,156 lbs
	MPLM (Raffaello) (20,340 lbs)	
	SEM-11, 12, and 15	
	MACH-1 (hitchhiker)	
	GAS (6)	
	SRMS s/n 303	

Notes:	Delayed 29 Nov 2001 (Russian Progress at ISS)
	Scrubbed 04 Dec 2001 (KSC weather)
	EVA1 10 Dec 2001 (Godwin and Tani)

Wakeup Calls:

06 Dec	"Soul Spirit" / "Put a Little Love in Your Life"
07 Dec	"God of Wonders"
08 Dec	"Wade Into the Water"
09 Dec	"It's A Grand Ol' Flag"
10 Dec	"Jumpin' at the Woodside"
11 Dec	"Let There Be Peace on Earth"
12 Dec	"Fly Me to the Moon"
13 Dec	"Here Comes the Sun"
14 Dec	"My Sweetheart" (Russian song)
15 Dec	"Where I Come From"
16 Dec	"I'll Be Home For Christmas"
17 Dec	"Please Come Home For Christmas"

The ribbons on the left side of the crew patch signify the Expedition Three crew and its American commander. The ribbons on the right depict the Expedition Four crew and its Russian commander. The white stars in the center represent the four Endeavour crew members. (NASA)

NASA rescheduled the 29 November 2001 launch of *Endeavour* for 4 December to allow sufficient time for the Expedition 3 crew to complete a spacewalk to clear an obstruction on the latching mechanism of a Progress supply vehicle. The launch team scrubbed the 4 December attempt after Charlie Precourt, flying a Shuttle Training Aircraft, detected rain over the launch site. *Endeavour* lifted off on 5 December to deliver supplies and the Expedition 4 crew to the ISS.

Dom Gorie and Mark Kelly brought *Endeavour* to a gentle linkup with the ISS as the spacecraft flew over the United Kingdom. The Expedition 3 crew ended their 117-day residency the following day when their custom Soyuz seat-liners were transferred to *Endeavour* for the trip home. The transfer of the Expedition 4 seat-liners to the Soyuz return vehicle attached to the station marked the official exchange of crews. Mark Kelly and Linda Godwin then used the SRMS to move Raffaello from the payload bay to Node 1 (Unity). The crews completed the transfer of more than 5,000 pounds of supplies from the middeck and Raffaello to the station. In turn, the crew packed Raffaello with items bound for Earth. Mission managers extended the mission to 12 days to allow the *Endeavour* crew to assist with additional maintenance tasks on the station. The single EVA had Linda Godwin and Dan Tani install insulation on the main solar arrays and retrieve an antenna cover for return to Earth.

To honor those lost on 11 September 2001, *Endeavour* carried an American flag recovered from the World Trade Center, a Marine Corps flag recovered from the Pentagon, an American flag that flew over the State Capital in Harrisburg, 23 shields (badges) of fallen NYPD officers, memorabilia from the FDNY, patches from the Port Authority of New York and New Jersey, and 6,000 American flags. The flags were later given to families that lost loved ones.

Flight controllers changed the departure to allow time for *Endeavour* to boost the ISS away from a derelict upper stage from a Soviet Kosmos-3M launch vehicle that would pass within 3 miles of the station. With the reboost, the station was more than 40 miles from the debris. Because the reboost used additional propellant, *Endeavour* only performed a quarter circle fly-around of the station.

After undocking, the crew deployed the STARSHINE-2 (student tracked atmospheric research satellite for heuristic international networking experiment) satellite from a canister on the MACH-1 hitchhiker in the payload bay. More than 30,000 students from 660 schools in 26 countries tracked STARSHINE-2 as it orbited the Earth for eight months. Landing at the Shuttle Landing Facility was accomplished on the first opportunity.

The three crews assemble in the Destiny laboratory. In the front, from the left, are Yuri Onufriyenko, Dan Bursch, Frank Culbertson, VladiMir Dezhurov, and Mikhail Tyurin. In the back are Carl Walz; Linda Godwin, Mark Kelly, Dom Gorie, and Dan Tani. (NASA)

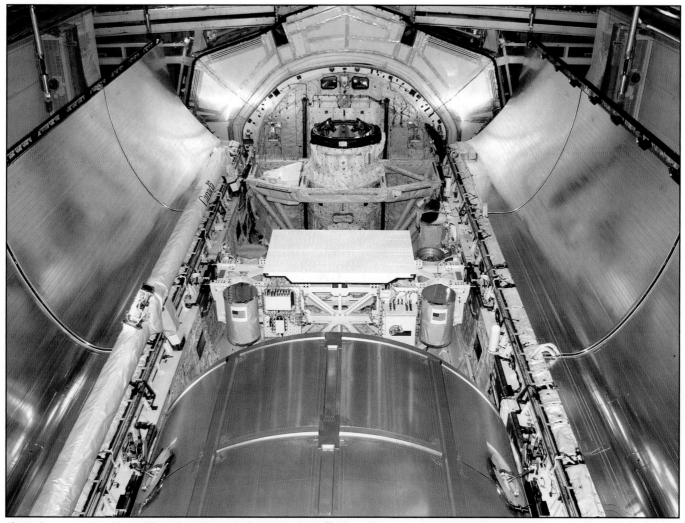

The Endeavour payload bay while in the OPF. At the bottom is the Raffaello multi-purpose logistics module. Above it is the Mach-1 GAS bridge that carried a variety of small payloads. In this view, from the left, are the SEM-11 container, the CAPL-3 electronics boxes, the hitchhiker avionics, and the SEM-15 container. The white square on top is the CAPL-3 radiator. The far (front) side included three GAS containers that carried STARSHINE-2, Collide-2, and G-761. Two other GAS containers were on the right payload bay wall and the external airlock and ODS is at the top. (NASA)

The crew of Liberty Star retrieving the frustum from the right solid rocket booster. Note the parachute lines heading to the large reels on the deck; the crews recovered the parachutes onto these reels before they lifted the frustum from the water. (NASA)

Endeavour touches down on the Shuttle Landing Facility. Typical of all delta-wing vehicles, the orbiter landed nose-high. The smoke from the tires was normal, a result of (in this case) a 201.4-knot touch-down speed and 1.30 fps sink rate, both well within the usual range. (NASA)

STS-109 (HST-SM3B)

Mission:	108	NSSDC ID:	2002-010A
Vehicle:	OV-102 (27)	ET-112 (SLWT)	SRB: BI111
Launch:	LC-39A	01 Mar 2002	11:22 UTC
Altitude:	312 nm	Inclination:	28.45 degrees
Landing:	KSC-33	12 Mar 2002	09:33 UTC
Landing Rev:	165	Mission Duration:	262 hrs 10 mins

Commander:	Scott D. "Scooter" Altman (3)
Pilot:	Duane G. "Digger" Carey (1)
MS1:	John M. Grunsfeld (4)
MS2:	Nancy J. Currie (4)
MS3:	Richard M. "Rick" Linnehan (3)
MS4:	James H. "Jim" Newman (4)
MS5:	Michael J. "Mike" Massimino (1)

Payloads:	Up: 27,564 lbs	Down: 6,409 lbs
	Hubble Servicing Mission (18,903 lbs)	
	SRMS s/n 301	

Notes:	Delayed 21 Feb 2002 (mission replanning)
	Delayed 28 Feb 2002 (KSC weather)
	EVA1 04 Mar 2002 (Grunsfeld and Linnehan)
	EVA2 05 Mar 2002 (Newman and Massimino)
	EVA3 06 Mar 2002 (Grunsfeld and Linnehan)
	EVA4 07 Mar 2002 (Newman and Massimino)
	EVA5 08 Mar 2002 (Grunsfeld and Linnehan)

Wakeup Calls:

01 Mar	"Blue Telescope"
02 Mar	"Theme from Mission: Impossible"
03 Mar	Five Variations on "Twinkle, Twinkle Little Star"
04 Mar	"Floating in the Bathtub" (children's song)
05 Mar	"Carmen Ohio" (Ohio State marching song)
06 Mar	"Sittin on Top of the World"
07 Mar	"Theme from Mission: Impossible II"
08 Mar	"Who Made Who"
09 Mar	"Fly Me to the Moon"
10 Mar	"Floating"
11 Mar	"Countdown"

The crew patch depicts the Hubble Space Telescope and Columbia over the North American continent. The STS designation is on the underside of the aperture door and the names of the crew members are around the edge. The SM3B emblem is at the right. (NASA)

This *Columbia* launch was scheduled for 28 February 2002, but forecasters were predicting a temperature of only 38°F (below minimums), so the mission management team rescheduled launch for 1 March. The countdown for that attempt proceeded smoothly with no unplanned holds. NASA considered this a night launch since liftoff was 25 minutes before sunrise.

This was the fourth servicing mission (SM3B) to the Hubble Space Telescope, the second part of a scheduled servicing that ended up becoming two missions after three gyros failed. The immediate problems had been addressed by STS-103 (SM3A), leaving this mission to work on planned upgrades to the telescope. The crew installed the advanced camera for surveys, rigid solar arrays, power control unit, and a cryocooler for the near infrared camera and multi-object spectrometer (NICMOS). Nancy Currie, assisted by Scott Altman, used the SRMS to grapple the telescope 46 hours after launch. However, when the crew attempted to open the internal airlock hatch, they reported it could not be unlatched. A closeout technician had noted a similar condition prior to launch. With a little effort the crew got the hatch working, although not quite the way it was supposed to. Nevertheless, two pairs of astronauts made five spacewalks on consecutive days. In typical NASA fashion, Duane Carey and Altman documented the activities with video and still images.

John Grunsfeld and Rick Linnehan made the first EVA on flight day 3. The crew members replaced one of the solar arrays and installed a multi-layer insulation tent for the NICMOS cooling system power feed. The following day Jim Newman and Mike Massimino replaced the other solar array and a reaction wheel assembly, and installed a new outer blanket. They also performed three get-ahead tasks, including installing a new outer layer blanket, two doorstop extensions, and an aft shroud latch repair kit.

While preparing for the third EVA, John Grunsfeld discovered water had leaked into his primary life support system. The crew resized the spare hard upper torso to fit Grunsfeld, causing a two-hour delay in the spacewalk. During the EVA, Grunsfeld and Rick Linnehan replaced the power control unit. Jim Newman and Mike Massimino conducted the fourth EVA to replace the original faint object camera with the advanced camera for surveys. The pair also installed the NICMOS electronics support module and finished a few final tasks from the power control unit installation. John Grunsfeld and Rick Linnehan used the fifth EVA to install the NICMOS cryogenic cooler system and radiator.

The weather at KSC was good and *Columbia* returned on the first opportunity. This was the last successful flight of *Columbia*.

The traditional preflight crew portrait. From the left are Mike Massimino, Rick Linnehan, Duane Carey, Scott Altman, Nancy Currie, John Grunsfeld, and Jim Newman. This was the fourth crew to visit Hubble on a servicing mission. (NASA)

From the top are the new solar array panels mounted on a Spacelab pallet, a carrier with the other instruments and hardware, and, at the bottom (back of the payload bay), the flight support system (FSS) that supported the telescope during the servicing. (NASA)

Jim Newman moves around the payload bay while working with Mike Massimino (out of frame) during the second EVA. Inside Columbia, Nancy Currie operated the SRMS arm from the aft flight deck. Part of the telescope can be seen to the left behind Newman. (NASA)

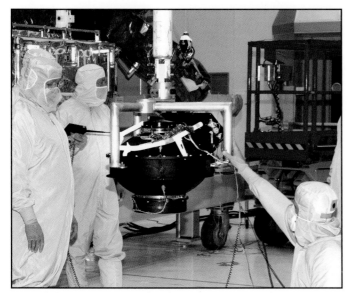

One of the six new gyros that had been installed during the third servicing mission (SM3A, STS-103) soon experienced a seven-minute telemetry dropout, so NASA decided to replace it during SM3B. This is the replacement gyro in the Vertical Processing Facility at KSC. (NASA)

The rotating service structure is rolled back from Columbia in preparation for launch. The American and Columbia flags (foreground) illustrate the brisk winds blowing at LC-39A that day. Each orbiter had a unique flag that was generally flown nearby. (NASA)

STS-110 (ISS-8A)

Mission:	109	NSSDC ID:	2002-018A		
Vehicle:	OV-104 (25)	ET-113 (SLWT)	SRB: BE112		
Launch:	LC-39B	08 Apr 2002	20:44 UTC		
Altitude:	219 nm	Inclination:	51.60 degrees		
Landing:	KSC-33	19 Apr 2002	16:28 UTC		
Landing Rev:	170	Mission Duration:	259 hrs 43 mins		

Commander:	Michael J. "Bloomer" Bloomfield (3)
Pilot:	Stephen N. "Steve" Frick (1)
MS1:	Rex J. Walheim (1)
MS2:	Ellen L. Ochoa (4)
MS3:	Lee M. E. Morin (1)
MS4:	Jerry L. Ross (7)
MS5:	Steven L. "Steve" Smith (4)

Payloads:	Up: 35,849 lbs	Down: 2,607 lbs
	S0 Truss (26,778 lbs)	
	Mobile Transporter (1,950 lbs)	
	SRMS s/n 202	

Notes:	Scrubbed 04 Apr 2002 (LH$_2$ drain line failure)
	EVA1 11 Apr 2002 (Smith and Walheim)
	EVA2 13 Apr 2002 (Ross and Morin)
	EVA3 14 Apr 2002 (Smith and Walheim)
	EVA4 16 Apr 2002 (Ross and Morin)

Wakeup Calls:

09 Apr	"The Best Years of Our Lives"
10 Apr	"Rapunzel Got a Mohawk"
11 Apr	University of California-Berkeley fight song
12 Apr	"Testify to Love"
13 Apr	"Voodoo Chile"
14 Apr	"All Star"
15 Apr	"Magic Carpet Ride"
16 Apr	"I Am an American"
17 Apr	"Noah"
18 Apr	"Somewhere Over the Rainbow"
19 Apr	"Message in a Bottle"

The crew patch was patterned after the cross-section of the S0 truss and encased the launch of Atlantis and a silhouette of the ISS as it appeared at the end of the mission. The successfully installed S0 segment was highlighted in gold. (NASA)

Engineers scrubbed the launch attempt on 4 April 2002 during ET loading when they detected a hydrogen leak in the mobile launch platform. Subsequent inspection revealed the 16-inch hydrogen vent line had a 0.125-inch-wide crack in a weld and the launch was delayed three days while technicians repaired the crack.

During rendezvous, the Ku-band radar acquired the ISS at 141,000 feet and tracked it until 630 feet when the crew changed to the communications mode. Mike Bloomfield and Steve Frick made three OMS and five RCS firings during rendezvous.

On flight day 3, the ISS crew lifted the S0 truss out of the payload bay using the SSRMS and attached it to the Destiny laboratory with the SRMS acting as a camera platform to provide a second angle. The S0 truss formed the backbone of the station to which the S1 and P1 truss segments were attached (during STS-112 and STS-113, respectively). Steve Smith and Rex Walheim made the first EVA to install the S0 truss. During the EVA they installed the port and starboard forward module-to-truss struts, connected the forward and port avionics umbilical trays, deployed the S0 aft laboratory tray, verified the power and data connections were functional, and installed the mobile transporter and Zenith trailing umbilical system.

Jerry Ross and Lee Morin used the second EVA to continue installing the S0 truss. They attached two aft mobile transporter system struts, removed and stowed two keels and two drag links, removed a thermal cover, and installed the airlock handrail.

The third EVA saw Steve Smith and Rex Walheim configuring the SSRMS onto the S0 truss. The pair installed SSRMS cabling, transferred cables from the laboratory to the S1 truss, removed the mobile transporter thermal covers, and brought tools from *Atlantis* to a temporary location on the S0 truss. All planned tasks except installing some MMOD shields and an airlock spur were completed. During the fourth EVA, Jerry Ross and Lee Morin performed a myriad of tasks. Among these were installing the airlock spur that had been deferred from the third EVA, attaching lights around the ISS, performing a photographic survey of the P6 radiator, and conducting a video survey of the mobile transporter. Like the first three spacewalks, the final EVA was conducted from Quest.

During docked operations, Mike Bloomfield and Steve Frick conducted three reboost maneuvers to raise the station orbit. After undocking, they made the customary fly-around.

The weather at KSC was acceptable and *Atlantis* returned on the first opportunity. The left air data probe did not deploy when the crew moved the switch but did after Frick wiggled the switch a few times. Once deployed, the probe operated normally.

The crew heads for the Astrovan to take them to LC-39B for the terminal countdown demonstration test. From left are Jerry Ross, Lee Morin, Steve Smith (rear), Rex Walheim, and Ellen Ochoa, Steve Frick, and Mike Bloomfield. Note the post-9/11 security team. (NASA)

Rex Walheim translates along the Destiny laboratory during the third EVA. The primary task for this spacewalk was to configure the SSRMS onto the S0 truss. (NASA)

Steve Smith (shown) and Rex Walheim also moved a lot of cables from Destiny to the new truss and transferred tools from Atlantis to a temporary location on the S0 truss. (NASA)

Lee Morin (left) and Jerry Ross check a procedures checklist on the aft flight deck of Atlantis. Note the four Sony video recorders at the top of the photo. (NASA)

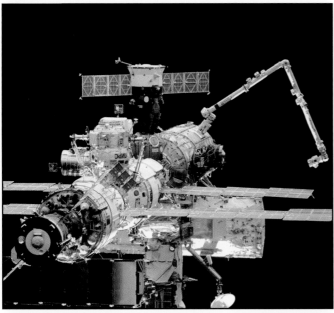

A view of the International Space Station with the newly installed S0 truss. S0 was the first segment of a truss structure that ultimately extended the station to the length of a football field. (NASA)

Ellen Ochoa and Dan Bursch used the SSRMS on the ISS to lift the S0 truss out of the payload bay and install it onto a temporary fixture on the Destiny laboratory. (NASA)

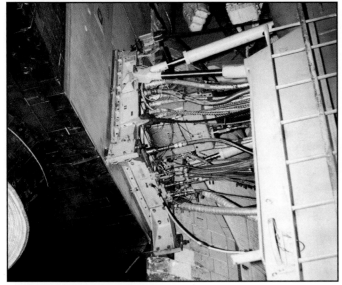

The liquid hydrogen T-0 umbilical showing the connections between the orbiter on the left and the tail service mast (TSM) on the mobile launch platform. The large hose in the center is the LH_2 connection. The smaller hoses carried various liquids and gases. (NASA)

Freedom Star brings the left solid rocket booster through the Canaveral Locks (and draw bridge) that separates Port Canaveral and the Banana River. Note the frustum on the aft deck. At this point the booster was towed in the "hip" position next to the ship. (NASA)

STS-111 (ISS-UF2)

Mission:	110		NSSDC ID:		2002-028A
Vehicle:	OV-105 (18)		ET-114 (SLWT)		SRB: BI113
Launch:	LC-39A		05 Jun 2002		21:23 UTC
Altitude:	214 nm		Inclination:		51.60 degrees
Landing:	EDW-22		19 Jun 2002		17:59 UTC
Landing Rev:	218		Mission Duration:		332 hrs 35 mins

Commander:	Kenneth D. "Taco" Cockrell (5)
Pilot:	Paul S. "Paco" Lockhart (1)
MS1:	Philippe Perrin (1)
MS2:	Franklin R. Chang-Diaz (7)
Up:	Valery Grigoryevich Korzun (2)
Up:	Peggy A. Whitson (1)
Up:	Sergei Yevgenyevich Treshchov (1)
Down:	Yuri Ivanovich Onufriyenko (2)
Down:	Carl E. Walz (4)
Down:	Daniel W. "Dan" Bursch (4)

Payloads:	Up: 36,082 lbs	Down: 6,342 lbs
	MPLM (Leonardo) (23,029 lbs)	
	Mobile Base System (3,181 lbs)	
	SRMS s/n 303	

Notes:	Scrubbed 30 May 2002 (KSC weather)
	Delayed 31 May 2002 (KSC weather)
	Delayed 03 Jun 2002 (OMS repair)
	EVA1 09 Jun 2002 (Chang-Diaz and Perrin)
	EVA1 11 Jun 2002 (Chang-Diaz and Perrin)
	EVA1 13 Jun 2002 (Chang-Diaz and Perrin)
	Orbiter returned to KSC on 29 Jun 2002 (N911NA)

Wakeup Calls:

06 Jun	"Gettin' Jiggy Wit It"
07 Jun	"American Woman"
08 Jun	"I Have a Dream"
09 Jun	"Drops of Jupiter"
10 Jun	"I Only Have Eyes for You" (American Graffiti)
11 Jun	"Mi PC"
12 Jun	"Chasing Sheep is Best Left to Shepherds"
13 Jun	"On the Road Again"
14 Jun	American National Anthem – "The Star-Spangled Banner"
15 Jun	"Hello to All the Children of the World"
16 Jun	"Where My Heart Will Take Me" (Star Trek: Enterprise)
17 Jun	"The Eyes of Texas"
18 Jun	"Sojourner"
19 Jun	"I Got You Babe"

The red, white, and blue ISS orbit represents the American, Costa Rican, French, and Russian flags while the Earth in the background shows Italy, which contributes the MPLM. The ten stars represent the ten crew members on-orbit during the flight. (NASA)

NASA planned this *Endeavour* launch for 30 May 2002, but adverse weather around LC-39A stopped the countdown during the T-9 minute hold. The launch was rescheduled to 31 May, but mission management slipped the attempt to 3 June based on the weather forecast. At a subsequent meeting, they decided a leaking OMS gaseous nitrogen regulator needed replaced and rescheduled the launch for 5 June. The countdown on 5 June proceeded smoothly. This was the last flight of a CNES astronaut (Philippe Perrin), the French Space Agency having transferred its astronaut corps to the European Space Agency (ESA).

Following an inspection of the active common berthing mechanism, Ken Cockrell used the SRMS to move Leonardo to Node 1 (Unity). The MPLM carried experiment racks and three stowage and resupply racks. On 12 June, the crews stowed 4,667 pounds of trash and hardware in Leonardo for return to Earth.

During the first EVA, Franklin Chang-Diaz and Philippe Perrin installed a power and data grapple fixture to the P6 truss that was later used to relocate the P6 truss to its final location. They also retrieved six micrometeoroid debris shields from the payload bay and stored them on PMA-1 pending their future installation on the Zvezda Service Module. An unscheduled task required the pair to inspect and photograph a failed control moment gyro on the Z1 truss.

The second EVA had Franklin Chang-Diaz and Philippe Perrin connect cables between the Canadarm2 mobile transporter and the mobile base system. They also deployed an auxiliary grapple fixture that could hold payloads as they are moved along the truss atop the mobile base system. During the third, and last, EVA, Chang-Diaz and Perrin replaced the wrist-roll joint on Canadarm2, restoring it to full use. The SRMS was used to maneuver the crew between the SSRMS and the payload bay during the various activities.

Endeavour undocked from the ISS on flight day 11 and flew one and a quarter laps around the station. Mission management waved-off two landing opportunities on flight day 12 due to low cloud cover, rain, and thundershowers at the Shuttle Landing Facility. While opening the payload bay doors after one attempt was waved off, a series of sensors on the doors began sending erroneous data, forcing the crew to open and close the doors under manual control. Mission control waved-off both KSC landing opportunities of the second extension day because of weather. The planned mission duration was 12 days plus 2 contingency days, but sufficient consumables existed for a third contingency day if needed. To avoid delaying the landing to the last possible day, the NASA moved the landing to Edwards on the first opportunity of the second extension day.

The Expedition 4 crew of Dan Bursch, Yuri Onufriyenko, and Carl Walz is at left and the Expedition 5 crew of Valery Korzun, Peggy Whitson, and Sergei Treshchov is at right. Philippe Perrin, Paul Lockhart, Ken Cockrell, and Franklin Chang-Diaz from STS-111 are at the bottom. (NASA)

Ken Cockrell and Paul Lockhart in a trainer at JSC. The orbiters were in transition between the original multifunction CRT display system (MCDS) and the new multifunction electronic display system (MEDS) glass cockpit, so there were mockups featuring both systems. (NASA)

Although unusual at the time, this would be come a routine photo after the Columbia accident when the ISS crews began taking detailed photos of each orbiter as it approached so that engineers on the ground could examine its thermal protection system for ascent damage. (NASA)

Wearing training versions of their pressure suits, the STS-111 and Expedition 5 crews are briefed by trainer Ken Trujillo in the Space Vehicle Mockup Facility at JSC. This building held a variety of space shuttle and ISS mockups, as well as various other training aids. (NASA)

A Boeing F-15A-13-MC (75-0043) from the 125th Fighter Wing of the Florida Air National Guard conducts a post-9/11 combat air patrol mission over LC-39A on 30 May 2002. Although not particularly unusual, it was not routine to have air cover during launches. (U.S. Air Force)

Endeavour is shown landing at Edwards AFB on the big screen in this overall view of the White Flight Control Room (WFCR) in the Mission Control Center at JSC. A generally similar Blue Fight Control Room (BFCR) monitored International Space Station operations. (NASA)

Philippe Perrin and Franklin Chang-Diaz practice EVA procedures while wearing training versions of the extravehicular mobility unit in the Neutral Buoyancy Laboratory at JSC. Note the safety divers that made sure the astronauts did not get caught on anything. (NASA)

STS-112 (ISS-9A)

Mission:	111	NSSDC ID:	2002-047A		
Vehicle:	OV-104 (26)	ET-115 (SLWT)	SRB: BI115		
Launch: Altitude:	LC-39B 220 nm	07 Oct 2002 Inclination:	19:46 UTC 51.60 degrees		
Landing: Landing Rev:	KSC-33 170	18 Oct 2002 Mission Duration:	15:45 UTC 259 hrs 58 mins		

Commander:	Jeffrey S. "Bones" Ashby (3)
Pilot:	Pamela A. "Pam" Melroy (2)
MS1:	Piers J. Sellers (1)
MS2:	Sandra H. "Sandy" Magnus (1)
MS3:	David A. Wolf (3)
MS4:	Fyodor Nikolayevich Yurchikhin (1)

Payloads:	Up: 37,441 lbs		Down: 1,839 lbs
	S1 Truss (29,543 lbs)		
	SRMS s/n 202		

Notes:	Delayed 02 Oct 2002 (JSC weather)
	First use of an "ET cam" to record ascent
	EVA1 10 Oct 2002 (Wolf and Sellers)
	EVA2 12 Oct 2002 (Wolf and Sellers)
	EVA3 14 Oct 2002 (Wolf and Sellers)

Wakeup Calls:

08 Oct	"Venus and Mars"
09 Oct	"The Best"
10 Oct	Medley of childhood songs
11 Oct	"Oh Thou Tupelo"
12 Oct	"Push It"
13 Oct	"Moscow Aviation Institute Hymn" – Aviation March
14 Oct	"You Gave Me The Answer"
15 Oct	"Only an Ocean Away"
16 Oct	"Prime Time"
17 Oct	"These are the Days"
18 Oct	"Someday Soon"

The crew patch depicts the ISS during departure, with the installed S1 truss segment outlined in red. A gold trail represents a portion of the rendezvous trajectory. The nine-pointed star represents the combined on-orbit team of six shuttle and three ISS crew members. (NASA)

This *Atlantis* launch was scheduled for 2 October 2002, but Hurricane Lili formed in the Atlantic on 21 September and forecasters were having difficulties determining her path. However, after Lili crossed Cuba as a category 2 storm on 1 October, NASA closed mission control in Houston and delayed the launch until 7 October. The countdown on that day went smoothly.

This was the first mission to carry an "ET cam" mounted on the external tank. The live video was near perfect until the booster separation motors coated the lens in residue, making it impossible to discern anything but general shapes. During ascent, the left bipod ramp shed a 7x12-inch, 0.3-pound piece of foam that caused a 4x3-inch dent in the ET attach ring on the left SRB. In response, space shuttle program manager Ron Dittemore ordered an investigation into the bipod foam loss issue. Nevertheless, the flight readiness review presentation for the next mission (STS-113) stated, "The ET is safe to fly with no new concerns (and no added risk)." It was a harbinger of things to come three months later.

Jeff Ashby and Pam Melroy docked to PMA-2 over central Asia. During the first EVA, David Wolf and Piers Sellers called a "stop" when the SSRMS elbow joint came near the right payload bay door. Initial calculations indicated the clearance was approximately 2 feet, but later analysis showed the clearance was possibly less than 2 inches. During the EVA, they mated the Zenith tray utilities, released the radiator beam launch locks, deployed the S-band antenna, released the crew and equipment translation aid (CETA) cart launch locks, installed the S1 nadir external television camera, and released the thermal radiator rotary joint drive-lock assembly launch locks.

David Wolf and Piers Sellers were on tap for the second EVA, installing 22 spool positioning devices, connecting the ammonia tank assembly umbilicals, installing the Destiny laboratory external television camera group, removing the radiator beam launch locks, affixing the S1/S3 line clamps, relocating the articulating portable foot restraint on the CETA cart, and installing the S1 thermal radiator rotary joint stinger bolts. The crew ran out of time before they could install the photovoltaic radiator spool positioning devices on two quick disconnects on the P6 truss.

The third EVA saw David Wolf and Piers Sellers retract the inertial unit assembly safing bolt, connect the S1/S1 fluid jumper, remove the port and starboard keel pins and drag links, install the thermal radiator rotary joint spool positioning devices, and reconfigure the squib firing unit.

The weather was good and *Atlantis* returned to the Shuttle Landing Facility on the first landing opportunity.

The crew poses in front of LC-39B during a tour prior to launch. From the left are Sandy Magnus, Jeff Ashby, Pam Melroy, David Wolf, Fyodor Yurchikhin, and Piers Sellers. At the time, Hurricane Lili was already threatening the Mission Control Center at JSC. (NASA)

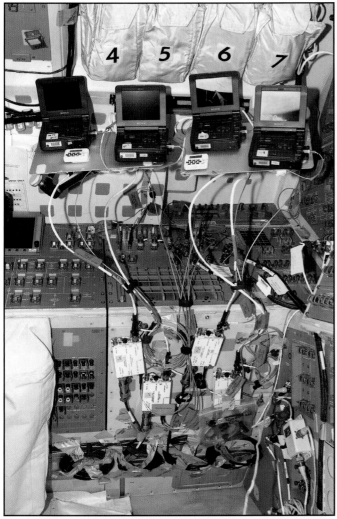

The aft flight deck of Atlantis as the crew set up the various video recorders that would be used during the mission. A bit later they would begin setting up several laptops and then unpack a menagerie of cameras and lenses. (NASA)

The Atlantis stack starts its journey from the VAB to LC-39B. Note each of the SRB forward skirts still has a yellow GSE (ground support equipment) door on it, not the final flight doors. Technicians would need to enter the forward skirts to configure the igniters prior to launch. (NASA)

Jeff Ashby prepares to taxi one of the Shuttle Training Aircraft at the Shuttle Landing Facility during a landing exercise. The aircraft were modified Gulfstream II business jets that had a set of orbiter displays and controls for the right seat pilot. (NASA)

The last main engine is installed in Atlantis after technicians welded and polished the flowliners to repair small cracks. All of the Atlantis engines were removed for inspection and repair after the anomaly was detected in the flowliners of other orbiters. (NASA)

STS-113 (ISS-11A)

Mission:	112	NSSDC ID:	2002-052A
Vehicle:	OV-105 (19)	ET-116 (SLWT)	SRB: BE114
Launch:	LC-39A	24 Nov 2002	00:50 UTC
Altitude:	214 nm	Inclination:	51.60 degrees
Landing:	KSC-33	07 Dec 2002	19:38 UTC
Landing Rev:	215	Mission Duration:	330 hrs 47 mins

Commander:	James D. "WxB" Wetherbee (6)
Pilot:	Paul S. "Paco" Lockhart (2)
MS1:	Michael E. "LA" López-Alegría (3)
MS2:	John B. Herrington (1)
Up:	Kenneth D. "Sox" Bowersox (5)
Up:	Nikolai Mikhailovich Budarin (3)
Up:	Donald R. "Don" Pettit (1)
Down:	Valery Grigoryevich Korzun (2)
Down:	Peggy A. Whitson (1)
Down:	Sergei Yevgenyevich Treshchov (1)

Payloads:	Up: 38,393 lbs	Down: 2,250 lbs
	P1 Truss (29,672 lbs)	
	SRMS s/n 201	

Notes:	Scrubbed 10 Nov 2002 (O$_2$ leak in payload bay)
	Delayed 17 Nov 2002 (SRMS damage)
	Scrubbed 22 Nov 2002 (TAL weather)
	EVA1 26 Nov 2002 (López-Alegría and Herrington)
	EVA2 28 Nov 2002 (López-Alegría and Herrington)
	EVA3 30 Nov 2002 (López-Alegría and Herrington)

Wakeup Calls:

24 Nov	"The Wind Ensemble"
25 Nov	"I Need You Like a Hole in My Head"
26 Nov	"Gimme All Your Lovin'"
27 Nov	"Copperhead Road"
28 Nov	"Figaro's Wedding" (*Trading Places*)
29 Nov	"Quiet Flame"
30 Nov	"Walkin' on Sunshine"
01 Dec	"Roll With It"
02 Dec	"Cheesty Proody" (Russian song)
03 Dec	"Flot" (Russian song)
04 Dec	"Asereje" ("The Ketchup Song")
05 Dec	"Hotel California"
06 Dec	Clips from *Groundhog Day* and *Jeopardy*

The *Endeavour* mission was scheduled for 10 November 2002, but an excessive oxygen concentration in the payload bay led to a scrub. Technicians replaced a leaking flexible oxygen hose and NASA tentatively rescheduled the launch for 17 November. However, during the repair activities in the Orbiter Processing Facility, a work platform hit the SRMS. As a result, the mission management team rescheduled launch for 22 November so the Canadian manufacturer could perform tests of the SRMS that ultimately showed its structural integrity was not compromised.

The launch team stopped the countdown on 22 November during the T-9 minute hold because of weather at the Zaragoza and Moron TAL sites. Launch proceeded smoothly the following day. This was the last space shuttle mission that included Russian cosmonauts (Nikolai Budarin, Valery Korzun, and Sergei Treshchov). Paul Lockhart replaced Christopher Loria as the pilot for this mission due to due to a severe back injury suffered in a bicycle accident.

The SRMS brought more excitement once on-orbit when data indicated the wrist-roll joint was experiencing issues. A quick test of the joint showed it was operating properly but the crew nevertheless performed a second test of the joint prior to cradling the arm to confirm the health of the joint. All seemed fine.

Flight day 3 saw Jim Wetherbee and Paul Lockhart dock with the ISS to begin seven days of station assembly and cargo transfers. *Endeavour* brought the Expedition 6 crew to replace the Expedition 5 crew that had been on the station for 185 days. Jim Wetherbee used the SRMS to unberth the P1 truss from the payload bay and position it over the port wing for handoff to Peggy Whitson operating the SSRMS, who then mated it to the S0 truss. *Endeavour* undocked on flight day 10 and did not perform the now-customary fly-around maneuver because of an earlier use of propellant for attitude control.

The crew closed the payload bay doors in preparation for two KSC landing opportunities, but mission control waved both off due to weather. The crew opened the payload bay doors to provide cooling for the orbiter. The crew did not close the payload bay doors for the first contingency day landing opportunities because of the forecasted weather and, as it happened, both KSC opportunities were waved off. The crew closed the payload bay doors in preparation for the first landing opportunity on the second contingency day, but again both opportunities were waved-off due to weather. The crew reopened the doors. The weather was acceptable for the first Shuttle Landing Facility opportunity on the third contingency day. This was the first time a mission ended on the fourth day of landing attempts.

The STS-113 crew (red shirts) are Jim Wetherbee, John Herrington, Michael López-Alegría, and Paul Lockhart. The Expedition 6 crew (left) are Ken Bowersox, Don Pettit, and Nikolai Budarin and the Expedition 5 crew (right) Valery Korzun, Peggy Whitson, and Sergei Treshchov. (NASA)

The crew patch depicts Endeavour docked to the ISS during the installation of the P1 truss with the gold astronaut symbol in the background. The seven stars represent the seven ascent crew members and the three stars symbolize the returning Expedition 5. (NASA)

The view of the pressurized mating adapter on the ISS just prior to docking. This was PMA-2 attached to the forward port of the Destiny laboratory. Ultimately, thirty-five missions docked to PMA-2 and two docked to PMA-3 (Zarya is permanently docked to PMA-1). (NASA)

Jim Wetherbee adds the STS-113 crew patch (actually, a sticker) to the growing collection of those representing space shuttle crews who had worked on the ISS. This location in the Unity module (Node 1) served as one of the traditional posting sites for the patches. (NASA)

A crawler-transporter has just delivered the Endeavour stack to LC-39A. The crawler is still under the mobile launch platform. Note the two tail service masts—the large gray boxes—next to the aft fuselage. These provided all gases, fluids, and power to the orbiter while it was on the launch pad. In addition, they provided the flow path for the liquid hydrogen and liquid oxygen that filled the ET; the propellants flowed through the orbiter on their way from the ground storage dewars to the ET. All of the fixed service structure umbilicals are in their retracted positions. (NASA)

STS-107

Mission:	113	NSSDC ID:	2003-003A
Vehicle:	OV-102 (28)	ET-93 (LWT)	SRB: BI116
Launch: Altitude:	LC-39A 156 nm	16 Jan 2003 Inclination:	15:39 UTC 39.00 degrees
Landing: Landing Rev:	(KSC-33) 255	(01 Feb 2003) Mission Duration:	(14:30) UTC (375 hrs 21 mins)

Commander:	Rick D. Husband (2)
Pilot:	William C. "Willie" McCool (1)
MS1:	David M. Brown (1)
MS2:	Kalpana Chawla (2)
MS3:	Michael P. Anderson (2)
MS4:	Laurel B. Clark (1)
MS5:	Ilan Ramon (1)

Payloads:	Up: 35,463 lbs	Down: 0 lbs
	Spacehab RDM (18,756 lbs)	
	SEM-14	
	FREESTAR (hitchhiker)	
	Extended duration orbiter pallet	
	No SRMS	

Notes:	Last flight of a lightweight external tank (LWT)
	Vehicle and crew lost during entry

Wakeup Calls (Blue Team / Red Team):

16 Jan	"EMA EMA" / "America, the Beautiful"
17 Jan	"Coming Back to Life" / "Space Truckin'"
18 Jan	"Cultural Exchange" / "Hatishma Koli"
19 Jan	"Fake Plastic Trees" / "Amazing Grace"
20 Jan	"Texan 60" / "God of Wonders"
21 Jan	"The Wedding Song" / "Prabhati"
22 Jan	"Hakuna Matata" (Blue Team)
22 Jan	"Ma ata osheh kesheata kam baboker? (Red Team)
23 Jan	"Burning Down the House" / "Kung Fu Fighting"
24 Jan	"Hotel California" / "The Prayer"
25 Jan	"I Say a Little Prayer" / "Drops of Jupiter"
26 Jan	"When Day is Done" / "Love of My Life"
27 Jan	"Slow Boat to Rio" / "Running to the Light"
28 Jan	"I Get Around" / "Up on the Roof"
29 Jan	"Imagine" / "Yaar ko hamne ja ba ja dekha"
30 Jan	"Silver Inches" / "Shalom lach eretz nehederet"
31 Jan	"If You've Been Delivered" / "Scotland the Brave"

The central element of the crew patch is the microgravity symbol, μg, flowing into the rays of the astronaut symbol. Six stars have five points, the seventh has six points, like a Star of David, for Ilan Ramon. The patch design was initiated by Laurel Clark and Kalpana Chawla. (NASA)

This *Columbia* countdown on 16 January 2003 proceeded smoothly with no major issues or unplanned holds. However, at approximately T+81.7 seconds, a large light-colored piece of debris originated from near the ET forward attach bipod. The debris appeared to move outboard and then fall aft along the left side of the fuselage, striking near the leading edge of the left wing. An assessment of this event performed during the mission concluded it was not a safety of flight issue.

Columbia was making the maiden flight of the Spacehab double research module along with the only existing extended duration orbiter pallet. The RDM was a new version of the commercially developed Spacehab that included avionics enhancements, more than 5kW of module power, and a new Ku-band communications capability that provided experiment-data downlink transmission rates up to 48 megabytes per second.

The primary payload consisted of 28 facilities (payloads) in the Spacehab module and on the middeck supporting more than 80 life science, Earth science, physical science, and commercial investigations from around the world. This included three roof-mounted experiments that focused on heat pipe design, as well as university and grade school-sponsored experiments. As was normal for science missions, the crew split into two teams to allow around-the-clock operations. The red team consisted of Rick Husband, Kalpana Chawla, Laurel Clark, and Ilan Ramon while the blue team included Willie McCool, David Brown, and Michael Anderson.

Initially, NASA had wanted to carry Triana, a deployable Earth-observing satellite first proposed by vice president Al Gore, on this mission. Political disagreements between Congress and the White House delayed Triana, which was ultimately replaced by the cobbled-together FREESTAR (fast reaction experiments enabling science, technology, applications and research) payload. This included a cross-bay carrier populated with CVX-2 (critical velocity of xenon), LPT (low-power transceiver), MEIDEX (Mediterranean Israeli dust experiment), SEM-14 (space experiment module), SOLCON-3 (measurement of solar constant), and SOLSE-2 (shuttle ozone limb sounding experiment). An additional secondary payload was the DoD RAMBO (ram burn observation) experiment, but the required ground observations were not completed because the scheduled OMS burns were not within the range of the military observation assets.

The crew closed the payload bay doors for the first landing opportunity at the Shuttle Landing Facility. Entry interface was at 13:44:09 universal coordinated time, 1 February 2003; *Columbia* never made it to KSC.

The crew in their on-orbit portrait in the Spacehab double module. Wearing red shirts to signify their shift, from the left, are Kalpana Chawla, Rick Husband, Laurel Clark, and Ilan Ramon. Wearing blue shirts are David Brown, Willie McCool, and Michael Anderson. (NASA)

By the time this photo was taken, flight controllers had lost radar, radio, and telemetry contact with Columbia but still did not understand that it was not just a technical glitch. Television news would soon report pieces of Columbia were raining down across eastern Texas. (NASA)

Ilan Ramon (left), Laurel Clark, and Michael Anderson, in one of the high fidelity trainers in the Space Vehicle Mockup Facility at JSC. The three, attired in training versions of their pressure suits, are seated on the middeck for an emergency egress exercise. (NASA)

It seemed like a picture-perfect launch on a beautiful winter day. The cameras, however told a different story, detecting a debris strike on the orbiter. Unfortunately, everybody would soon discover that Columbia was mortally wounded while going uphill. (NASA)

The Spacehab double module as seen from the aft flight deck of Columbia. Spacehab was a commercial venture and its name and logo could be replaced by whatever entity purchased the use of the equipment. In this case, NASA paid Spacehab for the flight. (NASA)

Columbia rolls out from the Vehicle Assembly Building on its way to LC-39A, as seen from the windows of the Launch Control Center. Originally the LCC had mechanical sunshades on the outside of the windows that many believed, incorrectly, were blast shields. (NASA)

COLUMBIA ACCIDENT

At 11:15 UTC on 16 January 2003, the Ice Team began a visual and photographic inspection of the launch pad and *Columbia* stack, finding nothing particularly noteworthy. Everything was within the launch commit criteria, as was the weather at 48 °F with 97 percent relative humidity and a 5-knot wind. Fifteen minutes later, the countdown clock came out of the planned T-9-minute hold and the main-engine start sequence commenced at 15:38:53. The final flight of *Columbia* began at 15:38:59.994, universal coordinated time, 16 January 2003.

The initial ascent was uneventful, but *Columbia* encountered a significant wind shear at T+57 seconds as the vehicle passed through 32,000 feet. This pushed the front of the stack to the right and increased the aerodynamic load on the ET bipod strut. Several post-accident studies showed the aerodynamic loads on the bipod, and the interacting aerodynamic loads between the ET and orbiter, were greater than normal but well within design limits. In fact, all loads were less than 70 percent of their design limit throughout ascent.

The day after launch, the Intercenter Photo Working Group noticed a debris strike on *Columbia* at T+81.9 seconds. The quick-look image analysis reported, "The debris appeared to move outward in a –Y [left] direction, then fell aft along the left orbiter fuselage and struck the underside (–Z) of the leading edge of the left wing." Based on the available imagery, the analysts determined one large and at least two smaller pieces of foam separated from the left bipod ramp at T+81.7 seconds. Further analysis determined the large piece was 21 to 27 inches long and 12 to 18 inches wide, tumbling at approximately 18 rpm. The wide range for these measurements was due to the relatively slow frame rate and poor resolution of the available cameras. Because of these uncertainties, the analysts were unable to determine how much the debris weighed, although later investigation concluded it was approximately 1.67 pounds. As it separated from the ET, the foam was traveling at the same 2,300 fps (1,568 mph) as the rest of the stack. The photo interpreters believed the debris impacted somewhere on the underside of the wing

approximately 0.161 second later. During that time, air resistance had caused the rather unaerodynamic and very light foam to slow to about 1,500 fps (1,022 mph), meaning the orbiter hit the foam (not the other way around) with a relative velocity of slightly more than 545 mph. Although there was a lot of initial confusion as to exactly where the foam hit, with most analysts assuming it was near the left main landing-gear door, subsequent analysis determined the foam impacted on the eighth reinforced carbon-carbon panel outboard from the fuselage on the left wing (RCC-8L).

On 1 February 2003, the crew completed their tasks and began guiding *Columbia* through the atmosphere on its way to the Shuttle Landing Facility. Unknown to the crew, or flight controllers, entry heating began melting the left wing through the hole in the reinforced carbon-carbon leading edge. Videos from ground observers showed *Columbia* was breaking up at 14:00:18 UTC while traveling approximately Mach 15 at 181,000 feet.

Just before 08:00 central time on 1 February, the residents of east Texas heard a low rumble generated by debris from *Columbia* traveling nearly 12,000 mph. Cattle stampeded in eastern Nacogdoches County as thousands of pieces of debris began impacting the ground along a swath 10 miles wide and 300 miles long across East Texas and into Louisiana. A fisherman on Toledo Bend reservoir saw a piece splash down in the water, while debris smashed the windshield of a woman driving near Lufkin. The heaviest parts flew the farthest. An 800-pound piece of a main engine hit the ground in Fort Polk, Louisiana, doing 1,400 mph and a 600-pound piece landed on the nearby golf course.

Ultimately, more than 25,000 people from 270 organizations expended more than 1.5 million manhours searching for debris. They covered 2.3 million acres, an area half again as large as Delaware, with some 700,000 acres being searched by foot. Searchers also used 37 helicopters, 7 fixed-wing aircraft, and various ultra-lights. Tragically, on 27 March 2003 a Bell 407 helicopter lost power and crashed into heavily wooded terrain near Broadus, Texas. The

Ultimately, more than 25,000 people from 270 organizations expended approximately 1.5 million manhours searching for debris, covering some 2.3 million acres. The majority of them needed fed and housed while they searched, creating a huge logistics challenge. (NASA)

Steve Altemus (left), the NASA debris reconstruction manager, briefs Hal Gehman (white shirt) and other members of the accident board. This is the "Columbia Hangar" at KSC where all of the debris was collected. The author has his back to the camera in the center. (CAIB)

accident killed Jules "Buzz" Mier Jr., a contract pilot, and Charles Krenek, a Texas Forest Service employee, and seriously injured Ronnie Dale (NASA/KSC), Richard Lange (USA/KSC), and Matt Tschacher (Forest Service). The air search was discontinued as the FAA and NTSB investigated the accident, but resumed on 9 April.

By 30 June 2003, teams had recovered 83,900 pieces of debris weighing approximately 84,900 pounds, representing 38 percent of the orbiter and its payload. Although anecdotal evidence indicated *Columbia* shed debris as it crossed California, Nevada, and New Mexico, the most westerly piece of confirmed debris was a single tile found in Littlefield, Texas. The most easterly debris, the turbopumps from the three main engines, was found near Fort Polk, nearly 500 miles away. By the end, FEMA had expended more than $305 million on the search, not including the wages of hundreds of civil servants, NASA contractors, and military personnel. It was the largest and most expensive debris recovery effort in history.

COLUMBIA ACCIDENT INVESTIGATION BOARD

Shortly after the planned landing time of 14:16 UTC, NASA began executing the contingency action plan for space flight operations, something developed in the wake of the *Challenger* accident 17 years earlier. Later that day, NASA administrator Sean O'Keefe named retired four-star admiral Harold "Hal" Gehman as the chairman of the International Space Station and Space Shuttle Mishap Interagency Investigation Board described in the plan. One of the first things Gehman did was change the name to the Columbia Accident Investigation Board (CAIB).

Almost immediately, pundits criticized the board for a perceived lack of independence. In response to the *Challenger* accident, Ronald Reagan had established a presidential commission headed by William Rogers, completely independent of NASA. For *Columbia*, George Bush left it to the NASA administrator to investigate his own agency, much as had been done after the AS-204 (Apollo 1) fire in 1967. By appearances, it was not an ideal response, but anybody who thought the board was not independent had never met Hal Gehman, who proved more than capable of keeping the CAIB out of the political fray that surrounded the accident.

Other members of the CAIB included John Barry, director of Plans and Programs at the Air Force Materiel Command; Duane Deal, commander of the 21st Space Wing; James Hallock, chief of the Department of Transportation Aviation Safety Division; Kenneth Hess, commander of the Air Force Safety Center; Scott Hubbard, director of the Ames Research Center; John Logsdon, director of the Space Policy Institute at George Washington University; Douglas

Osheroff, 1996 Nobel Laureate in Physics; Sally Ride, professor of space science at the University of California in San Diego; Roger Tetrault, retired chairman of McDermott International; Stephen Turcotte, commander of the Naval Safety Center; Steven Wallace, director of Accident Investigation for the FAA; and Sheila Widnall, professor of aeronautics and astronautics at MIT.

The board released its final report on 26 August 2003, severely criticizing many aspects of NASA in general and the Space Shuttle Program in particular. Ultimately, the accident board provided 29 recommendations to help ensure a safe return to flight in addition to many findings and observations that needed addressed.

The technical cause of the accident was rather straightforward: "The physical cause of the loss of *Columbia* and its crew was a breach in the thermal protection system on the leading edge of the left wing. The breach was initiated by a piece of insulating foam that separated from the left bipod ramp of the external tank and struck the wing in the vicinity of the lower half of reinforced carbon-carbon panel 8 at 81.9 seconds after launch. During entry, this breach in the thermal protection system allowed superheated air to penetrate the leading-edge insulation and progressively melt the aluminum structure of the left wing, resulting in a weakening of the structure until increasing aerodynamic forces caused loss of control, failure of the wing, and breakup of the orbiter."

The organizational and cultural causes were more nuanced. Of the 112 missions prior to STS-107, there was some evidence of foam loss from the external tank on 65 of the 79 missions for which usable imagery existed. This included 35 missions that had foam loss from the liquid hydrogen tank intertank flange, 25 from the intertank acreage, 6 from the left bipod, 3 from the thrust panels (popcorning), and 15 from various other areas of the tank. Of the 59 missions without a bipod ramp anomaly, 36 of them had experienced foam losses near the bipod. During the 22 years of the flight campaign before the *Columbia* accident, NASA had taken 17 distinct actions to eliminate debris from the external tank but it continued to shed foam. It should be noted that, in general, there was only a loose correlation between foam loss and orbiter damage. Many of the larger anomalies observed on the ET had not produced corresponding damage to the orbiter, highlighting the variability of transport mechanisms during ascent and contributing to the apparent lack of interest from management.

To better understand the complexity of this accident, the reader is encouraged to read the CAIB final report, as well as the various personal accounts of the events surrounding the accident, such as *Comm Check: The Final Flight of Shuttle Columbia* by Michael Cabbage and William Harwood.

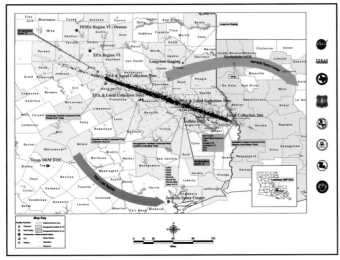

It was a huge debris field. The most westerly piece of confirmed debris was a single tile found in Littlefield, Texas. The most easterly debris, the turbopumps from the three main engines, was found near Fort Polk, Louisiana, nearly 500 miles to the east. (CAIB)

Investigators ultimately fired a representative piece of foam against a real RCC-8L panel that resulted in a large breach that easily could have caused the loss of Columbia. The tests were conducted at the Southwest Research Institute (SwRI) in San Antonio, Texas. (CAIB)

Return-to-Flight II

Interestingly, engineers had already identified corrective measures prior to both accidents, but NASA was in no rush to implement them because neither problem seemed critical. The CAIB had difficulty understanding, after the bitter lessons from *Challenger*, how the agency could have failed to notice it was continually normalizing technical anomalies. Each time an anomaly occurred, the flight readiness process declared it an "acceptable risk." Taken one at a time, each decision seemed correct. In retrospect, none of it made any sense.

As the investigation progressed, Sally Ride, who had also served on the Rogers Commission, opined there were "echoes of *Challenger*" in the *Columbia* accident. It became one of the enduring sound bites from the investigation. The foam debris was not the single cause of the *Columbia* accident, just as a leaking SRB joint seal was not the single cause of *Challenger*. Management failures and a distorted culture contributed to both accidents. George Santayana, a Spanish philosopher, wrote, "Those who cannot remember the past are condemned to repeat it." This certainly seems true for NASA, which never teaches its accidents and seldom even acknowledges them. This is in direct contrast to how the Navy handles its two nuclear submarine losses, which are front and center during early training for everybody entering the silent service. As the CAIB observed, "History is not a backdrop or a scene-setter. History is cause." As an agency, NASA still has not learned.

The CAIB noted many of the same systemic problems as the Rogers Commission had 17 years earlier. The investigation showed that, for all its cutting-edge technologies, "diving-catch" rescues, and imaginative plans for the future of space exploration, NASA had little understanding of the workings of its own organization. The echoes of *Challenger* had deadly implications for the crew of *Columbia*.

The fix for the physical cause of the *Columbia* accident, the bipod foam, was easy. After a series of wind tunnel tests, NASA and Lockheed simply deleted the bipod ramps entirely. Engineers took this opportunity to introduce a myriad of other, mostly minor, changes to the external tank. One thing they did not do was remove the protuberance air loads (PAL) ramps that protected the cable trays running the length of the tank. This would almost bite the program during the first return-to-flight mission, STS-114, and cause another long hiatus for the flight campaign.

Although the CAIB had many, often heated, internal discussions over the merits of continuing to fly space shuttle, in the end they believed the risks were worth the rewards, as long as they were approached with the full realization of the potential consequences. Ultimately, the CAIB recommended, "Prior to operating the shuttle beyond 2010, develop and conduct a vehicle recertification at the material, component, subsystem, and system levels." Essentially, this called for the same type of design certification review (DCR) the agency had undertaken after the *Challenger* accident. It was meant not to kill the program, but to make it better for the future. Somehow, that nuance was missed.

While the CAIB had been investigating the accident, the White House and NASA Headquarters had been formulating an alternate future that did not include space shuttle. On 14 January 2004, George Bush announced his Vision for Space Exploration, attempting to one-up the Space Exploration Initiative proposed by his father in 1989. The vision required NASA retire space shuttle after it completed assembly of the International Space Station, but no later than 2010. In its place would come a series of Ares launch vehicles derived from existing Saturn and space shuttle components

and a crew exploration vehicle (CEV, later named Orion) that resembled an overgrown Apollo capsule. According to the hype, the CEV would fly to the ISS beginning in 2008, return men to the moon by 2020, and carry humans to Mars at some unspecified time in the future. Somewhat later, NASA formed the Constellation Program (CxP) as an overarching organization to manage Ares and Orion. All of it ultimately came to naught and President Barack Obama canceled the program in 2010.

The primary rationale for retiring space shuttle appeared to be securing funding for the Vision for Space Exploration. However, the White House, and many within NASA, showed a great misunderstanding about exactly what the Space Shuttle Program paid for. The White House assumed the entire space shuttle budget would become available to what became the Constellation Program once the flight campaign was over. This ignored the significant portion of the budget that went to maintaining the infrastructure at the manned space centers and the industrial base, something that would still be needed, at least largely, by Constellation. Along with continuing to pay the standing army of civil servants, this resulted in far less funding than expected.

Impact on ISS

There were other nuances that were apparently unknown to, or ignored by, the White House planners. Most importantly, the Space Shuttle Program provided a significant amount of services for the International Space Station that would continue to be required. Kirk Shireman, the deputy ISS program manager, remembered, "all of our cargo—the food and the water and all that—that was flying up and down on shuttle, we [ISS] were not carrying any budget for that; the Space Shuttle Program had its own budget, so ISS basically got the transportation ... for free." With the impending retirement of space shuttle, the ISS Program "had a huge budget upper just for the transportation costs and ... we signed a $2.6 billion net contract with SpaceX and Orbital to fly cargo up through 2015. We'll have to sign another contract to carry us all the way out to 2020." Shireman concluded there were "very large impacts to the ISS Program as a result of space shuttle being retired." Even less money for Constellation.

Ignoring the funding implications, there were other issues. The ISS concept of operations included using a small number of orbital replacement units (ORU). Kirk Shireman explained, "The idea was ... if something failed ... we would unhook it, place the spare in, fly the old one home on a shuttle ... and repair it. Then we would fly it back up and it would be a new spare." This significantly reduced the cost of procuring and maintaining an inventory of spare parts. The retirement of space shuttle changed that. For one thing, many of the critical ORUs were too large to fit in any of the proposed commercial capsules and the ISS Program quickly realized that, since they were no longer going to be able to bring ORUs back to Earth for repair, they did not have enough spares. As a result, NASA needed to procure additional spares and launch them on space shuttle prior to the final mission. It all caused a massive headache for the ISS Program and consumed a great deal of treasure, which meant even less money for Constellation.

Then there was the issue of downmass. One of the initial promises by ISS was the ability to create things on-orbit and bring them back to Earth using space shuttle. The orbiters were unique among spacecraft in that they could carry a substantial payload, both weight and size, back to Earth. Without downmass, the results of scientific experiments, instruments needing repair, and products from prototype microgravity manufacturing efforts could only be returned if they fit within the limited confines of Soyuz along with whatever crew members were returning to Earth. Kirk Shireman began, "encouraging all our payload and science customers to do everything without having any or extremely limited downmass." Fortunately, as NASA began competing the commercial cargo contracts, SpaceX announced it could accommodate limited downmass on Dragon, "so that changed our thinking a little with respect to scientific utilization payloads." That also allowed NASA to back off slightly on the types and number of spare ORUs it procured since now it was possible for many smaller items to be returned, repaired, and reflown. In 2015, NASA awarded a commercial cargo contract to Sierra Nevada to fly Dreamchaser, which also will have a limited downmass capability.

Nevertheless, in the immediate aftermath of the *Columbia* accident, NASA needed to make other plans for supporting the space station. The Government Accountability Office (GAO) reported, "With the [space shuttle] fleet grounded, NASA is heavily dependent on its international partners, especially Russia, for operations and logistics support for the space station." One of the major problems was that at about 36,000 pounds, space shuttle could carry seven times more payload than the Russian Progress supply vehicle and thirty-five times the payload of Soyuz, along with four additional crew members. The GAO reported, "due to the limited payload capacity of the Russian space vehicles, on-orbit assembly has been halted. The program's priority has shifted from station construction to maintenance and safety, but these areas have also presented significant challenges and could further delay assembly of the core complete configuration. While some onboard research is planned, it will be curtailed by the limited payload capacity of the Russian vehicles." Despite the limitations, having the Russian vehicles available probably saved the space station. During the 29-month stand-down after the *Columbia* accident, Russia flew fourteen crew rotation and resupply missions to the ISS. Ultimately, space shuttle completed the assembly flights, and then turned the resupply and crew rotation functions back over to the Russians using Progress and Soyuz vehicles while America waited for the Commercial Resupply and Commercial Crew programs to mature. Eight years after the last flight of space shuttle, America still cannot launch humans into space.

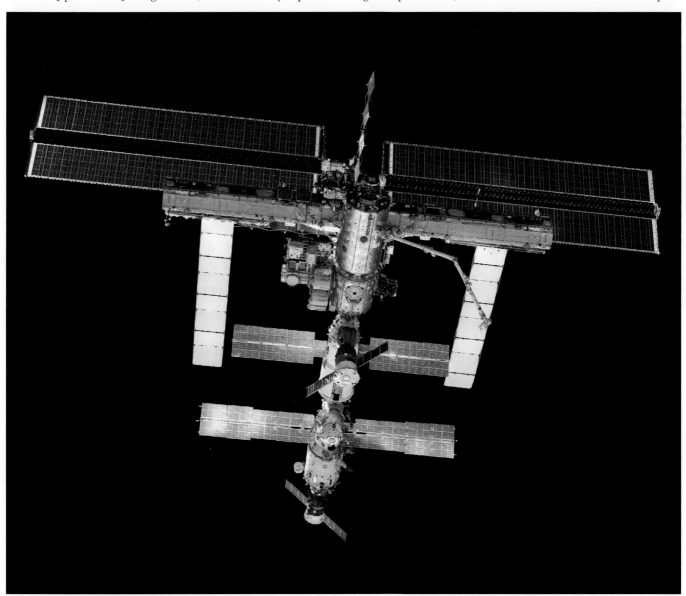

NASA was usually reluctant to admit it, but the International Space Station (seen here from STS-114) was designed to be assembled and serviced by the space shuttle fleet. No other vehicle was truly adequate for the task, although the Russian Soyuz and Progress did yeoman's work during the post-Columbia stand down and again after George W. Bush prematurely retired space shuttle. (NASA)

STS-114 (ISS-LF1)

Mission:	114		NSSDC ID:		2005-026A
Vehicle:	OV-103 (31)		ET-121 (SLWT)		SRB: BI125
Launch:	LC-39B		26 Jul 2005		14:39 UTC
Altitude:	191 nm		Inclination:		51.60 degrees
Landing:	EDW-22		09 Aug 2005		12:13 UTC
Landing Rev:	219		Mission Duration:		333 hrs 34 mins

Commander:	Eileen M. Collins (4)
Pilot:	James M. "Vegas" Kelly (2)
MS1:	Soichi Noguchi (1)
MS2:	Stephen K. "Steve" Robinson (3)
MS3:	Andrew S. W. "Andy" Thomas (4)
MS4:	Wendy B. Lawrence (4)
MS5:	Charles J. "Charlie" Camarda (1)

Payloads:	Up: 38,562 lbs	Down: 6,600 lbs
	MPLM (Raffaello) (18,127 lbs)	
	ICC-GD/ESP-2 (6,292 lbs)	
	LMC/CMG (3,355 lbs)	
	SRMS s/n 301	

Notes:	Tanking test 14 Apr 2005
	Tanking test 20 May 2005
	Scrubbed 13 Jul 2005 (LH$_2$ ECO sensor)
	First flight of the orbiter boom sensor system
	EVA1 30 Jul (Noguchi and Robinson)
	EVA2 01 Aug (Noguchi and Robinson)
	EVA3 03 Aug (Noguchi and Robinson)
	Orbiter returned to KSC on 21 Aug 2005 (N905NA)

Wakeup Calls:

26 Jul	"I Got You Babe"
27 Jul	"What a Wonderful World"
28 Jul	"Vertigo"
29 Jul	"Sanpo" ("Stroll")
30 Jul	"I'm Goin' Up"
31 Jul	"Walk of Life"
01 Aug	"Big Rock Candy Mountain"
02 Aug	"Where My Heart Will Take Me"
03 Aug	"Amarillo by Morning"
04 Aug	The Navy Hymn – "Anchors Aweigh"
05 Aug	The Air Force Hymn – "Wild Blue Yonder"
06 Aug	"The One and Only Flower in the World"
07 Aug	"Come On Eileen"
08 Aug	"Good Day Sunshine"

Against the background of the Earth at night, the orbit represents the ISS and the plume represents the broad spectrum of challenges for this mission. The blue orbiter rising above Earth includes the Columbia constellation of seven stars, honoring the crew of Columbia. (NASA)

NASA originally scheduled the return-to-flight mission after the *Columbia* accident for 13 July 2005, some 29 months after the launch of STS-107. During the countdown, a LH$_2$ engine cutoff (ECO) sensor provided faulty data, causing the launch to be scrubbed. Subsequent troubleshooting was unable to repeat the failure so engineers swapped the sensor wiring help isolate the source (orbiter or ET) of the failure should it occur again. The launch was rescheduled for 26 July. The ECO sensors performed as expected and *Discovery* lifted-off on time. This was the first flight that used Tyvek covers to protect the forward reaction control system jets from rain accumulation rather than the butcher paper that had been used since 1981. This was also the first mission to carry the orbiter boom sensor system (OBSS) on the right sill of the payload bay.

Needless to say, engineers were closely watching the ascent to determine if the external tank shed any debris. It did not take long. Approximately 127.1 seconds after liftoff, just 5.3 seconds after SRB separation, a large piece of foam separated from the ET protuberance air loads (PAL) ramp. Fortunately, the debris did not strike any part of the orbiter. However, 20 seconds later, a smaller piece of foam separated from the ET and struck the right wing. Based on the mass of the foam and the velocity at which it struck the wing, engineers were not concerned and scanning by the OBSS did not reveal any damage. Nevertheless, on 27 July, NASA announced it was grounding the orbiters, again, until it could resolve the foam problem.

On flight day 3, the crew performed the first R-bar pitch maneuver (RPM, often called a rendezvous pitch maneuver) as *Discovery* approached the ISS. The station crew used two digital cameras with telephoto lenses to image critical parts of the orbiter so that engineers on the ground could further evaluate the thermal protection system. The only areas of concern were two protruding Ames gap fillers and a damaged blanket beneath a window (W1).

Soichi Noguchi and Steve Robinson conducted three EVAs, all from the orbiter airlock. During the third EVA, the spacewalkers removed the two protruding gap fillers identified from the RPM photography. Engineers also considered having the crew remove the damaged thermal blanket located beneath W1, but wind tunnel tests at NASA Ames demonstrated the condition was safe for entry and mission management canceled plans for a fourth spacewalk.

Because of forecasted bad weather at the Shuttle Landing Facility, NASA waved off the first opportunity and, soon afterward, the second. The KSC weather had not improved substantially the following day, so the mission management team directed the crew to land at Edwards on the first opportunity.

Holding their helmets are James Kelly and Eileen Collins. Behind them are, from the left, Steve Robinson, Andy Thomas, Wendy Lawrence, Charlie Camarda, and Soichi Noguchi. Camarda and Noguchi were rookies while the rest were experienced astronauts. (NASA)

 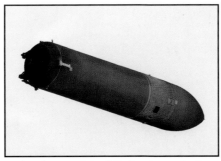 

Although the program had always attempted to take photos of the ET after it was jettisoned, this became a priority for each mission after Columbia. In addition to the automated umbilical well cameras (left), the crew rolled the orbiter and shot hand-held photos out the aft flight deck overhead windows (center). On STS-114, these revealed a large piece of the protuberance air loads (PAL) ramp was missing, although fortunately it missed hitting Discovery. Ground and airborne cameras also imaged the tank during ascent and as it reentered after being jettisoned. (NASA)

Beginning with this mission, each orbiter would perform an R-bar pitch maneuver (RPM) as it approached the ISS. The public affairs office believed the term "R-bar" (meaning Earth radius vector, an imaginary line connecting a target, in this case the ISS, to the center of the Earth) was too technical and usually called this the rendezvous pitch maneuver. The orbiter performed the 360-degree pirouette at 0.75-degree per second; while the bottom of the orbiter was facing the station, two ISS crew members took detailed photographs through the nadir windows on Zvezda. One used a digital camera with a 400 mm lens to photograph the acreage tiles on the belly while the other used an 800 mm lens to photograph the nose landing-gear doors, main landing-gear doors, ET doors, and the elevon coves. Although the entire maneuver took about eight minutes, the crew had only 93 seconds to take the pictures, long enough to photograph all of the critical areas twice. (NASA)

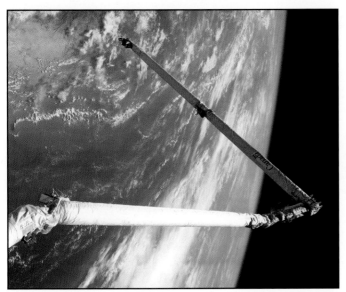 

This was the first mission to carry the orbiter boom sensor system (OBSS) that allowed the crew to inspect more of the orbiter TPS. Essentially, this was a 50-foot-long non-articulating extension to the SRMS arm that was carried on the starboard payload bay sill. (NASA)

Most every mission had banners signed by the employees, such as this one adorning the mobile launch platform as it carried the Discovery stack to LC-39B. The banners were left in each of the major facilities for several days so workers could sign them. (NASA)

STS-121 (ISS-ULF1.1)

Mission:	115	NSSDC ID:		2006-028A	
Vehicle:	OV-103 (32)	ET-119 (SLWT)		SRB: BI126	
Launch:	LC-39B	04 Jul 2006		18:38 UTC	
Altitude:	190 nm	Inclination:		51.60 degrees	
Landing:	KSC-15	17 Jul 2006		13:16 UTC	
Landing Rev:	202	Mission Duration:		306 hrs 38 mins	

Commander:	Steven W. "Steve" Lindsey (4)
Pilot:	Mark E. Kelly (2)
MS1:	Michael E. "Mike" Fossum (1)
MS2:	Lisa M. Nowak (1)
MS3:	Stephanie D. Wilson (1)
MS4:	Piers J. Sellers (2)
Up:	Thomas A. Reiter (2)

Payloads:	Up: 37,736 lbs	Down: 8,456 lbs
	MPLM (Leonardo) (20,999 lbs)	
	SRMS s/n 303	

Notes:	Piers Sellers replaced Carlos I. Noriega as MS4
	Scrubbed 01 Jul 2006 (KSC weather)
	Scrubbed 02 Jul 2006 (KSC weather)
	EVA1 08 Jul 2006 (Sellers and Fossum)
	EVA2 10 Jul 2006 (Sellers and Fossum)
	EVA3 12 Jul 2006 (Sellers and Fossum)

Wakeup Calls:

05 Jul	"Lift Every Voice and Sing"
06 Jul	"Daniel"
07 Jul	"Good Day Sunshine"
08 Jul	"God of Wonders"
09 Jul	"I Had a Dream"
10 Jul	"Clocks"
11 Jul	"All Star"
12 Jul	"I Believe I Can Fly"
13 Jul	"Theme from Charlie's Angels"
14 Jul	Aggie War Hymn (Texas A&M)
15 Jul	"Beautiful Day"
16 Jul	"Just Like Heaven"
17 Jul	"The Astronaut"

The crew patch depicts Discovery docked with the International Space Station in the foreground, overlaying the astronaut symbol with three gold columns and a gold star. The background shows the night-time Earth with a dawn breaking over the horizon. (NASA)

The stand-down after STS-114 lasted 11 months as the program struggled, again, to eliminate foam shedding from the external tank. This time, engineers completely eliminated the offending PAL ramps along with a myriad of smaller changes to the foam on the tank. The launch team scrubbed the first *Discovery* attempt on 1 July 2006 due to unacceptable weather around the launch site. A second attempt the following day was also scrubbed due to weather. The third attempt on 4 July proceeded smoothly.

This mission carried several new onboard cameras: each SRB contained a forward-looking camera located in the SRB-ET attach ring to view the orbiter wing leading edge and an aft-looking camera mounted in the forward skirt viewed most of the orbiter belly. The video from the new cameras was not telemetered to the ground but was recorded on the boosters for later retrieval. However, as with STS-114, a camera on the ET was broadcast live on NASA TV, allowing the public (and engineers) a spectacular view of launch. Immediately after tank separation, the crew used hand-held cameras to photograph the tank, supplementing the normal still and video cameras in the ET umbilicals. Unlike the previous two missions, engineers did not immediately see any foam debris during ascent.

Based on analysis of the imagery from the OBSS inspection and R-bar pitch maneuver, the mission management team identified six areas that required focused inspection. On flight day 4, the SSRMS unberthed the Leonardo MPLM from the payload bay and docked it to Node 1 (Unity). Later in the day, the crew began the focused inspections, marking the first operational use of the integrated sensor inspection system (ISIS) digital camera (IDC), which provided highly detailed imagery of the focused inspection areas.

Piers Sellers and Mike Fossum conducted the first EVA on flight day 5, installing the zenith integrated umbilical assembly blade blocker, rerouting the trailing umbilical system cable, and performing a loads test on the OBSS. During the second EVA Sellers and Fossum installed a pump on an external stowage platform and replaced the nadir trailing umbilical system reel assembly and nadir integrated umbilical assembly. After the EVA, George Bush called the crew, telling them that they represented the best of service and exploration, and thanking them for the job they were doing. Fossum and Sellers conducted the third EVA on flight day 9, testing a potential orbiter tile repair system and installing a grapple bar on the ISS.

On flight day 11 the crew unberthed Leonardo from the ISS, returned it to the payload bay, and conducted the late inspection prior to undocking to make sure there had not been any debris strikes. *Discovery* landed at KSC on the first opportunity.

Wearing shirts embroidered with a stylized version of the crew patch are, from the left, Mike Fossum, Piers Sellers, Lisa Nowak, Steve Lindsey, Mark Kelly, Stephanie Wilson, and Thomas Reiter. Sellers replaced Carlos Noriega due to an undisclosed medical condition. (NASA)

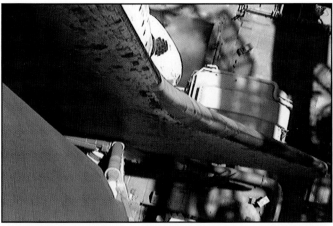

This was the first mission to carry cameras on the solid rocket boosters. This is the view from the camera mounted on the SRB forward skirt showing the orbiter belly; another camera looked forward from the SRB-ET attach ring at the bottom of each booster. (NASA)

This was the also first mission to carry the integrated sensor inspection system (ISIS) digital camera (IDC). This instrument on the end of the OBSS was used for focused inspections and provided high resolution monochrome images. Note the protruding gag filler at left. (NASA)

ET-119 being lowered into place between the two waiting solid rocket boosters in VAB High Bay 3. (Dennis R. Jenkins)

The Pegasus barge carrying ET-119 is being moved to the turn basin dock. The tank was rolled-out in Michoud for the first time on 28 May 2002 but this was a well-traveled tank, having been shipped back to Michoud twice for return-to-flight modifications. (Dennis R. Jenkins)

All three main engines are running but the sound suppression water has not yet started flowing. The external tank was more than four years old at this point, explaining its dark brown appearance. Note the broad black photo-reference band around the top of the left SRB. (NASA)

STS-115 (ISS-12A)

Mission:	116		NSSDC ID:		2006-036A
Vehicle:	OV-104 (27)		ET-118 (SLWT)		SRB: BI127
Launch:	LC-39B		09 Sep 2006		15:15 UTC
Altitude:	188 nm		Inclination:		51.60 degrees
Landing:	KSC-33		21 Sep 2006		10:22 UTC
Landing Rev:	186		Mission Duration:		283 hrs 07 mins

Commander:	Brent W. Jett, Jr. (4)
Pilot:	Christopher J. "Chris" Ferguson (1)
MS1:	Steven G. "Steve" MacLean (2)
MS2:	Daniel C. "Dan" Burbank (2)
MS3:	Joseph R. "Joe" Tanner (4)
MS4:	Heidemarie M. Stefanyshyn-Piper (1)

Payloads:	Up: 41,848 lbs	Down: 993 lbs
	P3/P4 Truss (35,677 lbs)	
	SRMS s/n 301	

Notes:	Scrubbed 27 Aug 2006 (KSC lightning)
	Rolled-back to VAB 29 Aug 2006 (Hurricane Ernesto)
	Rolled-back canceled halfway to VAB
	Delayed 06 Sep 2006 (fuel cell cooling pump)
	Scrubbed 08 Sep 2006 (LH$_2$ ECO sensor)
	EVA1 12 Sep 2006 (Tanner and Stefanyshyn-Piper)
	EVA2 13 Sep 2006 (Burbank and MacLean)
	EVA3 15 Sep 2006 (Tanner and Stefanyshyn-Piper)

Wakeup Calls:

10 Sep	"Moon River"
11 Sep	A solo cello performance
12 Sep	"My Friendly Epistle" (Ukrainian song)
13 Sep	"Takin' Care of Business"
14 Sep	"Wipe Out"
15 Sep	"Hotel California"
16 Sep	"Twelve Volt Man"
17 Sep	"Danger Zone"
18 Sep	"Rocky Mountain High"
19 Sep	"Ne Partez Pas Sans Mo"
	("Don't Leave Without Me")
20 Sep	"Beautiful Day"
21 Sep	"WWOZ"

The crew patch uses an ISS solar panel wing as the main element. As Atlantis launches toward the ISS, its trail depicts the astronaut symbol. The starburst, representing the power of the Sun, rises over the Earth and shines on the newly installed solar panel. (NASA)

The launch team scrubbed the *Atlantis* launch attempt on 27 August 2006 after lightning struck the LC-39B lightning mast and engineers observed a small power transient on the orbiter instrumentation, resulting in a 48-hour delay. In the meantime, Hurricane Ernesto had formed on 24 August in the eastern Caribbean and was heading up the east coast of Florida, so NASA decided to roll the stack back to the VAB. While rolling the stack back, updated weather forecasts predicted the storm would quickly dissipate, so mission management elected to return the stack to the launch pad. Ernesto passed central Florida as a weak tropical storm, but soon intensified and made landfall on the North Carolina coast just below hurricane strength.

The countdown began again on 3 September for a planned launch on 6 September. During the count, anomalies developed with the electrical supply to a fuel cell cooling pump and a Tyvek cover for the forward reaction control system module detached during a rainstorm. Engineers concluded neither presented a problem but still delayed launch until 8 September. While filling the ET, one of the LH$_2$ engine cutoff sensors indicated dry when it should have showed wet so the launch director scrubbed for 24 hours. The 8 September attempt was the 200th time an ET had been loaded at KSC. The count on 9 September proceeded smoothly with no unplanned holds.

On flight day 2, Chris Ferguson, Dan Burbank, and Steve MacLean used the SRMS and OBSS to perform the now-customary inspection of the orbiter while the rest of the crew prepared *Atlantis* for docking. After docking, Ferguson and Burbank used the SRMS to unberth the P3/P4 truss from the payload bay and maneuvered it for a handoff to the SSRMS. Oddly, this marked the first time a Canadian (MacLean) had operated the Canadarm2 (SSRMS) in space.

Joe Tanner and Heidemarie Stefanyshyn-Piper made the first EVA on flight day 4 to complete the installation of the P3/P4 truss. Also on flight day 4, the damage assessment team in Houston concluded that no focused inspections were required.

On flight day 5, Dan Burbank and Steve MacLean conducted the second EVA, with Maclean became only the second Canadian to perform a spacewalk, after Chris Hadfield. The third and final EVA came on flight day 7, by Joe Tanner and Heidemarie Stefanyshyn-Piper, mostly to remove the launch locks from the P3/P4 radiators.

Poor weather forecasts for the Shuttle Landing Facility delayed landing but the weather cleared sufficiently for *Atlantis* to come home on the first KSC landing opportunity on flight day 13. NASA considered this a night landing since it took place about 48 minutes before sunrise, marking the 21st night landing of the flight campaign.

The crew poses at the Shuttle Landing Facility after arriving in their T-38s. From the left are Steve MacLean, Dan Burbank, Chris Ferguson, Brent Jett, Heidemarie Stefanyshyn-Piper, and Joe Tanner. The T-38s allowed the rated pilots to maintain their proficiency. (NASA)

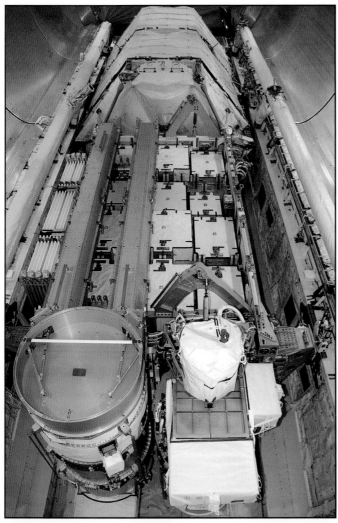

A full payload bay showing the P3/P4 truss. The hexagon object at the top is the solar alpha rotary joint (SARJ) while the integrated equipment assembly is in the middle. The round can at the bottom, and a similar one under the equipment next to it, contained the solar array wings. (NASA)

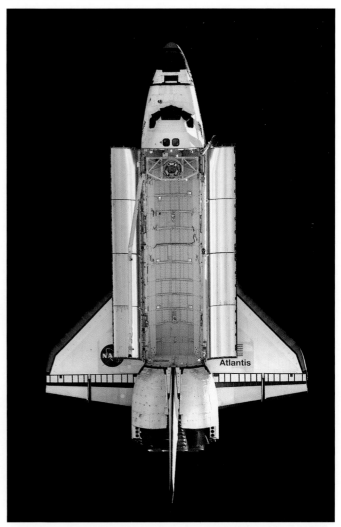

A mostly empty payload bay. The external airlock is at the front. The SRMS arm on the left is deployed while the OBSS arm on the right is stowed. Each payload bay door carried a full set of radiators but the forward radiators were not deployed on this mission. (NASA)

The hatch from the middeck of Atlantis into the external airlock. The hatch was hinged at the bottom and ground crews used a small block and tackle to open it since it was fairly heavy and the hinges were not designed to support it in 1-g. (Dennis R. Jenkins)

One of the crawler-transporters delivers MLP-2 to LC-39B. The crawler supported the MLP until it got to the pad then lowered it onto four large posts like the one shown here. There was also a smaller flip-down post on each side of the MLP for extra stability. (Dennis R. Jenkins)

STS-116 (ISS-12A.1)

Mission:	117		NSSDC ID:		2006-055A
Vehicle:	OV-103 (33)		ET-123 (SLWT)		SRB: BI128
Launch:	LC-39B		10 Dec 2006		01:48 UTC
Altitude:	192 nm		Inclination:		51.60 degrees
Landing:	KSC-15		22 Dec 2006		22:33: UTC
Landing Rev:	204		Mission Duration:		308 hrs 44 mins

Commander:	Mark L. "Roman" Polansky (2)
Pilot:	William A. "Bill" Oefelein (1)
MS1:	Nicholas J. M. Patrick (1)
MS2:	Robert L. "Beamer" Curbeam, Jr. (3)
MS3:	A. Christer Fuglesang (1)
MS4:	Joan E. Higginbotham (1)
Up:	Sunita P. L. "Suni" Williams (1)
Down:	Thomas A. Reiter (2)

Payloads:	Up: 35,690 lbs	Down: 806 lbs
	Spacehab LSM (11,903 lbs)	
	P5 Truss (4,031 lbs)	
	ICC-G (5,429 lbs)	
	SRMS s/n 303	

Notes:	Delayed 02 Dec 2006 (electrical transient)
	Scrubbed 07 Dec 2006 (KSC weather)
	Last LC-39B launch of the flight campaign
	EVA1 12 Dec 2006 (Curbeam and Fuglesang)
	EVA2 14 Dec 2006 (Curbeam and Fuglesang)
	EVA3 16 Dec 2006 (Curbeam and Williams)
	EVA4 18 Dec 2006 (Curbeam and Fuglesang)

Wakeup Calls:

10 Dec	"Here Comes the Sun"
11 Dec	"Beep Beep"
12 Dec	"Waterloo"
13 Dec	"Suavemente"
14 Dec	"Under Pressure"
15 Dec	"Low Rider"
16 Dec	"Fanfare for the Common Man"
17 Dec	"Beautiful Blue Danube"
18 Dec	"Good Vibrations"
19 Dec	"The Zamboni Song"
20 Dec	"Say You'll be Mine"
21 Dec	"The Road Less Traveled"
22 Dec	"Home for the Holidays"

The American and Swedish flags trail the orbiter, depicting the international composition of the crew. The seven stars of Ursa Major provide direction to the North Star, which is superimposed over the installation location of the P5 truss on ISS. (NASA)

During early pre-launch activities on 2 December 2006, *Discovery* experienced an electrical transient that exceeded the allowable voltage and duration. Engineers inspected all affected hardware and found no damage. The Air Force scrubbed the attempt on 7 December because the cloud cover over the Eastern Range violated range-safety requirements. The second attempt on 10 December proceeded smoothly with no unplanned holds. This was the last of 53 space shuttle launches from LC-39B, although two later launch-on-need (LON) missions briefly sat on the pad.

Nicholas Patrick used the SRMS to unberth the P5 truss segment on flight day 4. After slowly lifting it out of the payload bay, the SRMS handed off to the SSRMS operated by Joan Higginbotham. Afterward, Robert Curbeam and Christer Fuglesang conducted the first EVA, guiding P5 into its final position on the P4 truss. During the ISS P6 solar-array retraction, the array could not be fully retracted and stowed after multiple attempts. However, it retracted enough to allow controllers to activate the solar alpha rotary joint (SARJ) so the array could track the Sun.

On flight day 6, Robert Curbeam and Christer Fuglesang exited the Quest airlock for the second EVA. The crew members reconfigured the ISS power channel to reroute power from the P6 solar array to the P4 solar array installed by STS-115 and relocated the crew and equipment translation aid (CETA) cart. Flight controllers made several additional attempts to retract the P6 array, but none were successful. Robert Curbeam and Suni Williams conducted the third EVA on flight day 9, reconfiguring the main bus switching unit and moving the Zvezda Service Module debris panel from the payload bay to the ISS. The crew had additional time to troubleshoot the P6 array and managed to retract six additional bays of the solar array, leaving 11 bays extended.

The mission management team extended the mission one day and added a fourth EVA to finish retracting the P6 solar array. During the unplanned fourth EVA on flight day 10, Robert Curbeam and Christer Fuglesang finally successfully retracted the array.

Weather at KSC and Edwards concerned flight controllers, so NASA moved equipment and personnel to White Sands Space Harbor to support a possible landing, the first time this had been seriously contemplated since STS-3 in 1982. Because of the extra day at the ISS, consumables were running low and mission control activated Edwards, KSC, and White Sands. The orbiter would return to the site with the best weather, although nobody really wanted to land at White Sands. Fortunately, the weather cleared sufficiently to land at the Shuttle Landing Facility on the second opportunity.

This is the photo used for the STS-116 Space Flight Awareness poster distributed to the workforce. From the left are Joan Higginbotham, Nicholas Patrick, Bill Oefelein, Mark Polansky, Robert Curbeam, Christer Fuglesang, and Suni Williams. (NASA)

Christer Fuglesang prepares a meal at the galley on the middeck. Of interest is that this flight did not carry the middeck accommodations rack (MAR) that was normally to the left of the galley. Note the open door and curtain at the extreme left that led to the potty. (NASA)

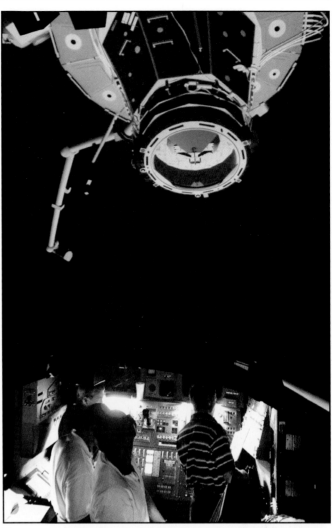

The crew participate in a simulation in the system engineering simulator in the Avionics Systems Laboratory at JSC. The facility included a complete aft flight deck and moving scenes of full-sized International Space Station components over a simulated Earth. (NASA)

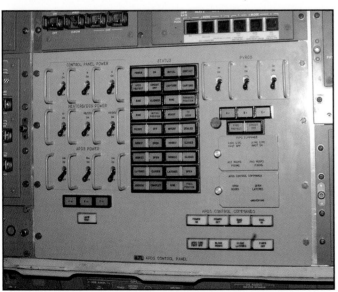

The APDS control panel on the aft flight deck. The Russian-provided panel looked completely different than any other control panel on the orbiter. Not only was it a different color (light green rather than gray), it used very different switches and indicators. (NASA)

The structural-carbon brakes installed after the Challenger accident had five rotors, six stators, and weighed 180 pounds. They had a maximum operating temperature of 2,100°F and could absorb up to 82 million foot-pounds of energy if needed. (Dennis R. Jenkins)

STS-117 (ISS-13A)

Mission:	118	NSSDC ID:	2007-024A	
Vehicle:	OV-104 (28)	ET-124 (SLWT)	SRB: BI129	
Launch:	LC-39A	08 Jun 2007	23:38 UTC	
Altitude:	181 nm	Inclination:	51.60 degrees	
Landing:	EDW-22	22 Jun 2007	19:51 UTC	
Landing Rev:	219	Mission Duration:	334 hrs 10 mins	

Commander:	Frederick W. "Rick" Sturckow (3)
Pilot:	Lee J. "Bru" Archambault (1)
MS1:	Patrick G. "Pat" Forrester (2)
MS2:	Steven R. "Swanny" Swanson (1)
MS3:	John D. "Danny" Olivas (1)
MS4:	James F. "JR" Reilly II (3)
Up:	Clayton C. "Clay" Anderson (1)
Down:	Sunita P. L. "Suni" Williams (1)

Payloads:	Up: 42,641 lbs	Down: 857 lbs
	S3/S4 Truss (35,677 lbs)	
	SRMS s/n 301	

Notes:	Delayed 16 Mar 2007 (ET hail damage)
	Rolled back to VAB 04 Mar 2007 (ET hail damage)
	EVA1 11 Jun 2007 (Reilly and Olivas)
	EVA2 13 Jun 2007 (Forrester and Swanson)
	EVA3 15 Jun 2007 (Reilly and Olivas)
	EVA4 17 Jun 2007 (Forrester and Swanson)
	Longest *Atlantis* mission
	Orbiter returned to KSC on 21 Aug 2005 (N905NA)

Wakeup Calls:

09 Jun	"Big Boy Toys"
10 Jun	"Riding the Sky"
11 Jun	"It Probably Always Will"
12 Jun	"What a Wonderful World"
13 Jun	"Questions 67 and 68"
14 Jun	"Indescribable"
15 Jun	"Radar Love"
16 Jun	University of Texas at El Paso Fight Song
17 Jun	Theme song from *Band of Brothers*
18 Jun	"Redeemer"
19 Jun	"Feelin' Stronger Every Day"
20 Jun	"If I Had $1,000,000"
21 Jun	"Makin' Good Time Coming Home"
22 Jun	The Marine Corps Hymn – "Halls of Montezuma"

Gold is used to highlight the portion of the ISS that will be installed by the crew. The two gold astronaut symbols, emanating from the "117" at the bottom of the patch, represent the concerted efforts of the shuttle and station programs toward the completion of the ISS. (NASA)

Hail from a thunderstorm on 26 February 2007 damaged the foam on the external tank from the liquid oxygen tank ogive to the aft interface hardware. The ground team removed the payload from *Atlantis* and stored it in the payload changeout room and then returned the stack to the VAB for repairs. They ultimately found more than 1,000 divots in the ET foam and 27 damaged tiles on *Atlantis*. After the repairs, the stack was rolled back to LC-39A and the launch on 8 June proceeded smoothly with no unplanned holds.

During the post-ascent survey on flight day 1, the crew reported a blanket on the left OMS pod was loose and extended upward about 5 inches. They downlinked the images to the damage assessment team in Houston and conducted a thorough inspection of the thermal protection system using the SRMS and OBSS.

On flight day 3, Rick Sturckow and Lee Archambault began chasing the ISS and then performed the R-bar pitch maneuver to allow the Expedition 15 crew to photograph the orbiter. After docking, Archambault and Pat Forrester unberthed the S3/S4 truss using the SRMS and handed it to the SSRMS for overnight parking. JR Reilly and Danny Olivas conducted the first EVA on flight day 4, releasing the launch restraints on the four boxes that housed the folded solar arrays and making preparations to activate the truss.

In Houston, the damage assessment team worried about the OMS pod. Based on a thermal analysis and the limited test data available for the graphite composite structure, the mission management team extended the mission two days to accommodate an EVA to repair the loose blanket.

On flight day 6, Pat Forrester and Steve Swanson conducted the second EVA to prepare the solar alpha rotary joints (SARJ) between the S3 and S4 trusses for rotation. By now, the mission management team had moved the OMS pod repair to the third EVA on flight day 8. Danny Olivas repaired the OMS pod while securely attached to the end of the SRMS, spending a little more than two hours on the task. Afterward, JR Reilly installed a vent for the oxygen generating system inside the Destiny laboratory. During the fourth EVA on flight day 10, Pat Forrester and Steve Swanson installed a camera on the S3 truss and removed the last six SARJ launch restraints.

Mission management waved-off both landing opportunities on flight day 14 because of bad weather at KSC and asked the crew to perform an orbit adjustment maneuver to allow an earlier Edwards opportunity on subsequent landing days if it was needed. The weather at the Shuttle Landing Facility was unacceptable for both opportunities on flight day 15, so mission management told the crew to land at Edwards. This was the longest mission for *Atlantis*.

The STS-117 and Expedition 15 crews in the Destiny laboratory. In the front, from the left, are Clay Anderson, Suni Williams, Fyodor Yurchikhin, and Oleg Kotov. In the back are Pat Forrester, Lee Archambault, Jim Reilly, Rick Sturckow, Steven Swanson, and Danny Olivas. (NASA)

Atlantis approaches the ISS. Docking occurred at 19:36 UTC on 10 June 2007. The docked Soyuz TMA-10 is at top. There was always a Soyuz docked at the ISS as an emergency crew return vehicle. (NASA)

The International Space Station as seen from Atlantis at the end of the mission. Lee Archambault was at the controls for the departure and fly-around, giving the crew a good look at the new configuration. (NASA)

Danny Olivas repairing a tear on an OMS pod blanket during the third EVA. During the post-ascent survey on flight day 1, the crew reported a blanket on the left OMS pod was loose and extended upward about 5 inches (left photo). They downlinked these initial images to the damage assessment team in Houston and conducted a more thorough inspection of the thermal protection system using the SRMS and OBSS the following day. The preliminary aero-thermal analysis in Houston indicated the entry thermal environment would result in localized temperatures that exceeded the certification limits of the OMS pod. Based on the thermal analysis, and the limited test data available for the graphite composite structure, the mission management team extended the mission two days to accommodate an EVA to repair the blanket (right photo). (NASA)

The first Shuttle Carrier Aircraft (N905NA) takes off from Edwards AFB with a full moon in the background. This was the last operational ferry flight by N905NA, although it would later deliver Discovery, Enterprise, and Endeavour to their final display sites. (NASA)

Atlantis in the Mate/Demate Device at Dryden. Note the cooling and purge units are still attached, evidenced by the trailers behind the orbiter and hoses connected to the T-0 umbilicals. Ground crews will soon install the tailcone needed for the ferry flight to KSC. (NASA)

STS-118 (ISS-13A.1)

Mission:	119	NSSDC ID:	2007-035A
Vehicle:	OV-105 (20)	ET-117 (SLWT)	SRB: BI130
Launch:	LC-39A	08 Aug 2007	22:37 UTC
Altitude:	186 nm	Inclination:	51.60 degrees
Landing:	KSC-15	21 Aug 2007	16:33 UTC
Landing Rev:	201	Mission Duration:	305 hrs 57 mins

Commander:	Scott J. Kelly (2)
Pilot:	Charles O. "Scorch" Hobaugh (2)
MS1:	Tracy E. "TC" Caldwell (1)
MS2:	Richard A. "Rick" Mastracchio (2)
MS3:	Dafydd R. "Dave" Williams (2)
MS4:	Barbara R. "Barby" Morgan (1)
MS5:	Benjamin A. "Alvin" Drew, Jr. (1)

Payloads:	Up: 37,390 lbs	Down: 316 lbs
	Spacehab LSM (11,580 lbs)	
	S5 Truss (4,030 lbs)	
	ICC-GD/ESP-3 (6,152 lbs)	
	SRMS s/n 201	

Notes: Clayton Anderson was replaced by Alvin Drew
First use of the SSPTS
Last flight of a Spacehab single module
EVA1 11 Aug 2007 (Mastracchio and Williams)
EVA2 13 Aug 2007 (Mastracchio and Williams)
EVA3 15 Aug 2007 (Mastracchio and Anderson)
EVA4 18 Aug 2007 (Williams and Anderson)

Wakeup Calls:

09 Aug	"Where My Heart Will Take Me"
10 Aug	"Mr. Blue Sky"
11 Aug	"Gravity"
12 Aug	"Up!"
13 Aug	"Outta Space"
14 Aug	"Happy Birthday Tracy"
15 Aug	"Good Morning World"
16 Aug	"Times Like These"
17 Aug	"Black Horse and the Cherry Tree"
18 Aug	"Learn to Fly"
19 Aug	"Teacher, Teacher"
20 Aug	"Flying"
21 Aug	"Homeward Bound"

The top of the gold astronaut symbol overlays the starboard S5 truss. The flame of knowledge represents the importance of education, and honors teachers everywhere. The seven white stars and the red maple leaf signify the American and Canadian crew members. (NASA)

Prior to the *Columbia* accident, NASA had manifested STS-118 as the 29th flight of *Columbia* and its first, and probably only, visit to the ISS. Technicians would have installed the *Endeavour* airlock in *Columbia* for the mission. After the accident, NASA changed STS-118 to *Endeavour*.

This mission included Barbara Morgan, who had trained as the backup to Christa McAuliffe as part of the Teacher in Space Project during 1985–86 before the *Challenger* accident. At the time, McAuliffe and Morgan were considered "space flight participants" and not mission specialists. After NASA canceled the Teacher in Space Project in 1990, Morgan continued to work with the agency until her selection "as a mission specialist and NASA's first educator astronaut," although she is generally considered a career astronaut and not part of a special class. Morgan completed two years of training in 2000 as a fully trained crew member that performed the same duties as any other astronaut.

As with all missions after the *Columbia* accident, much of flight day 2 was dedicated to inspecting the thermal protection system using the SRMS and OBSS. After reviewing the imagery, engineers in Houston directed the crew to perform a focused inspection on four tile locations and a frayed thermal barrier around one main landing-gear door. This inspection on flight day 5 revealed slight damage to several tiles; tests in Houston, including a full-temperature run in the arc jet facility, cleared the damage for entry. After the mission, NASA reported the damaged tiles had been removed in the OPF and engineers had found no evidence of heat-related damage to the orbiter structure beneath.

The mission delivered the S5 truss segment, an external stowage platform (ESP-3), and a replacement control moment gyroscope. The mission was also the final flight of a Spacehab single module. On flight day 3, *Endeavour* made the first use of the station-shuttle power transfer system (SSPTS) that allowed the orbiter to tap into the ISS power supply, thereby extending the amount of time it could remain docked to the station. NASA modified *Discovery* and *Endeavour* with the SSPTS, but *Atlantis* never received the proper hardware and continued to rely on her fuel cells while docked.

On flight day 11, Hurricane Dean was threatening Houston and the mission management team decided to shorten the mission one day. *Endeavour* undocked on flight day 12 and completed a late inspection of the RCC to ensure no damage from micrometeoroids or orbital debris. The weather at KSC was satisfactory so *Endeavour* deorbited on the first opportunity on flight day 14 and returned to the Shuttle Landing Facility.

Photoshopped in front of a photo of the launch pad are, from the left, Rick Mastracchio, Barbara Morgan, Charles Hobaugh, Scott Kelly, Tracy Caldwell, Dave Williams, and Alvin Drew. Clay Anderson was originally was slated to be launched to the ISS on this mission. (NASA)

Endeavour docked to PMA-2 attached to the forward port on the Destiny laboratory. The SRMS is deployed to the right and the SSRMS is on the left of the photo. Note the OBSS boom is stowed on the starboard payload bay sill and the Spacehab single module in the middle of the payload bay. The Ku-band antenna, in its communication mode, is deployed from the sill just behind the Endeavour name. (NASA)

Every crew practiced a myriad of emergency egress procedures at JSC. Techniques included using the escape pole (seen on the left side of the hatch opening), rappelling down Sky Genies, and using an airliner-like escape slide, shown here being shoved out the crew hatch. (NASA)

Somewhat more than twenty years after she was first selected to be a space flight participant as part of the Teacher in Space Project, Barbara Morgan, now a full-fledged astronaut, finally got to fly. Her initial selection was marred by the Challenger accident. (NASA)

STS-120 (ISS-10A)

Mission:	120	NSSDC ID:		2007-050A
Vehicle:	OV-103 (34)	ET-120 (SLWT)		SRB: BI131
Launch:	LC-39A	23 Oct 2007		15:38 UTC
Altitude:	188 nm	Inclination:		51.60 degrees
Landing:	KSC-33	07 Nov 2007		18:02 UTC
Landing Rev:	238	Mission Duration:		362 hrs 24 mins

Commander:	Pamela A. "Pam" Melroy (3)
Pilot:	George D. Zamka (1)
MS1:	Douglas H. "Wheels" Wheelock (1)
MS2:	Stephanie D. Wilson (2)
MS3:	Scott E. Parazynski (5)
MS4:	Paolo A. Nespoli (1)
Up:	Daniel M. Tani (2)
Down:	Clayton C. "Clay" Anderson (1)

Payloads:	Up: 40,872 lbs	Down: 1,577 lbs
	Node 2 (Harmony) (31,648 lbs)	
	SRMS s/n 202	

Notes: EVA1 26 Oct 2007 (Parazynski and Wheelock)
EVA2 28 Oct 2007 (Parazynski and Tani)
EVA3 30 Oct 2007 (Parazynski and Wheelock)
EVA4 03 Nov 2007 (Parazynski and Wheelock)

Wakeup Calls:

24 Oct	"Lord of the Dance"
25 Oct	"Dancing in the Moonlight"
26 Oct	"Rocket Man"
27 Oct	"Bellissime Stelle" ("Beautiful Stars")
28 Oct	"What a Wonderful World"
29 Oct	"One by One"
30 Oct	"Malaguena Salerosa"
31 Oct	"Nel Blu Dipinto di Blu"
01 Nov	"The Lion Sleeps Tonight"
02 Nov	"World"
03 Nov	"Theme from Star Wars"
04 Nov	"The Presence of the Lord"
05 Nov	"Roll Me Away"
06 Nov	"Space Truckin'"
07 Nov	"Chitty Chitty Bang Bang"

The star on the left represents the ISS; the red colored points represent the current location of the P6 solar array. During the mission, the crew will move P6 to the end of the port truss, represented by the gold points. The moon and Mars represent the future of NASA. (NASA)

During the final ice inspection on 23 October 2007, technicians noted clear ice with frost near the LH_2 umbilical. Ultimately, the mission management team approved a waiver since they expected the ice to fall off during SSME ignition. The countdown proceeded smoothly and *Discovery* was launched on the first attempt and the ice did, indeed, fall of as expected with no resulting damage.

On a lighter note, during the terminal countdown demonstration test (TCDT), astronaut Brian Duffy in jest called Pam Melroy "Pambo." In response, and beginning a thankfully short-lived tradition, the rest of the crew adopted "o" names ... "Zambo" Zamka, "Longbow" Parazynski, "Robeau" Wilson, "Flambo" Wheelock, Paolo "Rocky" Nespoli, and "Bo-Ichi" Tani. On another light note, the mission carried a Luke Skywalker Lightsaber to commemorate the 30th anniversary of Star Wars.

Pam Melroy and George Zamka rendezvoused with the ISS on flight day 3. Scott Parazynski and Douglas Wheelock conducted the first EVA the following day, removing a failed S-band antenna assembly from the ISS and stowing it in the payload bay, installing Node 2 (Harmony) on Node 1 (Unity), and preparing the P6 truss to move to the outboard location on the P5 truss.

Flight day 6 saw Scott Parazynski and Daniel Tani on the second EVA. Most significantly, they detached the P6 truss from the Z1 truss. The following day, the SSRMS handed off the P6 truss to the SRMS to allow the mobile transporter and SSRMS to translate to a different location on the ISS. The SRMS then handed the P6 truss back to the SSRMS. Meanwhile, the mission management team added an additional day of docked operations between the fourth and fifth EVAs to allow a longer spacewalk dedicated to inspecting a problematic starboard solar alpha rotary joint (SARJ).

Scott Parazynski and Douglas Wheelock conducted the third EVA on flight day 8, installing the P6 solar array. The crew successfully deployed one of the solar array wings but halted the other because of a tear in the solar array. Due to the anomaly on the starboard SARJ, mission management asked the spacewalkers to inspect the port SARJ, which appeared to be fine. On flight day 12, Parazynski and Wheelock made the fourth EVA, which included the first operational use of the OBSS as a work platform. In this case, the OBSS was attached to the SSRMS, not the SRMS.

The following day, the crew reported an debris strike to a center window; engineers on the ground concluded the damage was acceptable for entry. *Discovery* undocked from the ISS on flight day 14 and the crew closed the payload bay doors on flight day 16. *Discovery* returned to KSC on the first opportunity.

Looking in awe at the ISS, the crew poses for their Space Flight Awareness poster. From the left are Pam Melroy, George Zamka, Scott Parazynski, Doug Wheelock, Paolo Nespoli, Stephanie Wilson, and Daniel Tani. Clay Anderson was missing from the poster. (NASA)

A nice shot of ET-120 falling away from Discovery. Note the black scorching just in front of the intertank where the forward SRB separation motor exhaust impinges on the tank. The forward attach (bipod) is clearly visible as is the long 17-inch-diameter LO$_2$ feedline. (NASA)

An empty payload bay. The aft bulkhead shows a set of yellow handrails down the middle and the EVA winch to help a crew member close the payload bay doors, if needed, is at the top center of the bulkhead. A silver-insulated camera is on each side of the bulkhead. (NASA)

Scott Parazynski on the foot restraint at the end of the OBSS boom during the fourth EVA. This was the first operational use of the OBSS as a work platform. The OBSS is attached to the SSRMS at the mid-boom grapple fixture. Doug Wheelock is at the extreme left. (NASA)

Discovery lifts-off from LC-39A. Unusually, only a single bird appears in the photo; usually there are dozens or hundreds startled by the loud exhausts from the main engines and solid rocket boosters. This gives a good perspective of the layout of the launch pad, with the sound suppression water tower on the right and the fixed and rotating service structures on the left, adjacent to the mobile launch platform, that carried the space shuttle stack. The two pads were essentially identical. This was the sixty-seventh space shuttle launch from LC-39A. (NASA)

STS-122 (ISS-1E)

Mission:	121	NSSDC ID:	2008-005A
Vehicle:	OV-104 (29)	ET-125 (SLWT)	SRB: BI132
Launch: Altitude:	LC-39A 185 nm	07 Feb 2008 Inclination:	19:45 UTC 51.60 degrees
Landing: Landing Rev:	KSC-15 202	20 Feb 2008 Mission Duration:	14:08 UTC 306 hrs 22 mins

Commander:	Stephen N. "Steve" Frick (2)
Pilot:	Alan G. "Dex" Poindexter (1)
MS1:	Leland D. "Lee" Melvin (1)
MS2:	Rex J. Walheim (2)
MS3:	Hans W. Schlegel (2)
MS4:	Stanley G. "Stan" Love (1)
Up:	Léopold Eyharts (2)
Down:	Daniel M. "Dan" Tani (2)

Payloads:	Up: 40,296 lbs Down: 2,162 lbs Columbus Laboratory (26,627 lbs) ICC-L (4,548 lbs) SRMS s/n 301
Notes:	Scrubbed 06 Dec 2007 (LH$_2$ ECO sensor) Scrubbed 09 Dec 2007 (LH$_2$ ECO sensor) Tanking test 18 Dec 2007 (LH$_2$ only) EVA1 11 Feb 2008 (Walheim and Love) EVA2 13 Feb 2008 (Walheim and Schlegel) EVA3 15 Feb 2008 (Walheim and Love)

Wakeup Calls:

08 Feb	"The Book of Love"
09 Feb	"Tishomingo Blues"
10 Feb	"Maenner"
11 Feb	"Fly Like an Eagle"
12 Feb	"Dream Come True"
13 Feb	"Oysters and Pearls"
14 Feb	"Consider Yourself at Home"
15 Feb	"Marmor Stein und Eisen Bricht"
16 Feb	"I Believe I Can Fly"
17 Feb	"Hail Thee, Harvey Mudd"
18 Feb	"Over the Rainbow" / "What a Wonderful World"
19 Feb	"Always Look on the Bright Side" (*Spamalot*)
20 Feb	"Hail to the Spirit of Liberty"

The sailing ship denotes the travels of the early expeditions from the east to the west. A little more than 500 years after Columbus sailed to the new world, Atlantis will bring the European Columbus laboratory module to the International Space Station. (NASA)

NASA scrubbed the *Atlantis* launch attempt on 6 December 2007 during ET loading when three of four LH$_2$ engine cutoff (ECO) sensors failed. A similar anomaly caused the scrub of the second attempt on 9 December. The mission management team deferred the next launch attempt until after the first of the year while a troubleshooting team isolated the cause of the ECO failures. In the meantime, NASA conducted a dedicated tanking test on 18 December in an attempt to isolate the cause of the problem. Engineers finally believed they understood the problem and NASA rescheduled the launch for February. The countdown on 7 February proceeded smoothly with no unplanned holds.

On flight day 2 the crew unberthed the SRMS and OBSS and conducted the thermal protection system survey, taking slightly less than four hours to accomplish the task. The crew also photographed two OMS pod blankets that appeared to have minor damage. Steve Frick and Alan Poindexter began the rendezvous maneuvers on flight day 3, and performed the normal R-bar pitch maneuver as they approached the station. The mission delivered the European Columbus laboratory module to the station, along with the Biolab, EDR (European drawer rack), EPM (European physiology module), and FSL (fluid science laboratory) payloads. *Atlantis* also carried three green starting flags provided by NASCAR in recognition of the 50th running of the Daytona 500 on 17 February 2008.

On flight day 5, Rex Walheim and Stan Love conducted the first EVA. During the spacewalk, they retrieved the power data grapple fixture from the sidewall of the payload bay and installed it on the Columbus module. Lee Melvin then used the SSRMS to unberth Columbus from the payload bay and install it on the Node 2 (Harmony) starboard common berthing mechanism.

Rex Walheim and Hans Schlegel made the second EVA on flight day 7, successfully replacing the nitrogen tank assembly and securing the laboratory micrometeoroid and orbital debris (MMOD) shield. Later in the day, the mission management team added an additional day between the third EVA and undocking. After a day of rest, Rex Walheim and Stan Love completed the third on flight day 9. This EVA installed the external payload facility sun monitoring experiment (SOLAR) and the European technology exposure facility (EuTEF) on Columbus and transferred a failed control moment gyro (CMG) from the ISS to the orbiter for return to Earth. Steve Frick and Alan Poindexter undocked on flight day 12 and conducted a fly-around of the station. The weather at KSC was good and the deorbit burn occurred on the first landing opportunity.

In front of the Astrovan that transported crews from the crew quarters in the Operations and Checkout Building to the launch pad are, from the left, Léopold Eyharts, Stan Love, Hans Schlegel, Rex Walheim, Lee Melvin, Alan Poindexter, and Steve Frick. (NASA)

Atlantis approaching the ISS to dock at PMA-2, which was now attached to the forward port of Node 2 (Harmony). The ODS can be seen on top of the external airlock in the payload bay. By this point the Ku-band antenna was being used for communications, not as a radar. (NASA)

Steve Frick (right) and Alan Poindexter look out the starboard window (pilot's side) of Atlantis while it is docked to the ISS. The Ku-band antenna partially obscures the "A" in Atlantis. Note the various textures of the thermal protection system materials. (NASA)

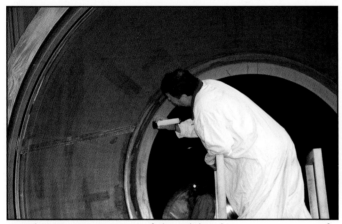

A technicians inspects one of the solid rocket motor segments for cracks in the propellant. Each segment was carefully inspected after it arrived at KSC to ensure it was not damaged during the train ride from Utah. The motors burned from the large center hole outward. (NASA)

After the boosters were cleaned and disassembled in Hangar AF, they were placed on rail cars for the trip back to Utah so they could be reloaded with propellant. Note the special covers that protected the segments during the trip across the country. (NASA)

The Pegasus barge transits from Port Canaveral through the Canaveral Locks (and drawbridge) into the Banana River carrying ET-125. A commercial tug is towing Pegasus since the SRB retrieval ships could not adequately control it in the shallow waters of the river. (NASA)

"Management Row" in Firing Room 4 at the KSC Launch Control Center. From the right are NASA administrator Mike Griffin, associate administrator Bill Gerstenmaier, and Wayne Hale. Space shuttle program manager Bill Parsons is standing at the back. (NASA)

STS-123 (ISS-1J/A)

Mission:	122	NSSDC ID:	2008-009A
Vehicle:	OV-105 (21)	ET-126 (SLWT)	SRB: BI133
Launch:	LC-39A	11 Mar 2008	06:28 UTC
Altitude:	186 nm	Inclination:	51.60 degrees
Landing:	KSC-15	27 Mar 2008	00:41 UTC
Landing Rev:	249	Mission Duration:	378 hrs 11 mins

Commander:	Dominic L. P. "Dom" Gorie (4)
Pilot:	Gregory H. "Box" Johnson (1)
MS1:	Robert L. "Bob" Behnken (1)
MS2:	Michael J. "Mike" Foreman (1)
MS3:	Richard M. "Rick" Linnehan (4)
MS4:	Takao Doi (2)
Up:	Garrett E. Reisman (1)
Down:	Léopold Eyharts (2)

Payloads:	Up: 38,915 lbs	Down: 4,891 lbs
	JEM-ELM/PS (18,377 lbs)	
	Spacelab pallet/Dextre (7,683 lbs)	
	SRMS s/n 201	

Notes:	Delayed 14 Feb 2008 (manifest)
	EVA1 14 Mar 2008 (Linnehan and Reisman)
	EVA2 15 Mar 2008 (Linnehan and Foreman)
	EVA3 17 Mar 2008 (Linnehan and Behnken)
	EVA4 20 Mar 2008 (Behnken and Foreman)
	EVA5 22 Mar 2008 (Behnken and Foreman)

Wakeup Calls:

11 Mar	"Linus & Lucy"
12 Mar	Music from *Godzilla Versus Space Godzilla*
13 Mar	"Saturday Night"
14 Mar	"Turn! Turn! Turn!"
15 Mar	"We're Going to be Friends"
16 Mar	"God of Wonders"
17 Mar	"Sharing the World"
18 Mar	"Hoshi Tsumugi no Uta" (Japanese song)
19 Mar	"Burning Love"
20 Mar	"Blue Sky"
21 Mar	"Enter Sandman"
22 Mar	"I Loved Her First"
23 Feb	"I Am Free"
24 Feb	"Furusato" (Japanese song)
25 Feb	"Con Te Partiro" (Italian song)
26 Feb	"Drops of Jupiter"

The crew patch depicts the space shuttle on-orbit with the crew names trailing behind. The ISS is shown in the configuration the crew will encounter when they arrive and the major components added by this mission are both illustrated. (NASA)

The original launch target date was 14 February 2008 but after the delay of STS-122, *Endeavour* launched on 11 March. The ET separation camera used new flash units located in the left umbilical well to better illuminate the tank after separation, which engineers thought, "were dramatic and effective."

The mission delivered the first part of the Japanese Experiment Module (JEM, Kibō), specifically, the experiment logistics module pressurized section (ELM-PS). *Endeavour* also carried the Canadian special purpose dexterous manipulator (SPDM, Dextre) robotics system on a modified Spacelab pallet. As originally designed, the pallets were non-deployable payloads, but several were modified with flight releasable grapple fixtures late in the program to allow them to be removed from the payload bay while on-orbit.

Endeavour successfully docked with the ISS on flight day 3. The following day, Rick Linnehan and Garrett Reisman conducted the first EVA to install Dextre after the SSRMS lifted the pallet out of the payload bay while being observed by cameras on the SRMS. A bit later, the SRMS successfully unberthed the ELM-PS from the payload bay and moved it to an interim location on Node 2 (Harmony). The pressurized section would be moved to its permanent location on the Kibō pressurized module after the next space shuttle mission delivered module it.

Rick Linnehan and Mike Foreman conducted the second EVA on flight day 6 to assemble the Dextre manipulator followed on flight day 8 by Linnehan and Bob Behnken installing a spare parts platform, cameras, and tool handling assembly on Dextre during the third EVA. The SSRMS then grappled the Spacelab pallet that had carried Dextre and returned it to the payload bay. Behnken and Foreman conducted the fourth EVA on flight day 11, replacing a failed remote power control module on the Z1 truss.

Operations during the next space shuttle mission presented a quandary for planners. The Kibō pressurized module (JEM-PM), the largest single ISS module, interfered with carrying the OBSS on *Discovery* during STS-124. The solution was to leave the *Endeavour* OBSS on ISS during this mission and have STS-124 retrieve it for the late thermal protection system inspections and return it to Earth. During the fifth EVA on flight day 13, Bob Behnken and Mike Foreman transferred the OBSS from *Endeavour* to the ISS.

Endeavour undocked from the ISS on flight day 15 after a 30-minute delay to latch the ISS solar panels. On flight day 17, flight controllers waved-off the first landing opportunity at KSC due to weather. Fortunately, things cleared up in Florida and *Endeavour* returned to the Shuttle Landing Facility on the second opportunity.

Awaiting the start of a training session in the Space Vehicle Mockup Facility at JSC are, from the left, Mike Foreman, Garrett Reisman, Bob Behnken, Takao Doi, Rick Linnehan, Greg Johnson, and Dom Gorie. *Endeavour* would bring Léopold Eyharts home from the ISS. (NASA)

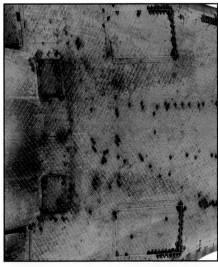

Three photos from the R-bar pitch maneuver as seen by the Expedition 16 crew aboard the ISS. At right is a shot of the nose landing-gear doors and forward ET attach point. In the middle, the ET umbilical doors and main landing-gear doors. At right, the SSME and OMS nozzles. (NASA)

Sitting in VAB High Bay 1, ET-126 shows the general configuration of a super lightweight tank. Most of the lighter, almost yellow, foam was shaved by hand to a specific shape, eliminating the "rind" that turned dark orange over time on the acreage foam. (NASA)

As designed, Spacelab pallets were supposed to remain in the payload bay, but late in the flight campaign several were modified with flight releasable grapple fixtures, trunnion scuff plates, and handrails so they could be removed from the payload bay while on-orbit. (NASA)

The Air Force Thunderbirds demonstration team fly over LC-39A in celebration of NASA's 50th anniversary. The aircraft had flown earlier at the Daytona 500, also commemorating its 50th anniversary, and agreed to fly over KSC on their way to their next show. (U.S. Air Force)

STS-124 (ISS-1J)

Mission:	123		NSSDC ID:		2008-027A
Vehicle:	OV-103 (35)		ET-128 (SLWT)		SRB: BI134
Launch:	LC-39A		31 May 2008		21:02 UTC
Altitude:	191 nm		Inclination:		51.60 degrees
Landing:	KSC-15		14 Jun 2008		15:16 UTC
Landing Rev:	217		Mission Duration:		330 hrs 14 mins

Commander:	Mark E. Kelly (3)
Pilot:	Kenneth T. "Hock" Ham (1)
MS1:	Karen L. Nyberg (1)
MS2:	Ronald J. "Ron" Garan, Jr. (1)
MS3:	Michael E. "Mike" Fossum (2)
MS4:	Akihiko "Aki" Hoshide (1)
Up:	Gregory E. Chamitoff (1)
Down:	Garret E. Reisman (1)

Payloads:	Up: 41,997 lbs	Down: 1,608 lbs
	JEM-PM (Kibō) (32,558 lbs)	
	SRMS s/n 202	

Notes:	The pad flame trench was damaged during launch
	Stephen Bowen was replaced by Ron Garan
	EVA1 03 Jun 2008 (Fossum and Garan)
	EVA2 05 Jun 2008 (Fossum and Garan)
	EVA3 08 Jun 2008 (Fossum and Garan)

Wakeup Calls:

01 Jun	"Your Wildest Dreams"
02 Jun	"Away from Home"
03 Jun	"Hold Me with the Robot Arm"
04 Jun	"Have You Ever"
05 Jun	"Fly Away"
06 Jun	"Bright as Yellow"
07 Jun	"Taking Off"
08 Jun	"Theme from The Mickey Mouse Club"
09 Jun	"The Spirit of Aggieland"
10 Jun	"All Because of You"
11 Jun	"Centerfield"
12 Jun	"Crystal Frontier"
13 Jun	"Baby, Won't You Please Come Home"
14 Jun	"Life on an Ocean Wave"

The Japanese flag is depicted on the JEM pressurized module and Kibō (Hope) is written in Japanese at the bottom of the patch. The view of the Sun shining down upon the Earth represents the increased "hope" that the entire world will benefit from JEM's discoveries. (NASA)

Despite several minor issues during processing and the subsequent countdown, *Discovery* launched on the first attempt on 31 May 2008. The east wall of the flame trench collapsed just after SRB ignition, resulting in a shower of bricks that severely damaged the perimeter fence 1,500 feet away.

NASA initially assigned Stephen Bowen to the crew but moved him to STS-126 to allow this mission to rotate an ISS crew member. Bowen would have performed the EVAs along with Mike Fossum, but Ron Garan took his place.

Montreal-born Gregory Chamitoff brought three bags of Fairmount sesame seed bagels, marking the first time this food had been in space; no word on the cream cheese. Also flying with the STS-124 crew was a Buzz Lightyear action figure, a yellow jersey from the Tour de France, the backup jersey Eli Manning took to the Super Bowl, and the last jersey that baseball great Craig Biggio wore in a game. Perhaps most importantly, the orbiter carried replacement parts in a middeck locker for a malfunctioning potty on the ISS. The crew had been using much less convenient waste-disposal methods until the new parts could be installed on the Zvezda Service Module.

On flight day 2, the crew performed the thermal protection system survey using only the SRMS since the OBSS could not be carried because of interference issues with the Japanese experiment module (JEM, Kibō) pressurized module (JEM-PM). A few wing leading edge upper carrier panels could not be captured in the field of view because of the limited reach of the SRMS, but these would be photographed during the R-bar pitch maneuver on flight day 3.

Mike Fossum and Ron Garan conducted the first EVA on flight day 4, servicing a solar alpha rotary joint (SARJ) and preparing the JEM-PM for installation. Using the SSRMS, Karen Nyberg and Akihiko Hoshide removed the JEM-PM from the payload bay and berthed it to Node 2 (Harmony). Later, the SSRMS handed the OBSS that had been left at the ISS by STS-123 to the SRMS. The crew would use the OBSS to take high-resolution imagery of five RCC panels on the right wing on flight day 8.

Mike Fossum and Ron Garan completed the second EVA on flight day 6. The following day, the crew moved the ELM-PS delivered by STS-123 to its permanent location on the JEM-PM. On flight day 9, Fossum and Garan conducted the final spacewalk, removing insulation and launch restraints from the JEM-RMS and collecting a sample of debris from the port SARJ.

Mark Kelly and Ken Ham undocked from the ISS on flight day 12 and performed a fly-around of the station. Mission control cleared *Discovery* to land at KSC on the first opportunity.

This cropped image was used in the "STS-124 and the Order of Discovery" poster that was distributed to the space shuttle workforce. From the left are Karen Nyberg, Ron Garan, Aki Hoshide, Mark Kelly, Greg Chamitoff, Mike Fossum, and Ken Ham. (NASA)

Ken Ham (left) and Mike Fossum on the aft flight deck after undocking from the ISS. Both are dressed a bit more formally than was normal for the crews, indicating they might have been preparing for a televised press conference or an educational classroom visit. (NASA)

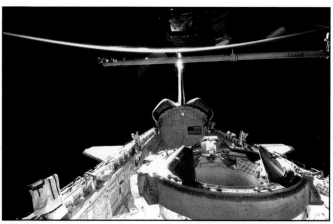

The view out of the aft flight deck windows at the orbiter docking system. The orbiter boom sensor system arm is at the top of the photo, being held by the SRMS that is slightly out of frame at the right. Of note is the reflection off the glass and a nice glow at the Earth's limb. (NASA)

Discovery going uphill. The solid rocket boosters burned so bright that they washed out any camera also capable of capturing natural light elements such as the orbiter. Some cameras used special filters to allow engineers to look at the exhaust plumes. (NASA)

Discovery touches down at the Shuttle Landing Facility. Mark Kelly set the orbiter down at 209 knots some 2,158 feet past the threshold. The total rollout was 9,354 feet. All about average for a landing. (NASA)

A worker examines damage to the east wall of the flame trench. Hot gases had penetrated the lining system, destroying more than 3,000 of the refractory bricks. The damage cost $2.5 million to repair. The "fire bricks" had been installed when the pad was built in 1965. (NASA)

STS-126 (ISS-ULF2)

Mission:	124	NSSDC ID:	2008-059A
Vehicle:	OV-105 (22)	ET-129 (SLWT)	SRB: BI136
Launch:	LC-39A	15 Nov 2008	00:56 UTC
Altitude:	193 nm	Inclination:	51.60 degrees
Landing:	EDW-04T	30 Nov 2008	21:26 UTC
Landing Rev:	250	Mission Duration:	380 hrs 29 mins

Commander:	Christopher J. "Chris" Ferguson (2)
Pilot:	Eric A. Boe (1)
MS1:	Donald R. "Don" Pettit (2)
MS2:	Stephen G. "Steve" Bowen (1)
MS3:	Heidemarie M. Stefanyshyn-Piper (2)
MS4:	Robert S. "Shane" Kimbrough (1)
Up:	Sandra H. "Sandy" Magnus (2)
Down:	Gregory E. Chamitoff (1)

Payloads:	Up: 39,471 lbs	Down: 19,436 lbs
	MPLM (Leonardo) (28,105 lbs)	
	LMC (MPESS) (3,296 lbs)	
	SRMS s/n 201	

Notes:	
	EVA1 18 Nov 2008 (Stefanyshyn-Piper and Bowen)
	EVA2 20 Nov 2008 (Stefanyshyn-Piper and Kimbrough)
	EVA3 22 Nov 2008 (Stefanyshyn-Piper and Bowen)
	EVA4 24 Nov 2008 (Bowen and Kimbrough)
	Only mission to land on EDW-04T (temporary)
	Orbiter returned to KSC on 12 Dec 2008 (N911NA)

Wakeup Calls:

15 Nov	"Shelter"
16 Nov	"Start Me Up"
17 Nov	"London Calling"
18 Nov	"City of Blinding Lights"
19 Nov	"Fanfare for the Common Man"
20 Nov	"Summertime"
21 Nov	"Unharness Your Horses, Boys" (Ukrainian song)
22 Nov	"You Are Here"
23 Nov	"Can't Take My Eyes Off of You"
24 Nov	"Can't Stop Loving You"
25 Nov	"Fever"
26 Nov	"North Sea Oil"
27 Nov	"Hold on Tight"
28 Nov	"In the Meantime"
29 Nov	"Twinkle, Twinkle, Little Star"
30 Nov	"The Rocky Theme – Gonna Fly Now"

Near the center of the crew patch, the constellation Orion reflects the goals of the manned spaceflight program, returning to the Moon and on to Mars, the red planet, which are also shown. The sunburst powers all these efforts through the solar arrays of the ISS. (NASA)

There was a last minute issue with the white room closeout door not being fully secured, *Endeavour* was launched on 14 November 2008 with no delays. The ascent imagery of the port T-0 umbilical area showed what engineers initially thought to be a missing FRSI blanket, so the damage assessment team requested additional imagery once on-orbit. During the RCC survey, the crew zoomed onto the port T-0 umbilical carrier panel and determined the FRSI was in place. There were no other possible concerns.

Chris Ferguson and Eric Boe rendezvoused and docked with the ISS on flight day 3. This mission coincided with the 10th anniversary of the International Space Station. In that time, 27 space shuttles, 30 Progress, 17 Soyuz, and 1 ATV had visited the outpost. Including the STS-126 crew, 167 people from 14 nations had docked at the station, which had orbited the Earth 57,509 times, for a total distance of more than 1,300 million miles.

On flight day 4, Don Pettit and Robert Kimbrough began maneuvering the Leonardo multi-purpose logistics module out of the payload bay and onto the nadir docking port on Node 2 (Harmony). Pettit, who served on the station for about five months during Expedition 6, led the transfer operations. The crew also began experiments on the orbiter middeck, including one that involved observing spiders and butterflies in space. Developed by school children in Colorado, Florida, and Texas, the experiment compared spider webs created in microgravity with those on Earth. The crew moved Leonardo back into the payload bay on flight day 12 in preparation to depart the station.

Unusually while on-orbit, the crew closed the star tracker doors to prevent contamination while they cleaned the starboard SARJ during the EVAs. Heidemarie Stefanyshyn-Piper and Steve Bowen conducted the first EVA on flight day 5, with Piper becoming the first female lead spacewalker. While she was preparing to begin work on the SARJ, Piper noticed a significant amount of grease in her tool bag. While she was removing the grease using a dry-wipe, one of her tool bags floated away. The bag floated aft and starboard of the ISS and did not pose a risk to the station or orbiter although it created a bit more space debris to be avoided in the future. After taking an inventory of the items in the lost bag, engineers determined Bowen had a duplicate set of tools and the two could share equipment, extending the EVA duration slightly but completing all of the assigned tasks.

Mission control waved-off both KSC landing opportunities on flight day 17 and decided to send *Endeavour* to Edwards on the first opportunity the following day.

Several variations of this Photoshopped crew portrait existed, some with the crew patch in various locations, others without. From the left are Sandy Magnus, Steve Bowen, Don Pettit, Chris Ferguson, Eric Boe, Shane Kimbrough, and Heidemarie Stefanyshyn-Piper. (NASA)

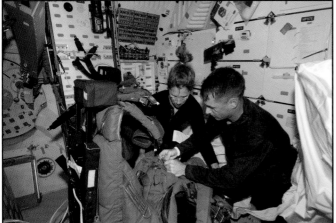

Once on-orbit, the crew quickly removed their S1035 advanced crew escape suits since they were bulky, hot, and somewhat uncomfortable. At top is Heidemarie Stefanyshyn-Piper giving a thumbs-up while Sandy Magnus and Shane Kimbrough work on their suits at bottom. (NASA)

Atlantis as STS-125 on LC-39A in the foreground and Endeavour as STS-400 on LC-39B in the background, with a rainbow between them. Endeavour would be launched as STS-126 after STS-125 encountered delays. NASA moved the Atlantis stack to LC-39A for launch. (NASA)

A slightly distorted wide-angle view of Endeavour making her way from OPF-2 to VAB High Bay 1. Endeavour was originally moved to LC-39B as the STS-400 launch-on-need (LON) rescue mission for STS-125. Every mission after the Columbia accident, except STS-135, had a LON mission being prepared. At first the rescue vehicle was physically moved to the launch pad; later the vehicle waited in the OPF or VAB. With both Atlantis and Endeavour on the pads, it was the eighteenth time that two orbiters were on the launch pads at the same time. (NASA)

STS-119 (ISS-15A)

Mission:	125	NSSDC ID:	2009-012A
Vehicle:	OV-103 (36)	ET-127 (SLWT)	SRB: BI135
Launch:	LC-39A	15 Mar 2009	23:44 UTC
Altitude:	189 nm	Inclination:	51.60 degrees
Landing:	KSC-15	28 Mar 2009	19:15 UTC
Landing Rev:	202	Mission Duration:	307 hrs 31 mins

Commander:	Lee J. "Bru" Archambault (2)
Pilot:	Dominic A. "Tony" Antonelli (1)
MS1:	Joseph M. "Joe" Acaba (1)
MS2:	Steven R. "Swanny" Swanson (2)
MS3:	Richard R. "Ricky" Arnold (1)
MS4:	John L. Phillips (3)
Up:	Koichi Wakata (3)
Down:	Sandra H. "Sandy" Magnus (2)

Payloads:	Up: 39,088 lbs	Down: 1,279 lbs
	S6 Truss (31,112 lbs)	
	SRMS s/n 202	

Notes:	Delayed 12 Feb 2009 (MPS control valves)
	Scrubbed 11 Mar 2009 (GUCP leak)
	EVA1 19 Mar 2009 (Swanson and Arnold)
	EVA2 21 Mar 2009 (Swanson and Acaba)
	EVA3 23 Mar 2009 (Arnold and Acaba)

Wakeup Calls:

16 Mar	"Free Bird"
17 Mar	"Radio Exercise"
18 Mar	"I Walk the Line"
19 Mar	"Que Bandera Bonita"
20 Mar	"Box of Rain"
21 Mar	"In a Little While"
22 Mar	"Alive Again"
23 Mar	"Ain't Nobody Here But Us Chickens"
24 Mar	"Andrew's Song"
25 Mar	"Dirty Water"
26 Mar	"Enter Sandman"
27 Mar	"Bright Side of the Road
28 Mar	"I Have a Dream""

The shape of the crew patch comes from the appearance of an ISS solar array viewed at an angle. The gold solar array of the ISS highlights the main payload. The seventeen white stars on the patch represent the crews of Apollo 1, Challenger, and Columbia. (NASA)

NASA originally scheduled this launch of *Discovery* for 12 February 2009, but postponed it on 3 February to allow additional time to investigate an MPS flow control valve anomaly that occurred on STS-126. These valves synchronized the flow of gaseous hydrogen between the ET and SSMEs. Engineers subsequently replaced the suspect valves with units that had less flight time. The launch team scrubbed the first attempt on 11 March after detecting excessive hydrogen concentrations around the ground umbilical carrier plate (GUCP) and rescheduled it for four days later. A similar leak would be seen during preparations for STS-127.

The revised launch date forced NASA to reduce the mission timeline because of the planned Soyuz TMA-14 launch on 26 March. To avoid any on-orbit conflicts, *Discovery* needed to undock no later than 25 March, eliminating a planned fourth EVA. During the second launch attempt, excessive hydrogen levels were again detected, but engineers adjusted the GUCP seal and the concentrations returned to allowable levels. Oddly, the ground controllers noted a free-tailed bat clinging to the ET during tanking, but engineers believed it would fly away before it could become debris and hit the orbiter. They were wrong and infrared cameras showed the bat was still clinging to the tank well after liftoff.

Flight day 5 saw Steven Swanson and Ricky Arnold exit the Quest airlock for the first EVA of the mission. Once the spacewalkers were in position, John Phillips and Koichi Wakata used the SSRMS to maneuver the S6 truss into its final position so that Swanson and Arnold could bolt the truss into place.

Steven Swanson and Joe Acaba conducted the second EVA on flight day 7, loosening bolts, installing foot restraints, and preparing tools so the STS-127 crew could more easily change out the P6 truss batteries. They also installed a second GPS antenna on Kibō, photographed the radiator panels extended from the P1 and S1 trusses, and reconfigured some cables on the Z1 truss.

On flight day 9, Ricky Arnold and Joe Acaba conducted the third EVA, assisting the SRMS to relocate the crew equipment translation aide cart from the P1 truss to the S1 truss, installing a new coupler on the CETA cart, and lubricating the SSRMS.

NASA waved-off the first KSC landing opportunity on flight day 14 because of weather near the Shuttle Landing Facility. The clouds cleared for the second opportunity. This was the first mission where *Discovery* carried a boundary layer transition experiment using a special tile on the bottom of the port wing. The experiment would be repeated on three additional *Discovery* flights, as well as one on *Endeavour*, albeit with a slightly different tile each time.

One of the variations of the photo used for the STS-119 Space Flight Awareness poster. Clockwise from lower left are Koichi Wakata, John Phillips, Steve Swanson, Lee Archambault, Tony Antonelli, Joe Acaba, and Ricky Arnold. (NASA)

The ISS as seen by the STS-119 crew on their way home. All of the solar array wings were now installed and the station appears mostly complete, with only a few smaller components left to be delivered before space shuttle was retired. (NASA)

KSC center director Bob Cabana watches the launch of Discovery from Firing Room 4 in the Launch Control Center. Before the Columbia accident, launches were conducted from Firing Rooms 1 and 3; beginning with STS-121, launches exclusively used the upgraded FR4. (NASA)

Looking back at the empty payload bay. The SRMS is still deployed at the top while the OBSS boom is stowed on the starboard payload bay sill at the bottom. Both arms say Canada on them. (NASA)

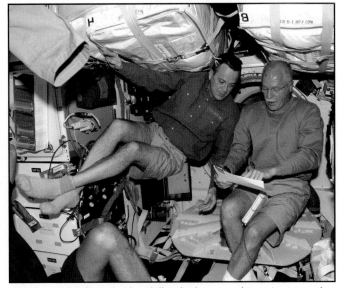

Ricky Arnold (left) and John Phillips look over a data printout as they float on the middeck of Discovery. Phillips appears to be sitting on the airlock hatch, but is in realty floating slightly above it. Note the large cargo bags secured to the ceiling above their heads. (NASA)

STS-125 (HST-SM4)

Mission:	126	NSSDC ID:	2009-025A
Vehicle:	OV-104 (30)	ET-130 (SLWT)	SRB: BI137
Launch:	LC-39A	11 May 2009	18:02 UTC
Altitude:	303 nm	Inclination:	28.45 degrees
Landing:	EDW-22	24 May 2009	15:40 UTC
Landing Rev:	197	Mission Duration:	309 hrs 39 mins

Commander:	Scott D. "Scooter" Altman (4)
Pilot:	Gregory C. "Ray-J" Johnson (1)
MS1:	Michael T. "Bueno" Good (1)
MS2:	K. Megan McArthur (1)
MS3:	John M. Grunsfeld (5)
MS4:	Michael J. "Mike" Massimino (2)
MS5:	Andrew J. "Drew" Feustel (1)

Payloads:	Up: 32,418 lbs	Down: 3,893 lbs
	Hubble Servicing Mission (22,076 lbs)	
	SRMS s/n 301	

Notes: Advanced 12 May 2009 (manifest)
EVA1 14 May 2009 (Grunsfeld and Feustel)
EVA2 15 May 2009 (Massimino and Good)
EVA3 16 May 2009 (Grunsfeld and Feustel)
EVA4 17 May 2009 (Massimino and Good)
EVA5 18 May 2009 (Grunsfeld and Feustel)
Orbiter returned to KSC on 02 Jun 2009 (N911NA)

Wakeup Calls:

12 May	"Kryptonite"
13 May	"Upside Down"
14 May	"Stickshifts and Safetybelts"
15 May	"God of Wonders"
16 May	"Hotel Cepollina"
17 May	"New York State of Mind"
18 May	"Sound of Your Voice"
19 May	"Lie in Our Graves"
20 May	"Theme from Star Trek"
21 May	"Cantina Band" (Star Wars)
22 May	"The Galaxy Song" (Monty Python)
23 May	"Where My Heart Will Take Me"
24 May	"The Ride of the Valkyries"

The overall composition of the universe is shown in blue and filled with planets, stars, and galaxies. The black background is indicative of the mysteries of dark-energy and dark-matter. The red border of the crew patch represents the red-shifted glow of the early universe. (NASA)

This *Atlantis* mission went through more gyrations than most during the planning phase, mostly because the revised flight rules issued after the *Columbia* accident dictated all flights go to the International Space Station. Obviously, this one would go to a much different orbit and, therefore, not have the services offered by the ISS such as the R-bar pitch maneuver damage assessment or the safe haven for the crew. NASA administrator Mike Griffin approved the mission on 31 October 2006. Ultimately, NASA scheduled the mission for 12 May 2009, but during the flight readiness review on 30 April 2009, NASA decided to advance the launch one day. The count on 11 May proceeded smoothly with no unplanned holds. This was the first *Atlantis* mission in 14 years not to visit a space station, the last being STS-66 in 1994.

Scott Altman and Greg Johnson rendezvoused with Hubble on flight day 3. The nominal height OMS-4 burn was so precise that two of the four planned midcourse corrections were not required. Communications between the orbiter and Hubble could not be established during the rendezvous because of a ground error. This resulted in the final Hubble roll maneuver not being performed, forcing Scott Altman and Greg Johnson to perform a manual fly-around to achieve the proper capture orientation for Megan McArthur to use the SRMS to grapple the telescope.

Atlantis carried two new instruments, the cosmic origins spectrograph (COS) and the wide field camera (WFC-3). The mission also replaced a fine guidance sensor, six gyroscopes, and two battery unit modules to allow the telescope to continue to function at least through 2014. The mission included five EVAs to service the telescope, with John Grunsfeld and Drew Feustel making three of them, and Mike Massimino and Mike Good making the other two.

Megan McArthur released the telescope from the SRMS on flight day 9. Mike Good and Mike Massimino were standing by to perform a spacewalk in the event that something went wrong during the deployment, but all went to plan.

After evaluating the weather, mission control elected to bring *Atlantis* home one revolution early to try to avoid the possibility of showers that would prevent a KSC landing. Nevertheless, mission control waved-off both opportunities on flight day 12 and both opportunities the following day. The weather at Shuttle Landing Facility had not improved on flight day 14 and mission management waved-off the first KSC opportunity. On the next revolution mission management worked both the KSC and Edwards opportunities until 15 minutes prior to the deorbit maneuver and decided to go to Edwards since the weather at KSC was still uncertain.

A photo from one of the dramatic crew posters the Space Shuttle Program began issuing toward the end of the flight campaign. From the left are Drew Feustel, Megan McArthur, Greg Johnson, Scott Altman, Mike Massimino, Mike Good, and John Grunsfeld. (NASA)

What appears to be a number of astronauts, reflected by the shiny surface on the exterior of the Hubble Space Telescope, is actually only two—John Grunsfeld (red stripes) and Drew Feustel during the first of five extravehicular activities to repair Hubble. (NASA)

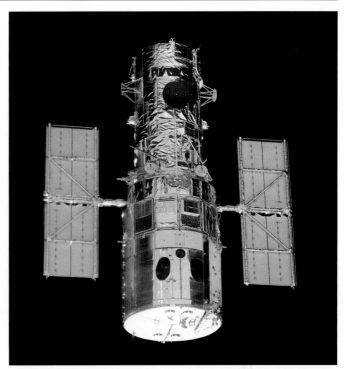

NASA initially canceled the fifth Hubble servicing mission following the Columbia accident. However, after spirited public discussion, NASA administrator Mike Griffin ultimately approved SM4. The telescope is still operating and NASA believes it could remain in service until 2030. (NASA)

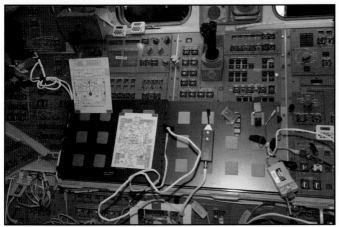

The aft flight deck shows a blue aluminum work surface covering part of the instrument panel. Given that Atlantis was not equipped with an orbiter docking system on this mission (see photo at right), covering its control panel did not matter much. (NASA)

A view from the end of the SRMS looking toward the forward bulkhead. The Ku-band antenna is deployed to the right and the OBSS is stowed on the starboard payload bay sill. Note the triangular tool stowage assembly on the side of the airlock truss structure. (NASA)

Mike Massimino collects something out the starboard tool stowage assembly (TSA) during the fourth EVA. This mission carried a TSA on each side of the external airlock truss. (NASA)

Perched in a foot restraint at the end of the SRMS arm, Drew Feustel picks up his tools during the first EVA. Part of John Grunsfeld's arm is visible at bottom frame. (NASA)

Mike Good is seen through an aft flight deck window during the second EVA. The photograph also (barely) captures a reflection of Megan McArthur in Atlantis. (NASA)

STS-127 (ISS-2J/A)

Mission:	127	NSSDC ID:	2009-038A
Vehicle:	OV-105 (23)	ET-131 (SLWT)	SRB: BI138
Launch:	LC-39A	15 Jul 2009	22:03 UTC
Altitude:	188 nm	Inclination:	51.60 degrees
Landing:	KSC-15	31 Jul 2009	14:49 UTC
Landing Rev:	249	Mission Duration:	376 hrs 46 mins

Commander:	Mark L. "Roman" Polansky (3)
Pilot:	Douglas G. "Doug" Hurley (1)
MS1:	Christopher J. "Chris" Cassidy (1)
MS2:	Julie Payette (2)
MS3:	Thomas H. "Tom" Marshburn (1)
MS4:	David A. Wolf (4)
Up:	Timothy L. "Tim" Kopra (1)
Down:	Koichi Wakata (3)

Payloads:	Up: 36,253 lbs	Down: 9,756 lbs
	JEM-EF (8,420 lbs)	
	JEM-ES (5,408 lbs)	
	ICC-VLD (8,699 lbs)	
	SRMS s/n 201	

Notes:
Scrubbed 13 Jun 2009 (GUCP leak)
Scrubbed 17 Jun 2009 (GUCP leak)
Delayed 11 Jul 2009 (KSC weather)
Scrubbed 12 Jul 2009 (KSC weather)
Scrubbed 14 Jul 2009 (KSC weather)
EVA1 18 Jul 2009 (Wolf and Kopra)
EVA2 20 Jul 2009 (Wolf and Marshburn)
EVA3 22 Jul 2009 (Wolf and Cassidy)
EVA4 24 Jul 2009 (Marshburn and Cassidy)
EVA5 27 Jul 2009 (Marshburn and Cassidy)

Wakeup Calls:

16 Jul	"These Are Days"
17 Jul	"Here Comes the Sun"
18 Jul	"Home"
19 Jul	"Learning to Fly"
20 Jul	"Theme from Thunderbirds" television show
21 Jul	"Life Is a Highway"
22 Jul	"Santa Monica"
23 Jul	"Tiny Dancer"
24 Jul	"Wish You Were Here"
25 Jul	"In Your Eyes"
26 Jul	"Dixit Dominus"
27 Jul	"On the Sunny Side of the Street"
28 Jul	"Proud to Be an American"
29 Jul	"Yellow"
30 Jul	"I Got You Babe"
31 Jul	"Beautiful Day"

The blue Earth without boundaries reminds us that we all share this world. In the center, the golden flight path of the space shuttle turns into the three distinctive rays of the astronaut symbol culminating in the star-like emblem characteristic of the Japanese Space Agency. (NASA)

The launch team scrubbed the *Endeavour* launch attempt on 13 June 2009 because of a gaseous hydrogen leak at the ET ground umbilical carrier plate (GUCP), similar to the fault that delayed STS-119 a couple of months earlier. NASA and the Air Force agreed to a second launch attempt on 17 June, only a day before the scheduled launch of the Lunar Reconnaissance Orbiter (LRO) on an Atlas V booster, which could not be moved.

NASA scrubbed that attempt when the launch team again detected leaking hydrogen from the GUCP. Engineers determined the ET carrier assembly and its seals were installed incorrectly. Due to conflicts with the LRO launch and a beta angle constraint, the next launch opportunity was 11 July. On 1 July, the launch team performed a tanking test to verify the modified GUCP seals no longer leaked. Because of lightning strikes near LC-39A, including one that hit the lightning mast on 10 July, NASA scrubbed the third launch attempt after instrumentation recorded energy levels that exceeded safety criteria. The launch team scrubbed an attempt on 12 July due to lightning and generally bad weather. The fifth attempt, on 14 July, was scrubbed due to yet more bad weather. A fuel cell anomaly during the count on 15 July almost scrubbed the launch, but engineers determined the unit would operate satisfactorily and *Endeavour* finally got off the ground.

When *Endeavour* docked with the ISS on flight day 3, it set a record for the most people in space in the same vehicle at 13. This was the last of three flight dedicated to the assembly of the Japanese Kibō laboratory complex and the main objective was to install the exposed facility (JEM-EF).

During the first EVA, on flight day 4, David Wolf and Tim Kopra prepared the attach points on the JEM-PM and Mark Polansky and Julie Payette then installed the JEM-EF using the SRMS along with Doug Hurley and Koichi Wakata using the SSRMS. *Endeavour* also brought up the exposed section (JEM-ES), a payload carrier that temporarily attached to the exposed facility. Three days after the astronauts installed the JEM-EF, they used the SRMS to unberth the JEM-ES from the payload bay and handed it to the SSRMS, which repositioned it near the Kibō complex.

Endeavour undocked from the ISS on flight day 14 and the crew deployed three satellites: CAPE/ANDE-2 (canister for all payload ejections/atmospheric neutral density experiment) and two DRAGONSat (dual radio frequency autonomous global positioning system on-orbit navigator satellite) experiments. Weather at the Shuttle Landing Facility was good and *Endeavour* returned on the first opportunity.

The official crew portrait. From the left are Dave Wolf, Chris Cassidy, Doug Hurley, Julie Payette, Mark Polansky, Tom Marshburn, and Tim Kopra. Although carrying a Japanese payload, the crew included Payette, who represented the Canadian Space Agency. (NASA)

An unusual view of the pad courtesy of the Ice Team that performs the final visual inspection of the vehicle just before launch. Note the vapors at the base of the orbiter; propellants are flowing in the vehicle. The crew cabin access arm is deployed against the crew hatch in case the crew needs to get out in a hurry. The red water "sausages" are evident in each SRB exhaust duct, something added after STS-1 to absorb some of the acoustical energy released by the boosters. The hydrogen vent arm is connected to the ground umbilical carrier plate on the ET at left. (NASA)

STS-128 (ISS-17A)

Mission:	128	NSSDC ID:	2009-045A	
Vehicle:	OV-103 (37)	ET-132 (SLWT)	SRB: BI139	
Launch:	LC-39A	29 Aug 2009	04:00 UTC	
Altitude:	192 nm	Inclination:	51.60 degrees	
Landing:	EDW-22	12 Sep 2009	00:55 UTC	
Landing Rev:	219	Mission Duration:	332 hrs 54 mins	

Commander: Frederick W. "Rick" Sturckow (4)
Pilot: Kevin A. Ford (1)
MS1: Patrick G. "Pat" Forrester (3)
MS2: José M. Hernández (1)
MS3: John D. "Danny" Olivas (2)
MS4: A. Christer Fuglesang (2)
Up: Nicole M. P. Stott (1)
Down: Timothy L. "Tim" Kopra (1)

Payloads: Up: 40,605 lbs   Down: 19,130 lbs
MPLM (Leonardo) (26,744 lbs)
LMC (MPESS) (3,920 lbs)
SRMS s/n 202

Notes: Scrubbed 25 Aug 2009 (KSC weather)
Scrubbed 26 Aug 2009 (MPS LH$_2$ valve)
Delayed 28 Aug 2009 (troubleshooting)
EVA1 01 Sep 2009 (Olivas and Stott)
EVA2 03 Sep 2009 (Olivas and Fuglesang)
EVA3 05 Sep 2009 (Olivas and Fuglesang)
Orbiter returned to KSC on 21 Sep 2009 (N911NA)
Last operational ferry flight

Wakeup Calls:

29 Aug	"Back in the Saddle Again"
30 Aug	"Made to Love"
31 Aug	"Mi Tierra"
01 Sep	"Indiana, Our Indiana"
02 Sep	"What a Wonderful World"
03 Sep	"There is a God"
04 Sep	"What a Wonderful World"
05 Sep	"El Hijo del Pueblo"
06 Sep	"Rocket"
07 Sep	"Only One"
08 Sep	"Beautiful Day"
09 Sep	"Sailing"
10 Sep	"Good Day Sunshine"
11 Sep	"Big Boy Toys"

Earth and the International Space Station wrap around the astronaut symbol, reminding us of the continuous human presence in space. The names of the STS-128 crew members border the crew patch in an unfurled banner that contains American and Swedish flags. (NASA)

The *Discovery* attempt on 25 August 2009 was scrubbed due to thunderstorms near LC-39A. Engineers canceled a second attempt the following day because the LH$_2$ inboard fill-and-drain valve did not respond as expected. After draining the ET, engineers cycled the valve and it appeared to operate normally. However, just before the start of tanking on 27 August, the mission management team delayed the launch another 24 hours to allow engineers to better understand the problem. There were no unplanned holds during the 29 August launch.

During flight day 4, the crew berthed the Leonardo MPLM to the nadir port on Node 2 (Harmony) using the SSRMS, then opened the hatch and began transferring cargo to the ISS.

The following day, Danny Olivas and Nicole Stott conducted the first EVA using the Quest airlock. While the spacewalk was taking place, other crew members transferred the C.O.L.B.E.R.T. (combined operational load bearing external resistance treadmill) exercise equipment and the Node 3 (Tranquility) air revitalization system rack to the ISS. Danny Olivas and Christer Fuglesang conducted the second EVA on flight day 7, installing a new ammonia tank assembly and fitting protective lens covers on the SSRMS wrist and elbow cameras. Two days later, Olivas and Fuglesang conducted the third EVA, installing two GPS antennas, deploying the S3 payload attach system, installing a new rate gyro assembly, and rerouting some Node 3 cables. From the inside, the ISS crew repaired the common berthing mechanism to ensure the capture of the Japanese H-II Transfer Vehicle (HTV) that was scheduled to dock at the ISS later that year.

Discovery undocked from the ISS on flight day 12 and the crew performed the normal ISS fly-around. Mission control delayed closing the payload bay doors since the weather at the Shuttle Landing Facility was forecast to be unacceptable for both opportunities on the first landing day. The weather did not look any better at KSC for the second day, so mission control elected to land at Edwards. *Discovery* landed on 11 September, marking the 54th and last space shuttle landing at the California site. The last operational ferry flight began on 20 September when N911NA and *Discovery* departed Edwards. The pair made fuel stops at Amarillo, Texas and NAS Fort Worth JRB before remaining overnight at Barksdale AFB. The vehicle arrived at the Shuttle Landing Facility on 21 September.

As part of the boundary layer transition experiment begun on STS-119, *Discovery* also carried two tiles covered with a catalytic coating engineers wanted to use on the Orion capsule. These instrumented tiles were located downstream of the BLT tile and suffered the full effects of the turbulent boundary layer during entry.

From the mission's safety poster. From the left are Pat Forrester, José Hernández, Christer Fuglesang, Rick Sturckow, Nicole Stott, Kevin Ford, and Danny Olivas. Stott was the last ISS crew member to ride on space shuttle (up on this mission; down on STS-129). (NASA)

The second Shuttle Carrier Aircraft (N911NA) departs Edwards AFB on the last operational ferry flight. The two 747s were essentially identical, but could be identified by the number of windows on the upper deck; two for N905NA and five for N911NA. (NASA)

The Mate/Demate Device at KSC was similar, but not identical, to the one at Dryden. The two were designed and built by different contractors to the same specification, and also made some slight concessions to the differing weather conditions at the two locations. (NASA)

After the tailcone was attached, workers raised the orbiter using two cranes in the MDD. Note the yellow "sling" that attached to the orbiter so the cranes could raise it without damaging the fragile TPS. (NASA)

Workers towed the 747 under the raised orbiter and then lowered the orbiter onto the attach fittings and secured them. The attach fittings were the same ones that normally attached to the external tank. (NASA)

Returning Discovery to KSC would be the last operational ferry flight, although there would later be ferry flights to deliver the orbiters to their display sites. Here Discovery is in the Mate/Demate Device at Dryden, being readied to be loaded onto the Shuttle Carrier Aircraft. (NASA)

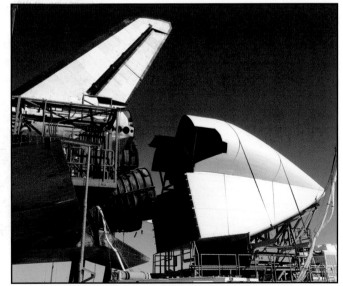

An essential part of the ferry preparations was installing the tailcone over the base area of the orbiter. Partly, this protected the main engine and OMS nozzles, but mostly it smoothed the airflow coming off the orbiter so it did not damage the tail of the 747. (NASA)

STS-129 (ISS-ULF3)

Mission:	129	NSSDC ID:	2009-062A
Vehicle:	OV-104 (31)	ET-133 (SLWT)	SRB: BI140
Launch:	LC-39A	16 Nov 2009	19:28 UTC
Altitude:	186 nm	Inclination:	51.60 degrees
Landing:	KSC-33	27 Nov 2009	14:44 UTC
Landing Rev:	171	Mission Duration:	259 hrs 17 mins

Commander:	Charles O. "Scorch" Hobaugh (3)
Pilot:	Barry E. "Butch" Wilmore (1)
MS1:	Leland D. "Lee" Melvin (2)
MS2:	Randolph J. "Komrade" Bresnik (1)
MS3:	Michael J. "Mike" Foreman (2)
MS4:	Robert L. "Bobby" Satcher, Jr. (1)
Down:	Nicole M. P. Stott (1)

Payloads:	Up: 38,893 lbs	Down: 1,176 lbs
	ELC-1 (14,101 lbs)	
	ELC-2 (13,528 lbs)	
	SRMS s/n 301	

Notes:	Last space shuttle crew rotation flight
	EVA1 19 Nov 2009 (Foreman and Satcher)
	EVA2 21 Nov 2009 (Foreman and Bresnik)
	EVA3 23 Nov 2009 (Satcher and Bresnik)

Wakeup Calls:

17 Nov	"I Can Only Imagine"
18 Nov	"Higher Ground"
19 Nov	"In Wonder"
20 Nov	"We Are Family"
21 Nov	"Voyage to Atlantis"
22 Nov	"Butterfly Kisses"
23 Nov	"Space Rise"
24 Nov	The Marine Corps Hymn – "Halls of Montezuma"
25 Nov	"Amazing Grace"
26 Nov	"Fly Me to the Moon"
27 Nov	"Home Sweet Home"

Atlantis is vividly silhouetted by the Sun highlighting how brightly the orbiters have performed over the past three decades. A space shuttle ascends on the astronaut symbol portrayed by the red, white, and blue swoosh bounded by the gold halo. (NASA)

The November 2009 window was bound by the launch of the mini-Russian research module (MRM-2, Poisk) on a Soyuz-U booster from Baikonur Cosmodrome on 10 November and Eastern Range launches of an Atlas V on 14 November and a Delta IV on 19 November. The Soyuz launched on time and the Atlas scrubbed, allowing *Atlantis* an opportunity. The 16 November 2009 countdown proceeded smoothly with no unplanned holds.

Robert Thirsk was initially scheduled to return from the ISS on this mission, but NASA worried possible delays could cause his stay to exceed the desired six months. Instead, Nicole Stott and Thirsk swapped rides, with Thirsk returning on Soyuz TMA-15 and Stott aboard STS-129. The marked the last ISS crew member to return on space shuttle. Randolph Bresnik carried a multicolored scarf worn by Amelia Earhart since his grandfather, Albert Louis Bresnik, had been her personal photographer from 1932 until she disappeared on 2 July 1937. Upon landing, the scarf was returned to The Ninety-Nines Museum of Women Pilots in Oklahoma City.

Charles Hobaugh and Barry Wilmore rendezvoused and docked with the ISS on flight day 3. The following day, Mike Foreman and Bobby Satcher conducted the first EVA, installing an S-band antenna and two space-to-ground antenna cable bundles on the Z1 truss, and performing a myriad of small tasks such as lubricating several items and securing wire bundles and MMOD shields.

The second spacewalk took place on flight day 6, with Mike Foreman and Randolph Bresnik installing the AIS (automatic identification system) and ARISS (amateur radio on the international space station) antennas on Columbus, among other tasks. On flight day 8, Bobby Satcher and Bresnik conducted the third and final EVA, installing a high-pressure gas tank on the Quest airlock and a MISSE experiment on ELC-2.

Charles Hobaugh and Barry Wilmore performed a 27-minute reboost maneuver on flight day 9. During docked operations, the crew transferred 2,211 pounds of material from the middeck to the ISS and 2,110 pounds from the ISS to *Atlantis*. In addition, the mission delivered 27,468 pounds of material from the payload bay and installed 1,100 pounds to return to Earth. The crew also transferred 1,389 pounds of water in 13 contingency water containers and 4 potable water reservoirs.

After undocking, Charles Hobaugh and Barry Wilmore performed the normal ISS fly-around and then conducted several separation burns in preparation for deorbit. The weather at KSC was good and *Atlantis* returned to the Shuttle Landing Facility on the first flight day 12 opportunity.

The flight crew speak after landing while Nicole Stott is examined by the flight docs after her 87-day stay aboard the International Space Station. From the left are Lee Melvin, Mike Foreman, Charlie Hobaugh, Bobby Satcher, Barry Wilmore, and Randy Bresnik. (NASA)

Bobby Satcher during the third EVA. Satcher and Randy Bresnik (out of frame) removed a pair of micrometeoroid and orbital debris shields from the Quest airlock during the 5-hour 42-minute spacewalk. (NASA)

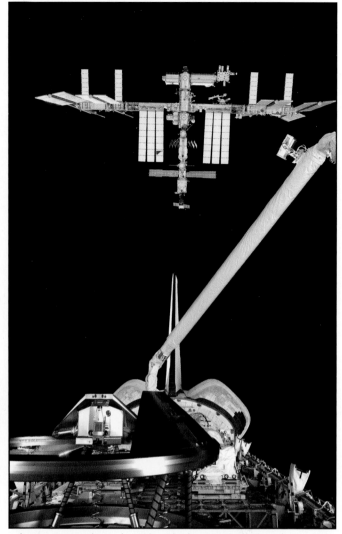

Atlantis approaching the ISS with the SRMS deployed. Note the extended docking ring on the orbiter docking system in the foreground, ready to mate with its counterpart on PMA-2. (NASA)

The cluttered flight deck of Atlantis during flight day 2. Somewhat unusually, the crew does not have the window shades installed. Note the laptop in the center of the overhead panel. Oddly, there does not seem to be a roll of gray tape anywhere in the photo. (NASA)

The future revealed; or not. The Ares I-X demonstrator launches from LC-39B on 28 October 2009 while Atlantis sits on LC-39A. The Constellation Program was the heir-apparent to space shuttle, but it was ultimately canceled before the flight campaign ended. (NASA)

STS-130 (ISS-20A)

Mission:	130		NSSDC ID:		2010-004A
Vehicle:	OV-105 (24)		ET-134 (SLWT)		SRB: BI141
Launch:	LC-39A		08 Feb 2010		09:14 UTC
Altitude:	190 nm		Inclination:		51.60 degrees
Landing:	KSC-15		22 Feb 2010		03:20 UTC
Landing Rev:	217		Mission Duration:		330 hrs 08 mins

Commander:	George D. "Zambo" Zamka (2)
Pilot:	Terry W. Virts, Jr. (1)
MS1:	Kathryn P. "Kay" Hire (2)
MS2:	Stephen K. "Steve" Robinson (4)
MS3:	Nicholas J. Patrick (2)
MS4:	Robert L. "Bob" Behnken (2)

Payloads:	Up: 40,956 lbs	Down: 1,262 lbs
	Node 3 (Tranquility) (33,367 lbs)	
	SRMS s/n 201	

Notes:	Scrubbed 07 Feb 2010 (KSC weather)
	EVA1 12 Feb 2010 (Behnken and Patrick)
	EVA2 14 Feb 2010 (Behnken and Patrick)
	EVA3 17 Feb 2010 (Behnken and Patrick)

Wakeup Calls:

08 Feb	"Give Me Your Eyes"
09 Feb	"Katmandu"
10 Feb	"Also Sprach Zarathustra"
11 Feb	"Beautiful Day"
12 Feb	"Theme from Firefly" television show
13 Feb	"Too Much Stuff"
14 Feb	"Forty Years On"
15 Feb	"Parabola"
16 Feb	"Window on the World"
17 Feb	"Oh Yeah"
18 Feb	"I'm Gonna Be (500 Miles)"
19 Feb	"In Wonder"
20 Feb	"The Distance"
21 Feb	The Marine Corps Hymn – "Halls of Montezuma"

The shape of the crew patch represents the Cupola. The image of Earth was the first photograph of our home planet taken from the moon by Lunar Orbiter I on 23 August 1966, symbolizing a future destination for manned space flight. (NASA)

Even prior to the start of tanking for the first launch attempt, mission management scrubbed the 7 February 2010 attempt because of unfavorable weather at the launch site. Weather forecasters initially predicted a 70 percent chance of favorable weather on 8 February, but that degraded to 30 percent in the hours before the planned launch due to low clouds. Nevertheless, *Endeavour* lifted-off while the ISS was over western Romania.

George Zamka and Terry Virts docked with the ISS on flight day 3. Based on the results of the RPM photography, the damage assessment team in Houston decided there were no requirements for a focused inspection. However, three areas required additional evaluation: a protruding flipper door seal, a protruding ceramic insert between windows 1 and 2, and some failed repair material on a cracked tile immediately aft of a window (W1). The engineers eventually satisfied themselves these were not problems and the mission management team cleared *Endeavour* for entry.

The mission delivered Node 3 (Tranquility) and its seven-window Cupola to the ISS, adding 2,600 cubic feet to the usable volume of the station. During the first EVA, Nicholas Patrick and Bob Behnken prepared Tranquility for its move out of the payload bay. From inside, Kay Hire and Terry Virts used the SRMS to position Tranquility on the port side of Node 1 (Unity) and secured it using 16 remotely controlled bolts. The spacewalkers then began connecting the new module to the ISS systems. Patrick and Behnken used their second EVA, on flight day 7, to further integrate Tranquility to the station and prepare for the relocation of the cupola.

On flight day 8, Kay Hire and Terry Virts used the SRMS to relocate the cupola from its transport location on the end of Tranquility to its permanent location on the side of the module. The crew then used the SSRMS to relocate a docking adapter from Node 2 (Harmony) to where the copula had been.

During the third EVA, Nicholas Patrick and Bob Behnken prepared the copula window covers to be opened and finished exterior work on Tranquility and the relocated docking port. As part of the dedication ceremony, the crew placed a moon rock from Apollo 11 inside the copula. Scott Parazynski, veteran of five space shuttle flights and seven spacewalks, had carried the same rock during his climb to the summit of Mount Everest.

On flight day 11 the crews received a congratulatory phone call from Barack Obama. The following day, they closed the hatches between the spacecraft and prepared for undocking. Despite not-so-favorable weather predictions, *Endeavour* returned to the Shuttle Landing Facility on the first opportunity during flight day 15.

The STS-130 Space Flight Awareness poster. From the left are Bob Behnken, Steve Robinson, George Zamka, Terry Virts, Kay Hire, and Nicholas Patrick. Note the different objects the crew is holding; various power tools, cameras, and a guitar. (NASA)

Liberty Star *towing the right solid rocket booster up the Banana River to Hangar AF. Liberty hipped her booster on the right side; Freedom Star hipped on the left side. (NASA)*

One of the space shuttle main engines on its way from the SSME Processing Facility to the OPF to be installed in Endeavour. It is being carried by a Hyster forklift. (NASA)

Like the payload bay doors, the OBSS (shown here) and SRMS arms did not really like 1-g conditions and had to be supported by large yellow ground support equipment. (NASA)

Endeavour, atop MLP-2, rolls-out of High Bay 1 on her way to LC-39A. At this point the orbiter still has the mounting pads for the payload bay door strongbacks installed (note the small yellow objects). The payload bay doors could not be opened in 1-g without support. (NASA)

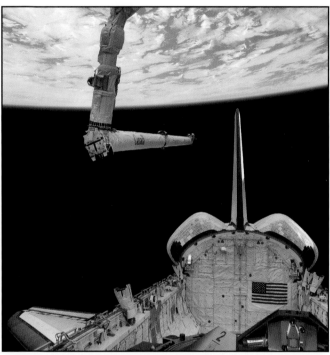

A nice shot of the SRMS holding the OBSS. The OBSS was the same length as the SRMS but lacked the articulated joints. The SRMS could grapple it on the end or in the middle of the boom as needed; the far end of the OBSS carried a variety of cameras and other sensors. (NASA)

The flight deck was cramped, especially when the crew were in pressure suits. Here are George Zamka (left) and Terry Virts going through the pre-entry checklists. Kay Hire and Steve Robinson were sitting behind them while the rest of the crew were on the middeck. (NASA)

STS-131 (ISS-19A)

Mission:	131	NSSDC ID:	2010-012A
Vehicle:	OV-103 (38)	ET-135 (SLWT)	SRB: BI142
Launch:	LC-39A	05 Apr 2010	10:21 UTC
Altitude:	190 nm	Inclination:	51.60 degrees
Landing:	KSC-15	20 Apr 2010	13:09 UTC
Landing Rev:	238	Mission Duration:	362 hrs 47 mins

Commander:	Alan G. "Dex" Poindexter (2)
Pilot:	James P. "Mash" Dutton, Jr. (1)
MS1:	Richard A. "Rick" Mastracchio (3)
MS2:	Dorothy M. "Dottie" Metcalf-Lindenburger (1)
MS3:	Stephanie D. Wilson (3)
MS4:	Naoko Yamazaki (1)
MS5:	Clayton C. "Clay" Anderson (2)

Payloads:	Up: 39,516 lbs	Down: 21,764 lbs
	MPLM (Leonardo) (18,263 lbs)	
	LMC (MPESS) (3,889 lbs)	
	SRMS s/n 202	

Notes:	Last seven-person crew
	Last rookie astronauts on space shuttle
	EVA1 09 Apr 2010 (Anderson and Mastracchio)
	EVA2 11 Apr 2010 (Anderson and Mastracchio)
	EVA3 13 Apr 2010 (Anderson and Mastracchio)
	Longest *Discovery* mission

Wakeup Calls:

05 Apr	"Find Us Faithful"
06 Apr	"I Will Rise"
07 Apr	"The Pigeon and a Boy"
08 Apr	"Defying Gravity"
09 Apr	"We Weren't Born to Follow"
10 Apr	"Stairway to the Stars"
11 Apr	"Because We Believe"
12 Apr	"Galileo"
13 Apr	"Miracle of Flight"
14 Apr	"The Earth in the Color of Lapis Lazuli"
15 Apr	"Theme from Stargate"
16 Apr	"Joy"
17 Apr	"What a Wonderful World"
18 Apr	American National Anthem – "The Star-Spangled Banner"
19 Apr	"On the Road Again"

The three gold bars of the astronaut symbol illustrate the 51.6-degree orbit and its elliptical wreath contains the orbit of the ISS. The star atop the astronaut symbol is the dawning Sun and the background star field contains seven stars, one for each crew member. (NASA)

Prior to the rollout from the vehicle assembly building, technicians noticed that birds had caused minor damage to the spray-on foam insulation on ET-135. NASA did not consider the damage significant, although technicians repaired the larger holes and top-coated the entire area after the vehicle was at the pad. The countdown on 5 April 2010 proceeded smoothly and *Discovery* was launched with no unplanned holds. This was the last mission that carried a seven-person crew and also the last that carried a rookie astronaut (in fact, three of them: James Dutton, Dottie Metcalf-Lindenburger, and Naoko Yamazaki).

Once on-orbit, the crew activated the Ku-band system, which failed its radar-mode self-test. When the crew switched to communications mode, the antenna tracked the TDRS satellite but never established a forward link and the return link was unusable. The Ku-band remained failed for the duration of the mission, forcing the crew to make a non-radar rendezvous with the ISS. The crew unberthed the SRMS and OBSS, and performed the RCC survey but, because of the failed Ku-band, could not downlink the imagery until after *Discovery* docked at the ISS. Alan Poindexter and James Dutton began the rendezvous using the "radar-fail procedure" that relied on the star trackers and crew optical alignment sight (COAS). After docking, Stephanie Wilson and Naoko Yamazaki removed the Leonardo MPLM from the payload bay and berthed it at the ISS.

On flight day 5, Clay Anderson and Rick Mastracchio conducted the first EVA, transferring a new ammonia tank assembly to the station for installation on a later spacewalk. James Dutton and Stephanie Wilson assisted the spacewalkers by operating the SSRMS. Anderson and Mastracchio performed the second EVA on flight day 7, replacing the ammonia tank assembly on the S1 truss and guiding the SSRMS to temporarily stow the old tank on the truss structure. During the third EVA the pair finished connecting the fluid lines to the ammonia tank and moved the old tank to the payload bay. On flight day 10, the ISS Program requested the crew perform an unplanned fourth EVA to fix an anomaly with a nitrogen tank assembly. After further consideration, mission management concluded the EVA could wait for a future mission.

Alan Poindexter and James Dutton undocked from the ISS on flight day 13. Mission management waved-off both KSC landing opportunities on flight day 15 because of a low ceiling at the Shuttle Landing Facility. Low clouds and precipitation caused the wave-off of the first opportunity the following day, but cleared sufficiently to allow *Discovery* to return on the second opportunity. This was the longest mission for *Discovery*.

The on-orbit portrait on the aft flight deck of Discovery. In the front, from the left, are Naoko Yamazaki, Alan Poindexter, and Stephanie Wilson. In the back are James Dutton, Clay Anderson, Dorothy Metcalf-Lindenburger, and Rick Mastracchio. (NASA)

Clay Anderson and Rick Mastracchio work near the ammonia tank assembly that is sitting on a lightweight MPESS carrier at the back of the payload bay. Interestingly, NASA removed the normal winch and handrails from the aft bulkhead to accommodate the LMC/ATA. (NASA)

The crew is using the SRMS and OBSS to conduct the detailed survey of the orbiter thermal protection system; note the OBSS crosses under Discovery and the sensor package on the end of the boom can be seen near the right wing leading edge. The SSRMS is at top. (NASA)

This was the third and last mission of the flight campaign with three female astronauts (STS-40 and STS-96 were the others). Combined with Expedition 23 flight engineer Tracy Caldwell-Dyson, this was also the first time four women had been in space at the same time. (NASA)

Discovery rolls past a fire engine and a crowd of employees at the Shuttle Landing Facility. The fire engine would chase the orbiter until it stopped rolling, just in case. The rest of the recovery convoy arrived a few minutes after the orbiter came to stop. (NASA)

STS-132 (ISS-ULF4)

Mission:	132	NSSDC ID:	2010-019A
Vehicle:	OV-104 (32)	ET-136 (SLWT)	SRB: BI143
Launch:	LC-39A	14 May 2010	18:20 UTC
Altitude:	194 nm	Inclination:	51.60 degrees
Landing:	KSC-33	26 May 2010	12:48 UTC
Landing Rev:	186	Mission Duration:	282 hrs 38 mins

Commander:	Kenneth T. "Hock" Ham (2)
Pilot:	Dominic A. "Tony" Antonelli (2)
MS1:	Garrett E. Reisman (2)
MS2:	Michael T. "Bueno" Good (2)
MS3:	Stephen G. "Steve" Bowen (2)
MS4:	Piers J. Sellers (3)

Payloads:	Up: 35,963 lbs	Down: 7,564 lbs
	MRM-1 (Rassvet) (17,688 lbs)	
	ICC-VLD (7,534 lbs)	
	SRMS s/n 301	

Notes: EVA1 17 May 2010 (Reisman and Bowen)
EVA2 19 May 2010 (Bowen and Good)
EVA3 21 May 2010 (Reisman and Good)

Wakeup Calls:

15 May	"You're My Home"
16 May	"Sweet Home Alabama"
17 May	"Alive Again"
18 May	"Macho Man"
19 May	"Start Me Up"
20 May	"Welcome to the Working Week"
21 May	"Traveling Light"
22 May	"Lord We Have Seen the Rising Sun"
23 May	"These Are Days"
24 May	"Theme to Wallace and Gromit"
25 May	"Empire State of Mind"
26 May	"Supermassive Black Hole"

NASA originally intended this to be the final flight of *Atlantis*, but ultimately the orbiter flew the last mission of the flight campaign, STS-135. The count on 14 May 2010 proceeded generally smoothly, with no unplanned holds.

Ken Ham and Tony Antonelli rendezvoused and docked with the ISS on flight day 3. During the R-bar pitch maneuver, three ISS crew members took 398 photographs of *Atlantis*: Oleg Valeriyevich Kotov used a 400 mm lens while Timothy Creamer and Soichi Noguchi used 800 mm lenses.

On flight day 4, Garrett Reisman and Steve Bowen conducted the first EVA. The pair installed a spare space-to-ground antenna, installed an enhanced tool platform for the special purpose dexterous manipulator (SPDM, Dextre), and released torque on six new batteries for the P6 truss. The crew members encountered several problems, including a slight gap between the space-to-ground antenna dish and its mounting pole. The spacewalkers loosened the bolts and used a higher torque setting, which narrowed the gap. Another problem occurred during the installation of the space-to-ground antenna when Bowen removed a cover from a connector and the primary computer detected an error and shut down. The cap closed the circuit for that connector, so when it was opened the sensor detected an error. The computer shutdown caused a two-minute loss of communications.

The next day, Ken Ham and Tony Antonelli used the SRMS to move the Russian-built Mini Research Module (MRM-1, Rassvet) from the payload bay to a point where the SSRMS grappled it and moved it to the nadir port on Zarya. Externally, the MRM-1 carried a spare elbow joint for the European robot arm, a spare radiator, a small payload airlock, and a portable work platform for maintaining the European robot arm.

Steve Bowen and Mike Good conducted the second EVA on flight day 6, mostly to replace the batteries on the P6 truss. Although the initial plan was to replace three batteries, the pair managed to replace four of them. Afterward, Bowen and Good worked more on the space-to-ground antenna and managed to close the dish-to-boom gap even further. On flight day 8, Mike Good and Garrett Reisman completed the third and final EVA. The pair connected a pair of ammonia jumpers on the P4/P5 truss segment before replacing the final two batteries on the P6 truss. Once that task was complete, they pair moved to the payload bay where they removed a grapple fixture and took it to the Quest airlock.

The weather at the Shuttle Landing Facility was acceptable for landing on the first opportunity on flight day 13.

Perhaps a bit overdressed for the traditional crew breakfast in the Operations and Checkout Building; notice the lack of any food. From the left are Piers Sellers, Steve Bowen, Tony Antonelli, Ken Ham, Garrett Reisman, and Mike Good. (NASA)

The crew patch features Atlantis flying off into the sunset as the end of the Space Shuttle Program approaches. However the Sun is also heralding the promise of a new day as it rises for the first time on a new ISS module, MRM-1 Rassvet, the Russian word for dawn. (NASA)

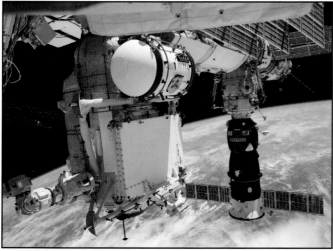

In the grasp of the SSRMS, the Mini-Research Module 1 (MRM-1) is attached to the earth-facing (nadir) port of Zarya (FGB). Named Rassvet, Russian for "dawn," the module was the last Russian hardware that space shuttle would carry. Note Progress M-05M in the background. (NASA)

Atlantis docked to PMA-2 attached to Node 2 (Harmony). Note the open star tracker doors in front of the windscreens. The European Columbus laboratory is to the right and the Japanese Kibō module is to the left of Node 2. (NASA)

ET-136 being lifted off its transporter (at the extreme right of the photo) on its way to its checkout cell in the Vehicle Assembly Building. The tank was completed in Michoud on 20 February 2010 and arrived at KSC aboard the Pegasus barge on 01 March 2010. (NASA)

After years of being largely ignored by the media and public, the last few missions of the flight campaign became very popular, despite a series of delays caused by external tank concerns. (NASA)

All three space shuttle main engines are up and running and the solid rocket boosters will ignite shortly. The SSMEs used liquid hydrogen and liquid oxygen propellants and the resulting exhaust was very pure steam. The solids were less environmentally friendly. (NASA)

STS-133 (ISS-ULF5)

Mission:	133	NSSDC ID:	2011-008A
Vehicle:	OV-103 (39)	ET-137 (SLWT)	SRB: BI144
Launch: Altitude:	LC-39A 202 nm	24 Feb 2011 Inclination:	21:53 UTC 51.60 degrees
Landing: Landing Rev:	KSC-15 202	09 Mar 2011 Mission Duration:	16:57 UTC 307 hrs 04 mins

Commander:	Steven W. "Steve" Lindsey (5)
Pilot:	Eric A. Boe (2)
MS1:	Nicole M. P. Stott (2)
MS2:	Benjamin A. "Alvin" Drew, Jr. (2)
MS3:	Michael R. "Mike" Barratt (2)
MS4:	Stephen G. "Steve" Bowen (3)

Payloads:	Up: 46,732 lbs	Down: 2,839 lbs
	PMM (Leonardo) (21,817 lbs)	
	ELC-4 (8,235 lbs)	
	Robonaut2 (300 lbs)	
	SRMS s/n 202	

Notes:	Delayed 01 Nov 2010 (OMS leak)
	Scrubbed 03 Nov 2010 (SSME electrical)
	Scrubbed 04 Nov 2010 (KSC weather)
	Scrubbed 05 Nov 2010 (GUCP leak)
	Tanking test 17 Dec 2010
	Rolled-back to VAB 22 Dec 2010 (ET stringers)
	EVA1 28 Feb 2011 (Bowen and Drew)
	EVA2 02 Mar 2011 (Bowen and Drew)
	Last flight of Discovery

Wakeup Calls:

25 Feb	"Through Heaven's Eyes"
26 Feb	"Woody's Roundup"
27 Feb	"Java Jive"
28 Feb	"Oh What a Beautiful Morning"
01 Mar	"Happy Together"
02 Mar	"The Speed of Sound"
03 Mar	"City of Blinding Lights"
04 Mar	"The Ritual / Ancient Battle / 2nd Kroykah"
05 Mar	"Ohio (Come Back to Texas)"
06 Mar	"Spaceship Superstar"
07 Mar	"Theme from Star Trek"
08 Mar	"Blue Sky"
09 Mar	"Coming Home"

This was the last crew patch designed by Robert McCall before his passing. Discovery is depicted ascending on a plume of flame as if it is just beginning a mission. However it is just the orbiter, without an ET or SRBs, signifying the completion of its operational life. (NASA)

The last *Discovery* mission suffered a series of delays due to problems with the ET and the payload. The launch, initially planned for September 2010, was pushed back to October, then to November. NASA slipped the planned launch from 1 November to 3 November after detecting helium and nitrogen leaks in the right OMS pod. Technicians replaced the quick-disconnect fittings. The 3 November launch was scrubbed to evaluate the condition of the backup controller for a main engine after it showed an irregular power drop. Recycling power seemed to cure the problem, but the mission management team slipped the launch a day to allow engineers more time to analyze the data. NASA scrubbed the 4 November attempt prior to tanking due to bad weather at KSC.

An attempt the next day was scrubbed after engineers detected a hydrogen leak in the ground umbilical carrier plate (GUCP) and the ground team noted a large crack in the insulation on the ET intertank. In the process of removing the cracked foam, technicians found two 9-inch cracks on either side of one of the intertank stringers. Engineers developed a repair plan, but technicians found two additional cracks on an adjacent stringer while they were preparing the original area for repair. This was the beginning of an intense effort to understand and correct the stringer cracking issues that affected ET-137 and ET-138, and were extended to ET-122 as a safety precaution.

NASA set the launch date for 24 February following the repairs to the ET. Late in the countdown, the range safety computer failed, forcing a hold at T-5 minutes. The Air Force restored the computer and the remainder of the countdown proceeded smoothly.

Steve Bowen was a late addition to the crew, replacing Tim Kopra following an injury while riding a bicycle on 15 January 2011. Bowen last flew on STS-132, making him the first (and so far, only) astronaut to fly on consecutive missions.

Steve Lindsey and Eric Boe docked to the ISS on flight day 3. Steve Bowen and Alvin Drew conducted the first EVA on flight day 5, installing a power cable linking Node 1 (Unity) and Node 3 (Tranquility), and moving an ammonia pump from its temporary location to ESP-2. On flight day 6, the crew transferred the newly modified Leonardo permanent multipurpose module (PMM) to the nadir port of Unity. The following day, Steve Bowen and Alvin Drew conducted the second EVA, removing thermal insulation from a platform and replacing a bracket on the Columbus module.

Steve Lindsey and Eric Boe undocked on flight day 12 and conducted the customary ISS fly-around. The weather at KSC was good and *Discovery* returned to the Shuttle Landing Facility on the first landing opportunity. This was the final flight of *Discovery*.

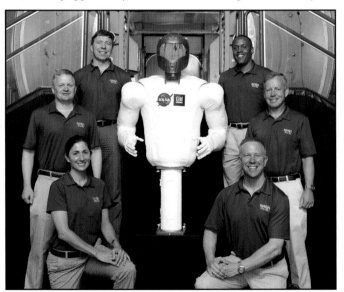

Robonaut2 is in the middle. Clockwise from lower right are Nicole Stott, Eric Boe, Mike Barratt, Alvin Drew, Steve Lindsey, and Tim Kopra. Unfortunately, Kopra was injured in a bicycle accident and was replaced by Steve Bowen about a month before launch. (NASA)

The History of the American Space Shuttle

The SSRMS moves Leonardo, now called a permanent multipurpose module (PMM) into place at the nadir (earth-facing) port on the Unity module. The most significant modification was adding more insulation, in this case scavenged from the unflown Donatello MPLM. (NASA)

Discovery shows the Leonardo permanent multipurpose module in the back of the payload bay with the ExPRESS logistics carrier (ELC-4) just ahead of it. The ELC-4 carried a spare heat rejection subsystem radiator for ISS and was installed on the S3 truss near ELC-2. (NASA)

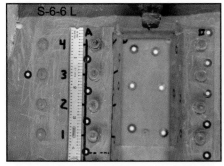

One of the stringer cracks is barely visible here, just to the right of the fasteners next to the ruler. The material used to manufacture the stringers was not hardened properly. (NASA)

Fixing the stringers required technicians to remove the sprayed-on foam insulation, a tedious manual process that required moving the stack back to the VAB. (NASA)

A technician replaces a seal on the hydrogen vent valve in the ground umbilical carrier plate. Leaking hydrogen is extremely explosive, so having a good seal was essential. (NASA)

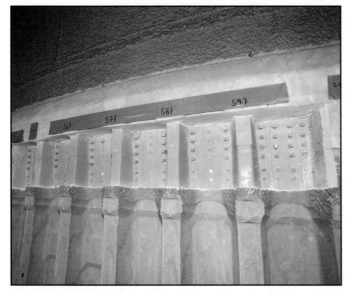

The corrugated look of the intertank was the result of using external stringers for strengthening. Like the rest of the ET, the stringers were covered with spray-on foam insulation. When workers noted a crack in the foam, they soon found cracks in the stringers. (NASA)

The hydrogen vent arm was connected to the ground umbilical carrier plate on the ET. There was a history of leaks at the GUCP, although usually they were easily corrected by changing various seals. This umbilical retracted at the same time the SRBs ignited. (NASA)

STS-134 (ISS-ULF6)

Mission:	134	NSSDC ID:	2011-020A
Vehicle:	OV-105 (25)	ET-122 (SLWT)	SRB: BI145
Launch:	LC-39A	16 May 2011	12:56 UTC
Altitude:	187 nm	Inclination:	51.60 degrees
Landing:	KSC-15	01 Jun 2011	06:35 UTC
Landing Rev:	248	Mission Duration:	377 hrs 38 mins

Commander:	Mark E. Kelly (4)
Pilot:	Gregory H. "Box" Johnson (2)
MS1:	Edward M. "Spanky" Fincke (3)
MS2:	Roberto Vittori (3)
MS3:	Andrew J. "Drew" Feustel (2)
MS4:	Gregory E. Chamitoff (2)

Payloads:	Up: 39,862 lbs	Down: 1,781 lbs
	AMS-02 (15,584 lbs)	
	ELC-3 (14,023 lbs)	
	SRMS s/n 201	

Notes:	Scrubbed 29 April 2011 (APU heaters)
	Scheduled to use ET-122; swapped with ET-138
	EVA1 20 May 2011 (Feustel and Chamitoff)
	EVA2 22 May 2011 (Feustel and Fincke)
	EVA3 25 May 2011 (Feustel and Fincke)
	EVA4 27 May 2011 (Chamitoff and Fincke)
	Last flight of *Endeavour*

Wakeup Calls:

16 May	"Beautiful Day"
17 May	"Drops of Jupiter"
18 May	"Luna"
19 May	"We All Do What We Can Do"
20 May	"In View"
21 May	"Il Mio Pensiero"
22 May	"Times Like These"
23 May	"Svegliarsi La Mattina (Woke Up This Morning)"
24 May	"Real World"
25 May	"Countdown"
26 May	"Fun, Fun, Fun"
27 May	"Will You Carry Me?"
28 May	"Galaxy Song"
29 May	"Slowness"
30 May	"Dreams You Give"
31 May	"Sunrise Number 1"

The space shuttle was supposed to have been retired after STS-133, but controversy over abandoning several ISS components, most notably the $1,500 million Alpha Magnetic Spectrometer (AMS-02), caused Congress to authorize, at first one, then two, additional missions. Initially, STS-134 was supposed to fly with the newer ET-138 while its corresponding STS-335 LON mission would use the ET-122 "Katrina tank," so named because it had been damaged during Hurricane Katrina and then repaired.

However, once it became clear that STS-135 would fly regardless, NASA swapped the tank assignments so that if STS-134 suffered damage from ET-122 foam shedding, STS-135 with the newer, and theoretically safer, ET-138 would be poised to rescue the *Endeavour* crew. Although ET-122 was an older tank that was not affected by the materials problem that had led to cracks in the intertank stringers on ET-137 and ET-138, engineers nevertheless applied the same repairs in an abundance of caution. NASA scrubbed the 29 April 2011 attempt when the auxiliary power unit heaters failed to activate but the count on 16 May proceeded smoothly. *Endeavour* docked at PMA-2 on flight day 3. The first task was to use the SRMS to unberth the ExPRESS logistics carrier (ELC-3) from the payload bay and transfer it to the SSRMS to attach it to the P3 truss. On flight day 4, Drew Feustel and Roberto Vittori lifted AMS-02 out of the payload bay using the SRMS and handed it to SSRMS operated by Gregory Chamitoff and Greg Johnson. The experiment was attached to the S2 truss and marked the completion of the American orbital segment of the ISS.

The first EVA came on flight day 5 when Drew Feustel and Greg Chamitoff installed a set of MISSE experiments, an ammonia jumper between the P3 and P6 trusses, and a cover on the starboard solar alpha rotary joint (SARJ). Feustel and Edward Fincke conducted the second EVA on flight day 7. During the EVA, the pair serviced the photovoltaic thermal control system, lubricated one of the SARJ joint, and installed a stowage beam on the S1 truss.

The ISS Program had decided having a 50-foot extension boom around might prove useful and asked the Space Shuttle Program to leave one of the OBSS booms at the station. The crew left the boom mated to the S1 truss, meaning they could not accomplish a late inspection before heading home. This was the second time *Endeavour* had left her OBSS at the station (the first being STS-123) and, it turned out, the OBSS was the final American component of the ISS.

The weather at KSC was good, and *Endeavour* returned to the Shuttle Landing Facility on the first opportunity on flight day 17. This was the final flight of *Endeavour*.

The shape of the crew patch is inspired by the international atomic symbol that represents the atom with orbiting electrons around the nucleus. The burst near the center refers to the big-bang theory and the origin of the universe, the subject of investigations by AMS-02. (NASA)

Mark Kelly is at bottom center. From the left are Greg Johnson, Edward Fincke, Gregory Chamitoff, Drew Feustel, and Roberto Vittori. Kelly almost missed the flight after his wife, Gabrielle Giffords, was seriously wounded during an assassination attempt in January 2011. (NASA)

Endeavour docked to the International Space Station during STS-134, as seen from the departing Soyuz TMA-20 on 23 May 2011. Dmitri Yuryevich Kondratyev backed the spacecraft away from the ISS about 600 feet and paused to give Paolo Nespoli the photo opportunity. It was the fulfillment, albeit much later than expected, of the original rationale for developing space shuttle—building a permanently man-tended orbital platform. In addition to Endeavour, there is a Soyuz, Progress, and ATV docked at the station. (NASA)

STS-135 (ISS-ULF7)

Mission:	135	NSSDC ID:	2011-031A		
Vehicle:	OV-104 (33)	ET-138 (SLWT)	SRB: BI146		
Launch:	LC-39A	08 Jul 2011	15:29 UTC		
Altitude:	212 nm	Inclination:	51.60 degrees		
Landing:	KSC-15	21 Jul 2011	09:57 UTC		
Landing Rev:	200	Mission Duration:	306 hrs 28 mins		

Commander: Christopher J. "Chris" Ferguson (3)
Pilot: Douglas G. "Doug" Hurley (2)
MS1: Sandra H. "Sandy" Magnus (3)
MS2: Rex J. Walheim (3)

Payloads: Up: 30,412 lbs   Down: 23,329 lbs
MPLM (Raffaello) (26,700 lbs)
LMC (MPESS) (2,918 lbs)
SRMS s/n 301

Notes: Scheduled to use ET-138; swapped with ET-122.
First four-person crew since STS-6 in 1983
Tanking test 15 Jun 2011
Last flight of Atlantis
End of space shuttle flight campaign

Wakeup Calls:

09 Jul	"Viva la Vida"
10 Jul	"Mr. Blue Sky"
11 Jul	"Tubthumping"
12 Jul	"More"
13 Jul	Message from Elton John / "Rocketman"
14 Jul	"Man on the Moon" (capella version)
15 Jul	Message from Paul McCartney / "Good Day Sunshine"
16 Jul	Message from Beyonce / "Run the World (Girls)"
17 Jul	"Celebration"
18 Jul	"Days Go By"
19 Jul	"Don't Panic"
20 Jul	"Fanfare for the Common Man"
21 Jul	"God Bless America"

The crew patch pays tribute to the entire NASA and contractor workforce that made possible all the accomplishments of the space shuttle. Omega, the last letter in the Greek alphabet, recognizes this mission as the last flight of the Space Shuttle Program. (NASA)

NASA initially planned for STS-134 to fly with the newer ET-138 and for STS-335 to use the nine-year-old ET-122 "Katrina tank." Mission management was concerned that ET-122 posed a slightly higher risk of foam shedding and therefore assigned it to the STS-335 launch-on-need mission that, at the time, appeared unlikely to fly. However, once it became clear that STS-135 would fly regardless, NASA swapped the tank assignments so that if STS-134 suffered damage from ET-122, STS-135 with the newer and theoretically safer ET-138 would be poised to rescue the *Endeavour* crew. During pre-launch operations, two lightning strikes occurred near *Atlantis*, but engineers were satisfied that neither affected the vehicle. Nearly one million spectators were in central Florida on 8 July 2011 to watch the final launch of the Space Shuttle Program.

Atlantis docked with the ISS on flight day 3, despite the failure of one general-purpose computer. The primary objective during flight day 4 was to move the Raffaello MPLM to the nadir port of Node 2 (Harmony); interestingly, this was the only time that *Atlantis* had carried an MPLM. Sandy Magnus and Doug Hurley used the SSRMS to remove Raffaello from the payload bay and install it on Harmony. This marked the only time two MPLMs were at the ISS at the same time (Leonardo, now called a PMM, had been left at the station by STS-133). Despite *Atlantis* not being equipped with the station-shuttle power transfer system (SSPTS), the crew saved enough power that mission management approved a one-day extension.

Because of a short training flow and a requirement to launch the orbiter with a reduced crew of four, NASA opted not to use the STS-135 crew for any EVAs during the docked operations, although the ISS crew performed several EVAs while *Atlantis* was present.

The next several days saw the ISS and *Atlantis* crews moving supplies from the orbiter to the ISS and performing a variety of maintenance tasks on the station. In all, they moved 2,097 pounds from the orbiter middeck to the ISS and 1,377 pounds from the ISS to the middeck. They also transferred 2,074 pounds of water.

On flight day 11, Sandy Magnus and Doug Hurley used the SSRMS to move Raffaello back into the payload bay. Chris Ferguson presented the ISS crew with a small American flag that had first flown 30 years earlier on STS-1. *Atlantis* undocked from the station early on flight day 12, marking the end of space shuttle visits to the orbiting outpost. Two days later, Doug Hurley fired the OMS engines for the deorbit maneuver for the first opportunity at KSC.

Atlantis landed 30 years, 3 months, 8 days, 21 hours, 57 minutes after the launch of *Columbia* on STS-1. This was the last flight of *Atlantis* and final mission of the Space Shuttle Program.

The beginning; and the end. The crews of STS-1 and STS-135 pose for a group photo at the Johnson Space Center. From the left are Doug Hurley, Bob Crippen, John Young, Chris Ferguson, Sandy Magnus, and Rex Walheim. (NASA/Houston Chronicle photo by Smiley N. Pool)

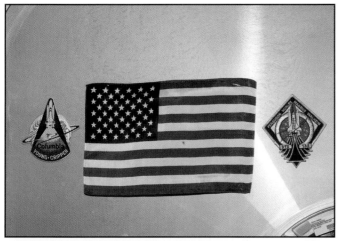

Chris Ferguson brought the ISS a small American flag that had flown on STS-1. In its presentation on the station, it is flanked by the crew patches from the first and last space shuttle missions. It would not have been possible to build the ISS without space shuttle. (NASA)

Red spray paint on Runway 15 at the Shuttle Landing Facility marks the location of the final wheels-stop. Green spray paint marked the nose wheels. Later, NASA would install a formal plaque by the side of the runway commemorating the location. (NASA)

Atlantis makes the last landing, trailing vortices off the wingtips that picked up some color from the xenon lights at the end of the runway. This was the 78th landing at KSC, the twenty-sixth night landing overall, and the twentieth night landing at KSC. (NASA)

The Atlantis stack lifts-off from LC-39A. This was the last external tank completed, although major components for several others had been manufactured but not assembled. There was one flight-qualified tank (ET-94) and three test tanks left at the end of the flight campaign. (NASA)

Not a good day to drive anywhere. This is the crowd on the new Max Brewer Bridge in Titusville waiting for launch; the city had closed the road in anticipation. Large crowds were present all over the Space Coast to watch the final mission of the flight campaign. (NASA)

RETIREMENT

At the time of the 2004 presidential announcement cancelling the Space Shuttle Program, NASA began planning what became known as transition and retirement (T&R). The T&R moniker was meant to represent the "transition" of the majority of assets (hardware and people) to what became the Constellation Program (CxP) and the "retirement" of the rest (primarily the orbiters and their related support equipment). Of course, it did not work that way and for the most part it became just a "retirement" exercise.

At the completion of their final mission, each orbiter underwent a somewhat streamlined version of the typical down-mission processing performed after every space shuttle flight. During this two-month process, the vehicle was thoroughly inspected for damage from its last mission, all payloads and personal items were removed, the galley and potty were shipped to JSC for cleaning, and the main engines were taken to the SSME Processing Facility for inspection and repair. In addition, technicians removed the forward reaction control system (FRCS) module and orbital maneuvering system (OMS) pods and transferred them to the Hypergolic Maintenance Facility for preliminary safing. This process was generally similar for all three orbiters.

For some time before the display site announcement, NASA and its contractors had been studying how to safe the orbiters for display. In addition to working internally, NASA sought recommendations from the Smithsonian and the National Museum of the United States Air Force on what hazards were acceptable for long-term public display. Obviously, the vehicles would be purged of all hazardous gases and liquids, such as hypergolic propellants, ammonia, and Freon. In addition, several hundred pieces of pyrotechnics would be removed from each vehicle. Ultimately, NASA decided all of the "softgoods" (seals, etc.) that had touched hypergols also would be removed to ensure the vehicle did not outgas any hazardous vapors (although this was contrary to how the Air Force handled F-16s and various missiles intended for long-term display). This involved gutting the FRCS and both OMS pods of essentially all plumbing and tankage; all that remains are a few pressurization tanks. For mostly political reasons, NASA sent the FRCS and OMS pods to the White

Sands Test Facility for this work rather than doing it at KSC, adding considerable time and expense to the process.

During 2010, NASA allowed internal organizations to request parts of the orbiters to support knowledge capture, future use, or educational display under a process known as STS-Last. Various organizations requested 452 pieces of hardware from *Discovery* as part of STS-Last. However, to preserve the integrity of the vehicle in the national collection, the Space Shuttle Program only agreed to 209 of these requests. The other parts were subsequently removed from the other two vehicles instead. These included a variety of valves and regulators, hydraulic pumps, power inverters, the wing leading edge detection system sensors and recorders, payload bay cameras, and the outer window panes (these were replaced by unflown spares). For the most part, all of these components were buried inside the vehicles and were not visible from the outside.

In addition, the nascent Space Launch System (SLS) requested components from the main feed lines and most of the large propellant valves from the main propulsion system in each orbiter to provide certified hardware for the initial SLS vehicles. For the most part, NASA declined to remove these parts from *Discovery* in order to keep the orbiter as intact as possible for the national collection, but removed them from the other two vehicles. None of the removals affected the outer mold line (excepting creating some empty holes in the T-0 and ET umbilicals), although the view inside the aft compartment of *Atlantis* and *Endeavour* is less than ideal.

SLS also requested all of the remaining SSMEs for use on the initial SLS vehicles. This left some rather large holes in the back of the orbiters, so the Space Shuttle Program requested Pratt & Whitney Rocketdyne manufacture nine replica shuttle main engines (RMSE). Contrary to their name, the RSMEs were not replicas nor were they engines. Rocketdyne found nine spare nozzles, all previously used for various tests, and refurbished them to a condition adequate for display. Mostly; each of the display sites subsequently had Guard-Lee cosmetically enhance the refurbished nozzles to present a more realistic appearance. Engineers attached these nozzles to support structures that mated with the normal SSME attachment points.

As part of the vehicle safing process, NASA removed the external airlocks from Atlantis and Endeavour for some "future program." As of 2018, they sit abandoned in the Space Station Processing Facility at KSC. Discovery retained her airlock but not the ODS. (NASA)

As part of the safing effort, the forward reaction control system module was removed from each orbiter, sent to the White Sands Test Facility, and completely gutted. The empty shells were then reinstalled on the vehicles and present a realistic external appearance. (NASA)

Perhaps the most invasive safing was to any of the systems that used hypergolic propellants. The auxiliary power units, their propellant tanks, and all propellant lines were removed from the vehicles. The orbital maneuvering system pods were completely gutted. Compare the OMS pod being removed from Discovery at right to one after it was safed at left; all of the tanks and lines have been removed. (NASA)

DISPLAY SITES

Although George Bush announced in 2004 that the Space Shuttle Program would end when ISS assembly was complete, it was not until 2007 that NASA began to seriously consider what to do with the orbiters once they were retired. Mike Griffin, the NASA administrator at the time, initially believed, without public or political input, the orbiters should be displayed at KSC, JSC, and MSFC. As with many things Griffin, others disagreed.

Traditionally, the National Air and Space Museum had received all flown manned space hardware and loaned the artifacts to other museums. This was codified in a 1967 memorandum of understanding between NASA and the Smithsonian that was renewed in 2008. However, applying this approach to the orbiters would have required NASA to absorb the entire cost of safing, preparing for display, and transporting the vehicles to their new homes, something the agency could not afford. Out of necessity, the orbiters would be handled differently than previous spacecraft.

Instead, NASA issued a request for information (RFI) on 17 December 2008 to "educational institutions, science museums, and other appropriate organizations" to gauge their interest in obtaining an orbiter or a space shuttle main engine. The agency asked interested parties to submit information concerning their accreditation, attendance, and financial condition, as well as about the population of the surrounding geographic area. In addition, each response needed to include information on the local infrastructure for transporting an orbiter from the delivery airport to the display site. It was the beginning of a long process that garnered a great deal of criticism – some justified, most not – from a variety of sources.

NASA received responses from 21 organizations by the 17 March 2009 deadline. An evaluation team at NASA Headquarters ultimately determined 11 of the 21 respondents met the essential requirements and had a credible proposal for financing, transporting, and displaying an orbiter.

In July 2009, about four months after NASA received the RFI responses, the Senate confirmed Charlie Bolden as the new NASA administrator to replace Mike Griffin. Bolden told the team, in contrast to the view held by Griffin, he believed the orbiters should be placed where the largest number of visitors could view them and promote interest in science, technology, and space exploration. Accordingly, he told them to revise the rankings to consider attendance, international visitors, and regional population.

In response to the new direction, NASA issued a second request for information on 15 January 2010. The agency received responses from 29 entities by the 19 February 2010 deadline, including many of the same organizations that had responded to the first request for information. The team concluded 13 institutions met the requirements.

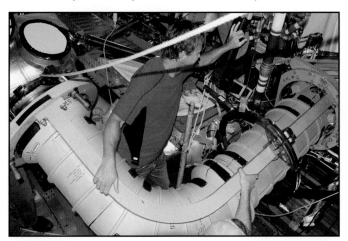

The Space Launch System (SLS) requested most of the main propulsion system plumbing from Atlantis and Endeavour as test articles. NASA generally tried not to remove many components from Discovery since she would become part of the national collection. (NASA)

SLS retained all 16 space shuttle main engines that remained at the end of the flight campaign. Since orbiters look rather odd without engines, NASA had Rocketdyne create "engines on a stick" using former test nozzles for the three orbiters. For the most part, they look correct. (NASA)

DISPLAY SITES

On 12 April 2011, a date well chosen to coincide with the 30th anniversary of the launch of STS-1, NASA administrator Charlie Bolden held a media event in front of Orbiter Processing Facility at KSC. Bolden announced that the three flown orbiters would be placed at the Smithsonian National Air and Space Museum, Kennedy Space Center Visitor Complex, and the California Science Center. *Enterprise*, already on display at the Smithsonian, would be moved to the Intrepid Sea, Air, & Space Museum.

While greeted with enthusiasm at the selected locations, the announcement was not well received by some members of Congress who represented states that did not receive an orbiter, particularly representatives from Houston, who called the decision a "Houston Shuttle Snub." Some raised concerns that NASA failed to follow the law, allowing politics to dictate the result, and called for an investigation. Responding to these concerns and public interest, the NASA Office of the Inspector General (OIG) examined the process for selecting the display sites. Ultimately, the OIG found no major issues with the decisions or the process used to make them, although the report noted the final scoring was flawed by a math error that would have led to the National Museum of the United States Air Force outscoring Intrepid. When asked if that would have swayed his decision, Charlie Bolden responded in the negative.

Each of the display sites, including the Smithsonian, prepared a logistics plan, transportation plan, and financial plan for NASA approval. These allowed the agency to ensure the vehicles would be displayed appropriately and provided insight into how the each organization would pay NASA for the display preparations and delivery. Complicating matters, each of the vehicles involved a different legal process to transfer title (or not, in the case of *Atlantis*).

Although not a part of the decision process, it is interesting to note the four recipients intended, at least initially, to display the vehicles very differently from each other, largely mimicking the four basic flight regimes flown by the orbiters. Intrepid intended to display *Enterprise* flaring for landing, complete with a tailcone as she appeared during the Approach and Landing Tests. The Smithsonian, never one for dramatic presentations, displays *Discovery* as she appeared on the runway after returning from STS-133. The KSC Visitor Complex configured *Atlantis* as she appeared on-orbit, positioned at a jaunty angle with the payload bay doors open. The California Science Center will display *Endeavour* as a vertical launch stack consisting of ET-94 and a set of steel solid rocket boosters. A complete mission in four widely separated locations.

DISPLAY PREPARATION

Once the vehicles were safed, United Space Alliance configured the orbiters as requested by each of recipient. For the most part, this involved identifying what items should be reinstalled in the orbiter and which should be left uninstalled for separate display. Most of the basic configuration was defined in the end-state subsystems requirements document, but many details were decided during multiple meetings with each recipient.

Even before Charlie Bolden announced the display sites, some in NASA had been looking at how to use the delivery flights as an opportunity to showcase the agency. One concept was for the shuttle carrier aircraft to make a grand tour of the United States during the ferry flights. This shuttle public outreach tour (SPOT, more often called shuttlepallooza) would be a 4-to-28-week ferry flight that would, potentially, have stopped in cities in every state. Oddly, NASA wanted the display sites to pay for the grand tour, something none of them were willing to do given its estimated $27 million cost. Also, the display sites were concerned about the increased risk of loss or damage to the SCA and orbiter during the tour, especially given the vehicles were irreplaceable artifacts and, essentially, uninsurable. A comment from one of the staff at the KSC Visitor Complex summed up the overall reaction: "It's a great idea ... with somebody else's vehicle." Eventually, the NASA executive council killed the shuttlepallooza idea and directed the final ferry flights to be along the most direct routes, although it authorized fly-overs of specific areas along the route and extensive fly-overs of the destination cities.

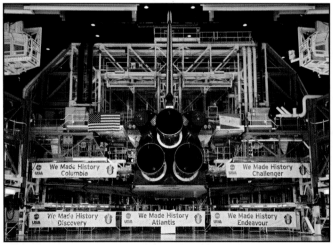

NASA decorated OPF-1 (left) with banners for the each of the five flight orbiters, including the two vehicles lost in the accidents. Atlantis was undergoing down-mission processing in the facility, and its flag can be seen to the right of the right OMS nozzle. Many employees took some time off to gather and watch the ceremony (right) including a speech from the crew aboard the ISS (note the TV screen). NASA administrator Charlie Bolden was joined on stage by STS-1 pilot Bob Crippen, astronaut Janet Kavandi, and KSC center director Bob Cabana. (NASA)

JAMES S. McDONNELL SPACE HANGAR

Discovery, the "vehicle of record" in the James S. McDonnell Space Hangar at the National Air and Space Museum's Steven F. Udvar-Hazy Center near the Dulles International Airport. At this point the workers from United Space Alliance were still finishing getting the orbiter ready for display; note the forward jacks and the man-lift in the background. Much to the delight of the visitors, the museum did not close that hangar during this period, allowing the public to see the work firsthand. (NASM/Dane A. Penland)

Enterprise was lifted off the 747 using the same cranes (and operators) as had unloaded Discovery and loaded Enterprise at Dulles. This greatly lessened the learning curve and the operation proceeded smoothly and quickly once the weather cooperated. (Dennis R. Jenkins)

Bay Crane lifted Enterprise onto a barge in Bergen Basin, at the west end of the John F. Kennedy International Airport. The operation went smoothly except for periodic interruptions by a Presidential helicopter detail that was operating nearby. (Dennis R. Jenkins)

INTREPID SEA, AIR, & SPACE MUSEUM

Enterprise had initially been delivered to the Smithsonian in 1985. After sitting in storage at Dulles for 18 years, it became the centerpiece of the space hangar at the new Steven F. Udvar-Hazy Center when it opened in 2003. Because of the long time in storage and on display, NASA worried the vehicle might no longer be flightworthy. Therefore, in 2010, NASA and United Space Alliance conducted a detailed inspection of the vehicle, finding no major concerns but a multitude of small things that needed fixed or modified before the vehicle could be ferried by the Shuttle Carrier Aircraft.

Actions taken by United Space Alliance to prepare *Enterprise* for its last ferry flight included reactivating the hydraulic system to retract the landing-gear, removing the plywood display-only OMS pods, installing the ferry-capable ALTA pods originally used for the Approach and Landing Tests, clearing the inside of the vehicle of any loose debris, and performing a careful weight and balance check. Interestingly, *Enterprise* is the only orbiter displayed with the set of OMS pods she was originally delivered with.

Once *Discovery* was offloaded from the SCA at Dulles, she was towed to the Udvar-Hazy Center where a carefully choreographed show saw *Enterprise* removed from the space hangar and replaced

with *Discovery* in front of thousands of spectators. *Enterprise* was then towed to Dulles and mated to the 747 for her ferry flight to New York. Mother Nature did not cooperate and the mated pair sat at Dulles for a few days waiting for acceptable weather.

At the John F. Kennedy International Airport, the team used a portion of apron where a former Trans World Airlines Hangar had been located; in fact, the hangar had still stood during the first site visit and was dismantled beginning in April 2011. The Port Authority of New York and New Jersey provided the site since it was not being used. A nearby infrared deicing tent provided a convenient place to let *Enterprise* and the SCA sit out a persistent weather system that hung over the airport and to store *Enterprise* prior to its move to the Intrepid. Once the weather cleared, *Enterprise* was moved to the Bergen Basin on the west side of the airport and loaded on a barge for its trip across Jamaica Bay, New York Bay, and up the Hudson River. A stop at the Weeks Marine facility in New Jersey allowed a crew to lower the landing-gear and ready the orbiter to be lifted onto Intrepid. Once the barge arrived at the museum, the Weeks 533 floating crane lifted *Enterprise* onto the aircraft carrier where it was secured. A temporary building was erected around the orbiter until a permanent facility can be designed and built at some point in the future.

Many private boats joined the procession across Jamaica Bay and up the Hudson, although the Coast Guard and NYPD kept them at a safe distance. Fire boats provided the typical New York welcome, although they did not spray water directly on the orbiter. (NASA)

Weeks Marine used the Weeks 533 crane to lift the orbiter onto Intrepid. This barge-mounted 500-ton capacity Clyde Iron Works model 52 is the largest revolving floating crane on the east coast of the United States. It made easy work of the lift. (NASA)

Since she would become the reference orbiter and part of the national collection, the Smithsonian wanted *Discovery* configured as closely as possible to her final mission and explicitly requested that any non-flight parts be clearly marked as such. In particular, Valerie Neal, the NASM curator, requested any non-flight thermal protection system elements be left "unmarked" (i.e., no attempt to imitate operational markings). In the end, this was impossible to do within the NASA and United Space Alliance configuration management system, and the eight "fake" (foam) tiles installed on the vehicle are clearly marked as "foam." The engineers and technicians at KSC understood the orbiter would be a part of the national collection and made every effort to ensure *Discovery* looked as operational as possible.

The Smithsonian elected to not reinstall the shuttle remote manipulator system (SRMS) arm and it is currently displayed next to *Discovery* at the Udvar-Hazy Center; the OBSS is installed on the starboard payload bay sill. This is the only authentic SRMS arm released to a museum (one was returned to Canada and one retained at JSC). In addition, *Discovery* is the only vehicle with a real external airlock installed, although it does not have an ODS attached and nobody can see it in any case since the payload bay doors are closed. *Discovery* has a complete flight deck and middeck, including a galley and potty. Initially, the escape pole and slide were delivered separately, but they were reinstalled into the orbiter during 2014. As delivered, the payload bay was configured largely as it had been after STS-133, but the Smithsonian subsequently allowed the California Science Center to remove two longeron bridges and passive latches to support the vertical display of *Endeavour*.

Delivering the orbiter, although the most straightforward of any of the display sites, was still a challenge for a program trying to go out of business. The magnitude of the task should not be underestimated. NASA negotiated with the Washington Metropolitan Airports Authority to use a large ramp area, called Apron W, between two active runways at Dulles International Airport. Because this apron was normally used for dicing airplanes, it could not be released to NASA until 15 April when the weather had warmed. Fortunately, this coincided nicely with when *Discovery* would be ready and the Smithsonian wanted to take delivery. Convincing the airport engineering organization to allow United Space Alliance to drill 200 holes in their concrete ramp to secure ground support equipment provided a lesson in diplomacy (USA filled the holes when the operation was finished).

Setting up the off-load area required 20 truckloads of equipment from KSC, in addition to renting forklifts, man-lifts, fences, light banks, office trailers, portable restrooms, and other equipment from

Discovery, and later Endeavour, flew a grand tour between KSC and their final landing sites. In this case, the FAA and Homeland Security cleared the SCA and its chase aircraft to make several circuits in the controlled airspace over Washington DC. (NASA)

local vendors. Then there were the two large cranes, the larger of which required 70 trucks to move on site. Although the procedure for loading and off-loading an orbiter from the SCA using cranes was well established, it had been almost 30 years since anybody had done it. The concept had been developed to support loading an orbiter after a TAL abort and had been used successfully to off-load and load *Enterprise* for the World's Fair in New Orleans, at the Marshall Space Flight Center for a series of tests, and for its delivery to the Smithsonian in 1985. A variation of it had also been used to load *Columbia* after it landed at White Sands at the end of STS-3. The NASA and United Space Alliance team approached the process carefully, not wanting to risk damaging a priceless artifact (and a spacecraft they had been taking care of for almost 30 years).

The Smithsonian planned a grand entrance for the new orbiter complete with the Marine Corps Marching Band (the director of the museum, Jack Dailey, was a retired four-star Marine). Inclement weather delayed the off-load of the orbiter, but NASA and United Space Alliance still managed to get it to the museum gate just before the start of the ceremonies. The late arrival of a guest of honor (Joe Engle) further delayed things a bit, but ultimately the show went off as planned and *Enterprise* and *Discovery* met for the first time as they were towed past each other.

Rain frustrated workers as they off-loaded Discovery from the SCA. The initial plan had been to accomplish the off-load during the day, but high winds forced the team to move the operation to late at night when things were calmer. (Dennis R. Jenkins)

Enterprise (left) and Discovery meet for the first time on the ramp behind the Steven F. Udvar-Hazy Center. Afterward, Enterprise was towed to Apron W at the Dulles airport and the waiting SCA while Discovery entered her new home in the space hangar. (NASA)

NASA gave Ivey's Construction the orbiter transporter system (OTS) to move Atlantis (and they later sold it to SpaceX for use with Falcon 9 first stages). For the most part, it was an easy drive, although it meant using this narrow off-ramp to avoid an overpass. (Dennis R. Jenkins)

Atlantis arrived at its new home just after dark and fireworks were part of the ceremony. Afterward, workers moved her into the partially completed building. It would take a few days to secure her to the attach hardware and take the wheels, tires, and brakes off. (Dennis R. Jenkins)

KENNEDY SPACE CENTER VISITOR COMPLEX

Possibly the easiest delivery, but one of the most difficult displays, was *Atlantis*. The KSC Visitor Complex was only 6 miles away from the Vehicle Assembly Building down fairly wide and well-maintained roads. A couple of stops for speeches and lunch were the only obstacles to delivering the orbiter. However, the display concept was imaginative and difficult to accomplish. The Visitor Complex displays *Atlantis* in an on-orbit configuration at a 43.21-degree angle, left wing down (the odd angle is a countdown: 4 … 3 … 2 … 1) inside a large new building on their campus.

As might be imagined, positioning an orbiter 20 feet in the air at a jaunty angle required some innovative work. Ivey's Construction and Beyel Brothers Crane and Rigging, supported by several former space shuttle engineers, developed a method of slowly raising the vehicle then tilting her into the final position. Since the vehicle is supported on the normal attach points, the ET umbilical doors were removed and the ferry doors installed to cover the umbilicals. In addition, *Atlantis* does not have wheels, tires, or brakes installed, mostly to save weight as part of the display. Before moving *Atlantis*, workers in the OPF installed a replica external airlock fabricated by Guard-Lee, Inc. of Apopka, Florida. The payload bay doors are

open, the replica SRMS is extended over one of the viewing ramps, and the real OBSS is on the starboard sill, but otherwise the payload bay is empty. An unfortunate oversight to an otherwise spectacular display was leaving the star tracker doors closed, although USA engineers offered to build clear plexiglass covers to protect the star trackers with the doors open.

The exhibit celebrates the workers, specifically at the Kennedy Space Center but more generally all of the centers and contractors, that contributed to space shuttle. Guard-Lee also built a full-size replica of the Hubble Space Telescope that hangs a short walk away. A variety of other artifacts, including a GOX beanie cap are nearby.

Perhaps the most moving aspect of the exhibit is called Forever Remembered. Designed by Luis Berrios at NASA and Daniel Gruenbaum at Delaware North, with assistance from Mike Ciannilli, the NASA project manager for the Space Shuttle Challenger Office, the tastefully rendered display highlights a major piece of debris from each orbiter in addition to personal mementos from the 14 crew members or their families. Joe Sembrat and his team at Conservation Solutions treated the *Challenger* and *Columbia* artifacts while Tom Wilkes and his artisans at Guard-Lee built the stands to support it while on display. It is likely the only location where pieces of debris will be displayed to the public on a routine basis.

Atlantis was attached to the large steel beams under her using a set of SCA ball mounts at the back and the normal bipod attachment at the front. Workers then jacked the beams, cribbed them with stands, then jacked some more until it was at the right height. (Dennis R. Jenkins)

The final display before opening day. The payload bay doors have thin wires connecting them to the overhead structure since they cannot support themselves in 1-g. The replica SRMS arm is deployed over the walkway, allowing a great view for the guests. (NASA)

CALIFORNIA SCIENCE CENTER

The California Science Center has the most ambitious plans of any of the display sites and is planning on displaying *Endeavour* indoors as a complete vertical launch stack. Unlike any other display stack, this one will be composed of all real flight hardware using ET-94 and a set of steel boosters donated by NASA (forward and aft skirts and many details) and Orbital ATK (the SRM segments). The external tank had been manufactured for use on *Columbia* science missions and was never used after the orbiter was lost on STS-107.

Getting to Los Angeles allowed the Science Center and NASA to show the flag throughout the California during the ferry flight. United Airlines graciously provided a location at LAX next to its maintenance hangar on the west side of the airport for the off-loading. The airline also allowed the California Science Center to store *Endeavour* inside for the few weeks it took to get ready for the move through the streets to Exposition Park. Moving a large object—the orbiter is 122 feet long, 78 feet wide, and 58 feet high—through 13 miles of city streets required a great deal of planning and coordination. Ultimately, it all worked well, with more than a million people lining the streets during the three days and three nights it took to move from the airport to Exposition Park. *Endeavour* is currently displayed in the temporary Samuel Oschin Pavilion while a permanent building is being designed and built.

After the Science Center received ET-94 from NASA, it was moved from the Michoud Assembly Facility near New Orleans, through the Panama Canal, to Marina del Rey where it was offloaded from its barge. It too was moved through the streets to Exposition Park, although this was somewhat easier since the tank was not as wide or tall, requiring fewer things be relocated.

The payload bay is configured loosely as the STS-118 mission that carried Barbara Morgan, including a Spacehab single module along with a replica external airlock and tunnel adapter fabricated by Guard-Lee. The *Endeavour* longeron bridge configuration did not allow placing Spacehab in the same location as STS-118, so the module is located 5 feet further aft than normal, but this does not distract from the experience. The Smithsonian provided two of the longeron bridges for the tunnel adapter from *Discovery* and Guard-Lee provided two non-structural replicas. An ICC-VLD is located in the aft portion of the payload bay, despite it not having been carried on STS-118 (which carried a somewhat similar ESP-3). The blankets covering two payload bays (8/9) were removed to expose the systems normally hidden from view. During preparations for its final display at the new Samuel Oschin Air and Space Center, workers reinstalled the middeck lockers, seats, and escape pole in *Endeavour*, although the galley and potty remain separately displayed.

ET-94 transiting the Miraflores Locks on the west side of the Panama Canal during its journey from Michoud to Los Angeles. Unlike the tanks that went to Vandenberg, ET-94 was on an open barge during the trip from Michoud to Marina del Rey. (Dennis R. Jenkins)

Initially, authorities were worried about potential vandalism while the orbiter was moving through the streets, but everybody that came out to watch was an enthusiastic supporter and there were no issues. The crowds were present at all hours during the move. (Dennis R. Jenkins)

A rendering of the final Endeavour stack in the Samuel Oschin Air and Space Center. Note the glass platform above the nose. Although not visible from this angle, the far payload bay door is open, revealing a Spacehab. (California Science Center Foundation/ZGF Architects)

ACRONYMS

ACBM	Active Common Berthing Mechanism	EES	Ejection Escape Suits (S1030A)
ACCESS	Assembly Concept for Construction of Erectable Space Structures	EI	Entry Interface
		ELM-PS	Experiment Logistics Module-Pressurized Section
ACES	Advanced Crew Escape Suit (S1035)	ELV	Expendable Launch Vehicle
ACTS	Advanced Communications Technology Satellite	EMU	Extravehicular Mobility Unit
AFB	Air Force Base	ESA	European Space Agency
AFFDL	Air Force Flight Dynamics Laboratory	ESP	External Stowage Platform
AFRSI	Advanced Flexible Reusable Surface Insulation	ET	External Tank
AFS	Air Force Station	EURECA	European Retrievable Carrier
AIAA	American Institute of Aeronautics and Astronautics	EVA	Extravehicular Activity
		EXPRESS	Expedite the Processing of Experiments to the Space Station
ALT	Approach and Landing Tests		
AMOS	Air Force Maui Optical Site Calibration	FDNY	Fire Department of New York
AOA	Abort-Once-Around	FGB	Functional Cargo Block (Zarya)
APU	Auxiliary Power Unit	FIB	Flexible Insulation Blankets (formerly AFRSI)
ASEM	Assembly of Station by EVA Methods	FRCS	Forward Reaction Control System
ASI	Italian Space Agency (Agenzia Spaziale Italian)	FREESTAR	Fast Reaction Experiments Enabling Science, Technology, Applications, and Research
ASSC	Alternate Space Shuttle Concepts		
ATK	Alliant Techsystems	FRF	Flight Readiness Firing
ATO	Abort to Orbit	FRGF	Flight-Releasable Grapple Fixture
		FRSI	Felt Reusable Surface Insulation
BFS	Backup Flight System (orbiter)	FSS	Fixed Service Structure
BoB	Bureau of the Budget (now OMB)	FY	Fiscal Year
CAIB	Columbia Accident Investigation Board	GAO	Government Accountability Office
CAPCOM	Capsule Communicator	GAS	Get-Away Special
CBM	Common Berthing Mechanism	GPC	General-Purpose Computer
CDR	Commander	GPS	Global Positioning System
CEIT	Crew Equipment Interface Test	GUCP	Ground Umbilical Carrier Plate
CETA	Crew and Equipment Translation Aid		
CFES	Continuous Flow Electrophoresis System	HAC	Heading Alignment Cone
CIRRIS	Cryogenic Infrared Radiance Instrumentation for Shuttle	HOST	Hubble Orbiting Systems Test
		HPU	Hydraulic Power Unit
CLOUDS	Cloud Logic to Optimize use of Defense Systems	HRSI	High-Temperature Reusable Surface Insulation
CMG	Control Moment Gyro	HST	Hubble Space Telescope
CNES	French Space Agency (Centre National d'Etudes Spatiales)	HUD	Heads-Up Display
		HUT	Hard Upper Torso
COAS	Crew Optical Alignment Sight		
CSA	Canadian Space Agency	ICC	Integrated Cargo Carrier
CSCS	Contingency Shuttle Crew Support	IECM	Induced Environment Contamination Monitor
		IML	International Microgravity Laboratory
DARA	German Space Agency (Deutsche Agentur für Raumfahrtangelegenheiten)	IMU	Inertial Measurement Unit
		ISS	International Space Station
DCR	Design Certification Review	IUS	Inertial Upper Stage
DFI	Developmental Flight Instrumentation	IVA	Intravehicular Activity
DFVLR	German Test and Research Institute for Aviation and Space Flight		
		JAXA	Japan Aerospace Exploration Agency
DLR	German Aerospace Center (Deutsches Zentrum für Luft- und Raumfahrt)	JEM	Japanese Experiment Module (Kibō)
		JEM-EF	JEM Exposed Facility
DoD	Department of Defense	JEM-ES	JEM Exposed Section
DSCS	Defense Satellite Communications System	JEM-PM	JEM Pressurized Module
DSP	Defense Support Program	JEM-RMS	JEM Remote Manipulator
		JPL	Jet Propulsion Laboratory
EASE	Experimental Assembly of Structures Through EVA	JRB	Joint Reserve Base
		JSC	Johnson Space Center, Texas
ECAL	East Coast Abort Landing		
ECO	Engine Cutoff	KSC	Kennedy Space Center, Florida
EDO	Extended Duration Orbiter		

L/D	Lift-to-Drag Ratio		PLT	Pilot
lbf	Pounds-Force		PMA	Pressurized Mating Adapter
LC	Launch Complex		PMM	Permanent Multipurpose Module
LCC	Launch Control Center		POCC	Payload Operations Control Center
LDEF	Long Duration Exposure Facility		PSx	Payload Specialist
LDM	Logistics Double Module			
LEH	Launch-Entry Helmet		RCC	Reinforced Carbon-Carbon
LES	Launch-Entry Suit (S1032)		RCRS	Regenerative Carbon Dioxide Removal System
LH$_2$	Liquid Hydrogen		RCS	Reaction Control System
LiOH	Lithium Hydroxide		RDM	Research Double Module
LMS	Life and Microgravity Spacelab		RFI	Request for Information
LO$_2$	Liquid Oxygen		RFP	Request for Proposals
LON	Launch-on-Need		RME	Radiation Monitoring Experiment
LRSI	Low-Temperature Reusable Surface Insulation		RPM	R-Bar Pitch Maneuver
LUT	Launch Umbilical Tower		RSRM	Redesigned/Reusable Solid Rocket Motor
LWT	Lightweight Tank		RTHU	Roll-To-Heads-Up
			RTLS	Return-To-Launch-Site (abort)
MAF	Michoud Assembly Facility, Louisiana		RTV	Room-Temperature Vulcanizing Silicone
MCDS	Multifunction CRT Display System			
MECO	Main Engine Cutoff		SAFER	Simplified Aid for EVA Rescue
MEDS	Multifunction Electronic Display System		SARJ	Solar Alpha Rotary Joint
MEEP	Mir Environmental Effects Payload		SCA	Shuttle Carrier Aircraft
MIT	Massachusetts Institute of Technology		SDS	Satellite Data System
MLI	Multi-Layer Insulation		SEADS	Shuttle Entry Air Data System
MLP	Mobile Launch Platform		SFP	Space Flight Participant
MLR	Monodisperse Latex Reactor		SFU	Space Flyer Unit
MMH	Monomethyl Hydrazine		SILTS	Shuttle Infrared Leeside Temperature Sensing
MMOD	Micrometeoroid and Orbital Debris		SLDP	Spacelab Logistics Double Pallet
MMU	Manned Maneuvering Unit		SLF	Shuttle Landing Facility
MPESS	Mission Peculiar Experiment Support Structure		SLP	Spacelab Logistics Pallet
MPLM	Multi-Purpose Logistics Module		SLS	Space Launch System
MPS	Main Propulsion System		SLS	Spacelab Life Sciences
MPTA	Main Propulsion Test Article		SLWT	Super Lightweight Tank
MSE	Manned Spaceflight Engineer		SM	Servicing Mission (Hubble Space Telescope)
MSFC	Marshall Space Flight Center, Alabama		SOFI	Spray-on Foam Insulation
MSL	Microgravity Science Laboratory		SPARTAN	Shuttle Pointed Autonomous Research Tool for Astronomy
MSx	Mission Specialist		SPAS	Shuttle Pallet Satellite
			SRB	Solid Rocket Booster
N$_2$H$_4$	Hydrazine		SRL	Space Radar Laboratory
N$_2$O$_4$	Nitrogen Tetroxide		SRM	Solid Rocket Motor
NACA	National Advisory Committee for Aeronautics		SRMS	Shuttle Remote Manipulator System
NASA	National Aeronautics and Space Administration		SRTM	Shuttle Radar Topography Mission
NASDA	National Space Development Agency of Japan		SSBUV	Shuttle Solar Backscatter Ultraviolet
NASM	National Air and Space Museum		SSME	Space Shuttle Main Engine
NGA	National Geospatial-Intelligence Agency		SSOR	Space-to-Space Orbiter Radio
NIMA	National Imagery and Mapping Agency		SSPTS	Station-Shuttle Power Transfer System
NRO	National Reconnaissance Office		SSRMS	Space Station Remote Manipulator System
NSTL	National Space Transportation Laboratory		SSSR	Space-to-Space Station Radio
NYPD	New York Police Department		STS	Space Transportation System
			SWT	Standard-Weight Tank
OARE	Orbital Acceleration Research Experiment			
OBSS	Orbiter Boom Sensor System		T&R	Transition and Retirement
ODS	Orbiter Docking System		TAEM	Terminal Area Energy Management
OEX	Orbiter Experiments		TAL	Trans-Oceanic Abort Landing
OFT	Orbital Flight Tests		TCDT	Terminal Countdown Demonstration Test
OIG	Office of the Inspector General		TDRS	Tracking and Data Relay Satellite
OMB	Office of Management and Budget		TPS	Thermal Protection System
OMDP	Orbiter Maintenance and Down Period		TSS	Tethered Satellite System
OMM	Orbiter Major Modification			
OMS	Orbital Maneuvering System		UARS	Upper Atmosphere Research Satellite
OPF	Orbiter Processing Facility		USA	United Space Alliance
OSTA	Office of Space and Terrestrial Applications		USBI	United Space Boosters, Inc.
OV	Orbiter Vehicle		USML	United States Microgravity Laboratory
			USMP	United States Microgravity Payload
PAL	Protuberance Air Loads			
PAM-D	Payload Assist Module-Delta		VAB	Vehicle Assembly Building
PAM-S	Payload Assist Module-Special			
PCR	Payload Changeout Room		WSF	Wake Shield Facility
PLSS	Primary Life Support System			

Endeavour *as STS-134.* (NASA)